A T Dore

ENGINEERING FORMULAS

Kurt Gieck

Fourth Edition

McGraw-Hill Book Company
New York St. Louis San Francisco Montreal Toronto

Library of Congress Catalog Card Number 82-20337
ISBN 0-07-023219-9

First published in the English language under the title
A COLLECTION OF TECHNICAL FORMULAE
and copyright © 1967 by Kurt Gieck, Heilbronn/N, West Germany.
Second English edition copyright © 1974 by Kurt Gieck, Heilbronn/N, West Germany (the 47th in a series),
Fourth English edition copyright © 1979 by Kurt Gieck, Heilbronn/N, West Germany, (the 58th in a series)
Fifth English edition copyright © 1982 by Kurt Gieck, Heilbronn/N, West Germany (the 64th in a series).

First American edition
published by McGraw-Hill, Inc. in 1971

Second American edition
published by McGraw-Hill, Inc. in 1976

Third American edition
published by McGraw-Hill, Inc. in 1979
English translation by J. Walters, B. Sc., Fareham

Fourth American edition
published by McGraw-Hill, Inc. in 1983
English translation by J. Walters, B. Sc., Fareham

Printed in Germany

Preface

The purpose of this collection of technical formulae is to provide a brief, clear and handy guide to the more important technical and mathematical formulae.

Since the book has been printed on one side of the page only, the facing pages are available for additional notes.

Each separate subject has been associated with a capital letter. The various formulae have been grouped under corresponding small letters and numbered serially. This method enables the formulae used in any particular calculation to be indicated.

Preface

to the 2nd edition

"Engineering Formulas" has been revised to include S.I. Units and in addition, several sections have been extended and brought up to date.

Preface

to the enlarged and revised 3rd edition

In the 3rd edition, the SI base quantity "*n*" and the base unit "mol" have been introduced.

The sections on Hydraulics and Heat have been completely revised and those on Arithmetic and Chemistry enlarged.

Included for the first time are pages on Fourier series, Permutations and Combinations, and in line with technical committees of I. S. O., the unit N/mm^2 is used for mechanical strength and stress.

Preface

to the enlarged and revised 4th edition

The sections on Strength and Machine Parts have been revised and considerably enlarged. The section known previously as Optics has been renamed Radiation Physics and formulae on ionizing radiation have been included because of its growing importance in connection with environmental pollution.

Since the pocket calculator is now commonplace, the tables of Circular Functions, Natural Logarithms and Degrees to Radians have been omitted.

Reference to BS, DIN and VDE

BS · British Standards Institution
(Address: 2 Park St, LONDON W 1 A 2 BS

DIN · Deutsches Institut für Normung
(Address: D-1000 BERLIN 30, Postfach 1107)

VDI · Verein Deutscher Ingenieure
(Address: D-4000 DUESSELDORF 1, Postfach 1139).

Method of Presentation and Use of Units

Most of the equations clearly reveal the physical relationships which they describe and are valid regardless of the system of units employed, provided that they are consistent.

Some of the equations are empirical in origin and the units quoted must be used in the formula to obtain the correct result, these are mainly to be found in sections Q and R.

It is intended that the Stroud notation is used when evaluating the formulae i.e. both the quantity and the unit is substituted for a given symbol and the subsequent calculation involves manipulation of numbers and units together.

For example, taking equation I 23: $t = \frac{s}{v}$

if s (distance) = 2·8 metres
v (speed) = 8 metres/second

then $t = \frac{2{\cdot}8 \text{ metres} \times \text{second}}{8 \text{ metres}}$

hence t = 0·35 seconds (time)
cancelling the unit 'metres'

It is clear that t should have the units of time; if it does not, then it is obvious that an error has been made and the working should be checked. As a help, in many cases, the anticipated units are quoted using the abbreviation "EU", **E**xample-**U**nit.

When the numerical values and the units are included in the calculations, their equivalents or definitions are best written so that they are dimensionless and have the value of 1·0. In this form they are sometimes called "Unity Brackets" and their use can be illustrated in three ways:

with consistent units,

equation a 6

$$1 \text{ km} = 10^3 \text{ m} \qquad \text{becomes} \qquad 1 = \left[\frac{1 \text{ km}}{10^3 \text{ m}}\right]$$

equation a 62

$$12 \text{ in} = 1 \text{ ft} \qquad \text{becomes} \qquad 1 = \left[\frac{1 \text{ ft}}{12 \text{ in}}\right]$$

equation a 90

$$778{\cdot}6 \text{ ft lbf} = 1 \text{ Btu} \qquad \text{becomes} \qquad 1 = \left[\frac{778{\cdot}6 \text{ ft lbf}}{1 \text{ Btu}}\right]$$

for example, to convert 14·7 lbf/in^2 to lbf/ft^2

$$14{\cdot}7 \frac{\text{lbf}}{\text{in}^2} = 14{\cdot}7 \frac{\text{lbf}}{\text{in}^2}\left[\frac{12 \text{ in}}{1 \text{ ft}}\right]^2 = 14{\cdot}7 \times 144 \frac{\text{lbf}}{\text{ft}^2} = 2117 \frac{\text{lbf}}{\text{ft}^2}$$

in the conversion between different systems of units,

equation a 37

$$1 \text{ kgf} = 9{\cdot}81 \text{ N} \qquad \text{becomes} \qquad 1 = \left[\frac{9{\cdot}81 \text{ N}}{1 \text{ kgf}}\right]$$

equation a 65

$$1 \text{ m} = 3{\cdot}281 \text{ ft} \qquad \text{becomes} \qquad 1 = \left[\frac{1 \text{ m}}{3{\cdot}281 \text{ ft}}\right]$$

equation a 108

$$1 \text{ Btu/lb} = 0{\cdot}556 \text{ kcal/kg} \qquad \text{becomes} \qquad 1 = \left[\frac{0{\cdot}556 \text{ kcal lb}}{1 \text{ kg Btu}}\right]$$

For example, to convert 1000 kgf/cm^2 to S.I. units,

$$1000 \frac{\text{kgf}}{\text{cm}^2} = 1000 \frac{\text{kgf}}{\text{cm}^2}\left[\frac{9{\cdot}81 \text{ N}}{1 \text{ kgf}}\right]\left[\frac{10^4 \text{ cm}^2}{1 \text{ m}^2}\right]\left[\frac{1 \text{ MN}}{10^6 \text{ N}}\right]$$

$$= 98{\cdot}1 \frac{\text{MN}}{\text{m}^2}$$

in the use of definitions:

1 lbf is the force required to accelerate a mass of 1 lb at the rate of 32·174 ft/s².

$$1\ \text{lbf} = 1\ \text{lb} \times 32{\cdot}174\ \frac{\text{ft}}{\text{s}^2} \quad \text{becomes} \quad 1 = \left[\frac{32{\cdot}174\ \text{lb ft}}{\text{s}^2\ \text{lbf}}\right]$$

Similarly, the Newton is defined by the equation

$$1\ \text{N} = 1\ \text{kg} \times \frac{1\ \text{m}}{\text{s}^2} \quad \text{which becomes} \quad 1 = \left[\frac{\text{N s}^2}{\text{kg m}}\right]$$

$$\text{and } 1\ \text{kgf} = 1\ \text{kg} \times 9{\cdot}81\ \frac{\text{m}}{\text{s}^2} \quad \text{becomes} \quad 1 = \left[\frac{9{\cdot}81\ \text{kg m}}{\text{kgf s}^2}\right]$$

For example, to find the force in S.I. units required to accelerate a mass of 3 lb at the rate of 2·5 ft/s², proceed as follows:

$F = m\,a$, equation m 1.

$$F = 3\ \text{lb} \times 2{\cdot}5\ \frac{\text{ft}}{\text{s}^2} \left[\frac{0{\cdot}4536\ \text{kg}}{1\ \text{lb}}\right]\left[\frac{1\ \text{m}}{3{\cdot}281\ \text{ft}}\right]\left[\frac{\text{N s}^2}{\text{kg m}}\right]$$

$$= \frac{3 \times 2{\cdot}5 \times 0{\cdot}4536}{3{\cdot}281}\ \text{N} = 1{\cdot}036\ \text{N}$$

which is a unit of force.

Base Quantities and Base Units of the International System of Measurement

base quantity		base unit	
name	symbol (*italic letters*)	name	symbol (vertical letters)
length	l	metre	m
mass	m	kilogram	kg
time	t	second	s
electric current	I	ampere	A
absolute temperature	T	kelvin	K
amount of		candela	cd
substance	n	mole	mol
light intensity	I_V	candela	cd

(Old units are put in [ ] brackets)

LIST OF SYMBOLS

Space and time

- α, β, γ angles
- Ω solid angle
- b, B breadth
- d, D diameter (diagonal)
- h, H height
- l, L length
- p pitch
- r, R radius
- s distance covered, perimeter
- t thickness
- u, U circumference
- A area, cross section
- A_m generated surface
- A_o surface area
- V volume
- t time
- v velocity, linear
- ω velocity, angular
- a acceleration, linear
- α acceleration, angular
- g acceleration, gravitational

Periodical and related phenomens

- T period
- f frequency
- n rotational speed
- ω angular frequency
- λ wavelength
- c velocity of light

Mechanics

- m mass
- ϱ density
- F force, direct force
- f, σ direct stress
- q, τ shear stress
- p normal pressure
- ε extension, strain
- E modulus of elasticity (Young's modulus)
- G modulus of rigidity (shear modulus)
- M bending moment
- T torsional moment, torque
- Z modulus of section
- Q shear force, shear load
- V vertical reaction
- W weight or load, work
- w uniformly distributed load
- I moment of inertia, second moment of area
- I_p polar moment of inertia
- J torsion constant
- Z modulus of section
- μ coefficient of sliding friction
- μ_o coefficient of static friction
- μ_q coefficient of friction of a radial bearing
- μ_l coefficient of friction of a longitudinal bearing
- f coefficient of rolling friction
- η dynamic viscosity
- ν kinematic viscosity
- P power
- η efficiency

Heat

T absolute temperature
t temperature
α linear coefficient of expansion
γ cubic coefficient of expansion
Φ heat current or flow
φ density of heat flow
q quantity of heat per unit mass
Q quantity of heat
c_p specific heat at constant pressure
c_v specific heat at constant volume
γ ratio of c_p to c_v
R gas constant
λ thermal conductivity
α heat transfer coefficient
k coefficient of heat transmission
C radiation constant
v specific volume

Electricity and magnetism

I current
J current density
V, U voltage
U_q source voltage
R resistance
G conductance
Q quantity of electricity (charge)
C capacitance
D dielectric displacement
E electric field strength
Φ magnetic flux
B magnetic induction
L inductance
H magn. field strength
Θ circulation (magnetic potential)
V magnetic voltage
R_m magnetic resistance
Λ magnetic conductance
δ length of air gap
α temperature coefficient of resistance
γ conductivity
ϱ resistivity
ε permittivity, dielectric constant
ε_0 absolute permittivity
ε_r relative permittivity
N number of turns
μ permeability
μ_0 absolute permeability
μ_r relative permeability
p number of pairs of poles
z number of conductors
Q quality, figure of merit
δ loss angle
Z impedance
X reactance
P_s apparent power
P_q reactive power
C_M moment constant

Light and related electromagnetic radiations

I_e radiant intensity
I_v luminous intensity
Φ_e radiant power, radiant flux
Φ_v luminous flux
Q_e radiant energy
Q_v quantity of light
E_e irrediance
E_v illuminance
H_e radiant exposure
H_v light exposure
L_e radiance
L_v luminance
c velocity of light
n refractive index
f focal length
D refractive power

UNITS

A 1

Decimal multiples and fractions of units

da =	deca =	10^{1}	d =	deci =	10^{-1}
h =	hecto =	10^{2}	c =	centi =	10^{-2}
k =	kilo =	10^{3}	m =	milli =	10^{-3}
M =	mega =	10^{6}	µ =	micro =	10^{-6}
G =	giga =	10^{9}	n =	nano =	10^{-9}
T =	tera =	10^{12}	p =	pico =	10^{-12}
P =	peta =	10^{15}	f =	femto =	10^{-15}
E =	exa =	10^{18}	a =	atto =	10^{-18}

Units of length

		m	µm	mm	cm	dm	km
a 1	1 m =	1	10^{6}	10^{3}	10^{2}	10	10^{-3}
a 2	1 µm =	10^{-6}	1	10^{-3}	10^{-4}	10^{-5}	10^{-9}
a 3	1 mm =	10^{-3}	10^{3}	1	10^{-1}	10^{-2}	10^{-6}
a 4	1 cm =	10^{-2}	10^{4}	10	1	10^{-1}	10^{-5}
a 5	1 dm =	10^{-1}	10^{5}	10^{2}	10	1	10^{-4}
a 6	1 km =	10^{3}	10^{9}	10^{6}	10^{5}	10^{4}	1

Units of length (continued)

		mm	µm	nm	[Å]	pm	[m Å]
a 7	1 mm =	1	10^{3}	10^{6}	10^{7}	10^{9}	10^{10}
a 8	1 µm =	10^{-3}	1	10^{3}	10^{4}	10^{6}	10^{7}
a 9	1 nm =	10^{-6}	10^{-3}	1	10	10^{3}	10^{4}
a 10	[1 Å] =	10^{-7}	10^{-4}	10^{-1}	1	10^{2}	10^{3}
a 11	1 pm =	10^{-9}	10^{-6}	10^{-3}	10^{-2}	1	10
a 12	[1 mÅ] =	10^{-10}	10^{-7}	10^{-4}	10^{-3}	10^{-1}	1

Units of area

		m^2	$µm^2$	mm^2	cm^2	dm^2	km^2
a 13	1 m^2 =	1	10^{12}	10^{6}	10^{4}	10^{2}	10^{-6}
a 14	1 $µm^2$ =	10^{-12}	1	10^{-6}	10^{-8}	10^{-10}	10^{-18}
a 15	1 mm^2 =	10^{-6}	10^{6}	1	10^{-2}	10^{-4}	10^{-12}
a 16	1 cm^2 =	10^{-4}	10^{8}	10^{2}	1	10^{-2}	10^{-10}
a 17	1 dm^2 =	10^{-2}	10^{10}	10^{4}	10^{2}	1	10^{-8}
a 18	1 km^2 =	10^{6}	10^{18}	10^{12}	10^{10}	10^{8}	1

Å = Ångström | 1 mÅ = 1 XE = 1 X-unit

UNITS

A 2

Units of volume

		m^3	mm^3	cm^3	$dm^3 = l$ [2]	km^3
a 19	1 m^3 =	1	10^9	10^6	10^3	10^{-9}
a 20	1 mm^3 =	10^{-9}	1	10^{-3}	10^{-6}	10^{-18}
a 21	1 cm^3 =	10^{-6}	10^3	1	10^{-3}	10^{-15}
a 22	1 dm^3 = 1 l =	10^{-3}	10^6	10^3	1	10^{-12}
a 23	1 km^3 =	10^9	10^{18}	10^{15}	10^{12}	1

Units of mass

		kg	mg	g	dt	t = Mg
a 24	1 kg =	1	10^6	10^3	10^{-2}	10^{-3}
a 25	1 mg =	10^{-6}	1	10^{-3}	10^{-8}	10^{-9}
a 26	1 g =	10^{-3}	10^3	1	10^{-5}	10^{-6}
a 27	1 dt =	10^2	10^8	10^5	1	10^{-1}
a 28	1 t = 1 Mg =	10^3	10^9	10^6	10	1

Units of time

		s	ns	µs	ms	min
a 29	1 s =	1	10^9	10^6	10^3	$16{\cdot}66 \times 10^{-3}$
a 30	1 ns =	10^{-9}	1	10^{-3}	10^{-6}	$16{\cdot}66 \times 10^{-12}$
a 31	1 µs =	10^{-6}	10^3	1	10^{-3}	$16{\cdot}66 \times 10^{-9}$
a 32	1 ms =	10^{-3}	10^6	10^3	1	$16{\cdot}66 \times 10^{-6}$
a 33	1 min =	60	60×10^9	60×10^6	60×10^3	1
a 34	1 h =	3600	$3{\cdot}6 \times 10^{12}$	$3{\cdot}6 \times 10^9$	$3{\cdot}6 \times 10^6$	60
a 35	1 d =	$86{\cdot}4 \times 10^3$	$86{\cdot}4 \times 10^{12}$	$86{\cdot}4 \times 10^9$	$86{\cdot}4 \times 10^6$	1440

Units of force (gravitational force also)

		N [1]	kN	MN	[kgf]	[dyn]
a 36	1 N =	1	10^{-3}	10^{-6}	$0{\cdot}102$	10^5
a 37	1 kN =	10^3	1	10^{-3}	$0{\cdot}102 \times 10^3$	10^8
a 38	1 MN =	10^6	10^3	1	$0{\cdot}102 \times 10^6$	10^{11}

[1] 1 N = 1 kg m/s^2

[2] 1 l = 1 liter

Units of pressure

		Pa=N/m²	N/mm²	bar	[kgf/cm²]	[torr]
a 39	1 Pa=1 N/m² =	1	10^{-6}	10^{-5}	$1\cdot02\times10^{-5}$	0·0075
a 40	1 N/mm² =	10^{6}	1	10	10·2	$7\cdot5\times10^{3}$
a 41	1 bar =	10^{5}	0·1	1	1·02	750
a 42	[1 kgf/cm²=1 at] =	98100	$9\cdot81\times10^{-2}$	0·981	1	736
a 43	[1 torr] [1] =	133	$0\cdot133\times10^{-3}$	$1\cdot33\times10^{-3}$	$1\cdot36\times10^{-3}$	1

Units of work

		J	kW h	[kgf m]	[kcal]	[hp h]
a 44	1 J [2] =	1	$0\cdot278\times10^{-6}$	0·102	$0\cdot239\times10^{-3}$	$0\cdot373\times10^{-6}$
a 45	1 kW h =	$3\cdot60\times10^{6}$	1	367×10^{3}	860	1·34
a 46	[1 kgf m] =	9·81	$2\cdot72\times10^{-6}$	1	$2\cdot345\times10^{-3}$	$3\cdot65\times10^{-6}$
a 47	[1 kcal] =	4186·8	$1\cdot16\times10^{-3}$	426·9	1	$1\cdot56\times10^{-3}$
a 48	[1 hp h] =	$2\cdot69\times10^{6}$	0·746	$0\cdot28\times10^{6}$	641	1

Units of power

		W	kW	[kgf m/s]	[kcal/h]	[hp]
a 49	1 W [3] =	1	10^{-3}	0·102	0·860	$1\cdot34\times10^{-3}$
a 50	1 kW =	1000	1	102	860	1·34
a 51	[1 kgf m/s] =	9·81	$9\cdot81\times10^{-3}$	1	8·43	$13\cdot2\times10^{-3}$
a 52	[1 kcal/h] =	1·16	$1\cdot16\times10^{-3}$	0·119	1	$1\cdot55\times10^{-3}$
a 53	[1 hp] =	746	0·746	76	643	1

Unit of mass for jewels

a 54 1 carat = 200 mg = $0\cdot2\times10^{-3}$ kg = 1/5000 kg

Unit of fineness for precious metals

a 55 24 carat ≙ 1000·00 ‰ | 18 carat ≙ 750·00 ‰

a 56 14 carat ≙ 583·33 ‰ | 8 carat ≙ 333·33 ‰

Units of temperatur

a 57 $T = \left(\frac{t}{^\circ C} + 273\cdot15\right) K = \frac{5}{9}\,\frac{T_R}{\text{Rank}}\; K$

a 58 $T_R = \left(\frac{t_F}{^\circ F} + 459\cdot67\right) \text{Rank} = \frac{9}{5}\,\frac{T}{K}\; \text{Rank}$

a 59 $t = \frac{5}{9}\left(\frac{t_F}{^\circ F} - 32\right) {}^\circ C = \left(\frac{T}{K} - 273\cdot15\right) {}^\circ C$

a 60 $t_F = \left(\frac{9}{5}\,\frac{t}{^\circ C} + 32\right) {}^\circ F = \left(\frac{T_R}{\text{Rank}} - 459\cdot67\right) {}^\circ F$

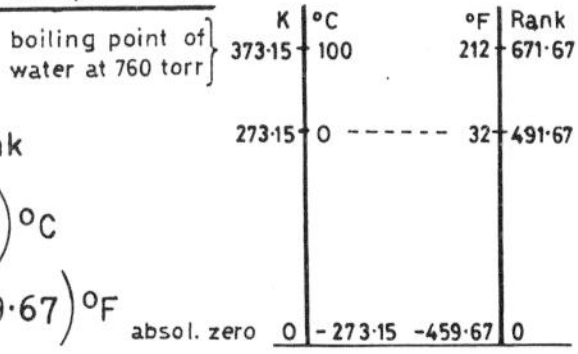

T, T_R, t and t_F are the temperatures in the scales for Kelvin, Rankine, Celsius, Fahrenheit

[1] 1 torr = 1/760 atm = 1·333 22 mbar = 1 mm Hg at t = 0 °C

[2] 1 J = 1 N m = 1 W s | [3] 1 W = 1 J/s = 1 N m/s

Conversion, Anglo-American to metric units

Units of length

		in	ft	yd	mm	m	km
a 61	1 in =	1	0·08333	0·02778	25·4	0·0254	–
a 62	1 ft =	12	1	0·3333	304·8	0·3048	–
a 63	1 yd =	36	3	1	914·4	0·9144	–
a 64	1 mm =	0·03937	3281×10^{-6}	1094×10^{-6}	1	0·001	10^{-6}
a 65	1 m =	39·37	3·281	1·094	1000	1	0·001
a 66	1 km =	39370	3281	1094	10^6	1000	1

Units of area

		sq in	sq ft	sq yd	cm^2	dm^2	m^2
a 67	1 sq in =	1	$6{\cdot}944 \times 10^{-3}$	$0{\cdot}772 \times 10^{-3}$	6·452	0·06452	$64{\cdot}5 \times 10^{-5}$
a 68	1 sq ft =	144	1	0·1111	929	9·29	0·0929
a 69	1 sq yd =	1296	9	1	8361	83·61	0·8361
a 70	1 cm^2 =	0·155	$1{\cdot}076 \times 10^{-3}$	$1{\cdot}197 \times 10^{-4}$	1	0·01	0·0001
a 71	1 dm^2 =	15·5	0·1076	0·01196	100	1	0·01
a 72	1 m^2 =	1550	10·76	1·196	10000	100	1

Units of volume

		cu in	cu ft	cu yd	cm^3	dm^3	m^3
a 73	1 cu in =	1	$5{\cdot}786 \times 10^{-4}$	$2{\cdot}144 \times 10^{-5}$	16·39	0·01639	$1{\cdot}64 \times 10^{-5}$
a 74	1 cu ft =	1728	1	0·037	28316	28·32	0·0283
a 75	1 cu yd =	46656	27	1	764555	764·55	0·7646
a 76	1 cm^3 =	0·06102	3532×10^{-8}	$1{\cdot}31 \times 10^{-6}$	1	0·001	10^{-6}
a 77	1 dm^3 =	61·02	0·03532	0·00131	1000	1	0·001
a 78	1 m^3 =	61023	35·32	1·307	10^6	1000	1

Units of mass

		dram	oz	lb	g	kg	Mg
a 79	1 dram =	1	0·0625	0·003906	1·772	0·00177	$1{\cdot}77 \times 10^{-6}$
a 80	1 oz =	16	1	0·0625	28·35	0·02832	$28{\cdot}3 \times 10^{-6}$
a 81	1 lb =	256	16	1	453·6	0·4531	$4{\cdot}53 \times 10^{-4}$
a 82	1 g =	0·5643	0;03527	0·002205	1	0;001	10^{-6}
a 83	1 kg =	564·3	35·27	2·205	1000	1	0·001
a 84	1 Mg =	$564{\cdot}4 \times 10^3$	35270	2205	10^6	1000	1

continued A 5

continued from A 4

Units of work

		ft lb	kgf m	J = W s	kW h	kcal	Btu
a 85	1 ft lb =	1	0·1383	1·356	$376{\cdot}8\times10^{-9}$	324×10^{-6}	$1{\cdot}286\times10^{-3}$
a 86	1 kgf m =	7·233	1	9·807	$2{\cdot}725\times10^{-6}$	$2{\cdot}344\times10^{-3}$	$9{\cdot}301\times10^{-3}$
a 87	1J=1W s=	0·7376	0·102	1	$277{\cdot}8\times10^{-9}$	239×10^{-6}	$948{\cdot}4\times10^{-6}$
a 88	1 kW h =	$2{\cdot}655\times10^{6}$	$367{\cdot}1\times10^{3}$	$3{\cdot}6\times10^{6}$	1	860	3413
a 89	1 kcal =	$3{\cdot}087\times10^{3}$	426·9	4187	$1{\cdot}163\times10^{-3}$	1	3·968
a 90	1 Btu =	778·6	107·6	1055	293×10^{-6}	0·252	1

Units of power

		hp	kgf m/s	J/s=W	kW	kcal/s	BTU/s
a 91	1 hp =	1	76·04	745·7	0·7457	0·1782	0·7073
a 92	1kgf m/s=	$13{\cdot}15\times10^{-3}$	1	9·807	$9{\cdot}807\times10^{-3}$	$2{\cdot}344\times10^{-3}$	$9{\cdot}296\times10^{-3}$
a 93	1J/s=1W =	$1{\cdot}341\times10^{-3}$	0·102	1	10^{-3}	239×10^{-6}	$948{\cdot}4\times10^{-6}$
a 94	1 kW =	1·341	102	1000	1	0·239	0·9484
a 95	1kcal/s=	5·614	426·9	4187	4·187	1	3·968
a 96	1 Btu/s=	1·415	107·6	1055	1·055	0·252	1

Other units

a 97	1 mil = 10^{-3} in	=	0·0254 mm
a 98	1 sq mil = 10^{-6} sq in	=	645·2 μm^2
a 99	1 yard = 3 ft	=	0·914 m
a100	1 English mile = 1760 yds	=	1609 m
a101	1 Nautical mile	=	1852 m
a102	1 Geographical mile	=	7420 m
a103	1 long ton = 2240 lb	=	1·016 Mg
a104	1 short ton (US) = 2000 lb	=	0·9072 Mg
a105	1 long ton = 2240 lbf	=	9·96 MN
a106	1 short ton (US) = 2000 lbf	=	9·00 MN
a107	1 Imp. gallon (Imperial gallon)	=	4·546 dm^3
a108	1 US gallon	=	3·785 dm^3
a109	1 BTU/ft^3 = 9·547 kcal/m^3	=	39·964 kJ/m^3
a110	1 BTU/lb = 0·556 kcal/kg	=	2·327 kJ/kg
a111	1 lbf/ft^2 = 4·882 kgf/m^2	=	47·8924 N/m^2
a112	1 lbf/in^2 (p.s.i.) = 0·0703 kgf/cm^2	=	0·6896 N/cm^2
a113	1 chain = 22 yds	=	20·11 m
a114	1 Hundredweight (GB) (cwt) = 112 lbf	=	498 kN
a115	1 Quarter (GB) = 28 lbf	=	124·5 kN
a116	1 Stone (GB) = 14 lbf	=	62·3 kN

AREAS

B 1

square

b 1 $A = a^2$

b 2 $a = \sqrt{A}$

b 3 $d = a\sqrt{2}$

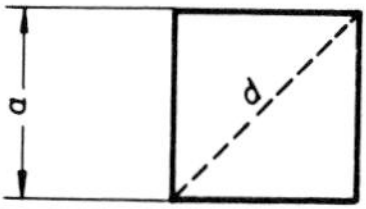

rectangle

b 4 $A = a\,b$

b 5 $d = \sqrt{a^2 + b^2}$

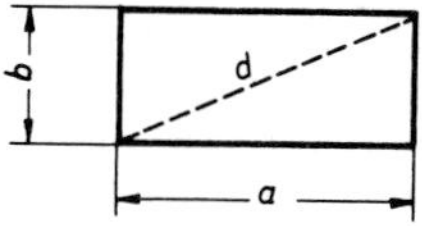

parallelogram

b 6 $A = a\,h = a\,b\sin\alpha$

b 7 $d_1 = \sqrt{(a + h\cot\alpha)^2 + h^2}$

b 8 $d_2 = \sqrt{(a - h\cot\alpha)^2 + h^2}$

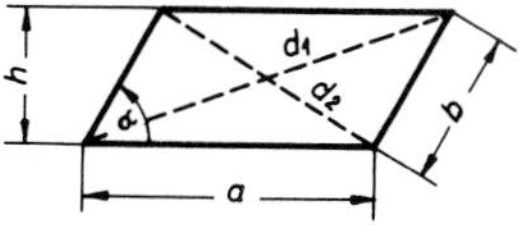

trapezium

b 9 $A = \dfrac{a + b}{2}\,h = m\,h$

b 10 $m = \dfrac{a + b}{2}$

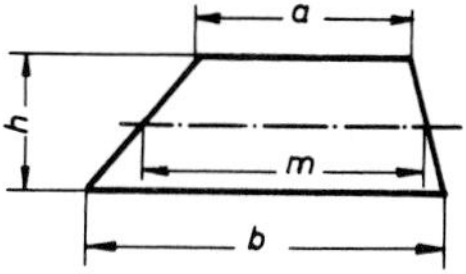

triangle

b 11 $A = \dfrac{a\,h}{2} = \varrho\,s$

b 12 $= \sqrt{s(s-a)(s-b)(s-c)}$

b 13 $s = \dfrac{a + b + c}{2}$

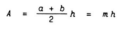

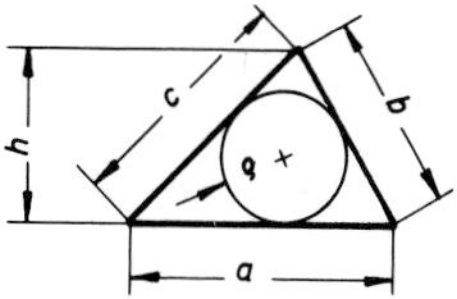

AREAS

equilateral triangle

b 14 $A = \frac{a^2}{4}\sqrt{3}$

b 15 $h = \frac{a}{2}\sqrt{3}$

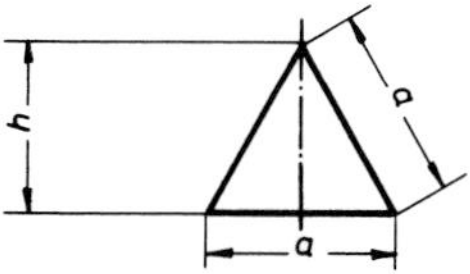

pentagon

b 16 $A = \frac{5}{8} r^2 \sqrt{10 + 2\sqrt{5}}$

b 17 $a = \frac{1}{2} r \sqrt{10 - 2\sqrt{5}}$

b 18 $\varrho = \frac{1}{4} r \sqrt{6 + 2\sqrt{5}}$

construction:
$\overline{AB} = 0{\cdot}5\, r$, $\overline{BC} = \overline{BD}$, $\overline{CD} = \overline{CE}$

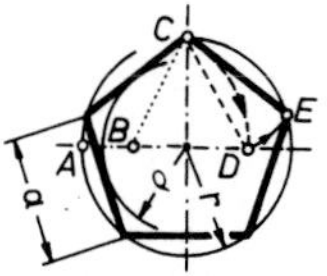

hexagon

b 19 $A = \frac{3}{2} a^2 \sqrt{3}$

b 20 $d = 2\,a$

b 21 $= \frac{2}{\sqrt{3}} s \approx 1{\cdot}155\, s$

b 22 $s = \frac{\sqrt{3}}{2} d \approx 0{\cdot}866\, d$

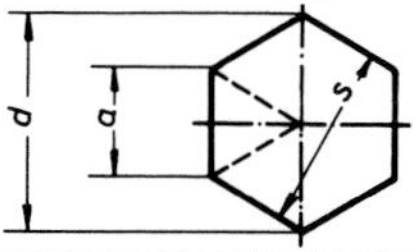

octagon

b 23 $A = 2\,a\,s \approx 0{\cdot}83\, s^2$

b 24 $= 2\, s \sqrt{d^2 - s^2}$

b 25 $a = s \times \tan 22{\cdot}5^\circ \approx 0{\cdot}415\, s$

b 26 $s = d \times \cos 22{\cdot}5^\circ \approx 0{\cdot}924\, d$

b 27 $d = \frac{s}{\cos 22{\cdot}5^\circ} \approx 1{\cdot}083\, s$

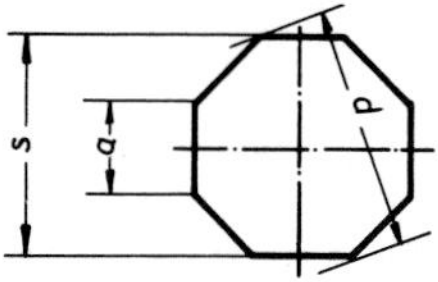

polygon

b 28 $A = A_1 + A_2 + A_3$

b 29 $= \frac{a\,h_1 + b\,h_2 + b\,h_3}{2}$

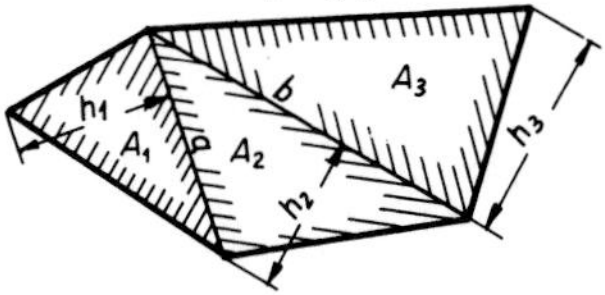

circle

b 30 $A = \frac{\pi}{4} d^2 = \pi r^2$

b 31 $\approx 0{\cdot}785\, d^2$

b 32 $U = 2 \pi r = \pi d$

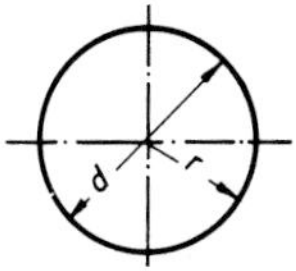

annulus

b 33 $A = \frac{\pi}{4} (D^2 - d^2)$

b 34 $= \pi (d + b) b$

b 35 $b = \frac{D - d}{2}$

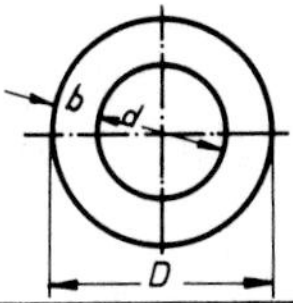

sector of a circle

b 36 $A = \frac{\pi}{360^\circ} r^2 \alpha = \frac{\hat{\alpha}}{2} r^2$

b 37 $= \frac{b\, r}{2}$

b 38 $b = \frac{\pi}{180^\circ} r \alpha$

b 39 $\hat{\alpha} = \frac{\pi}{180^\circ} \alpha$ ($\hat{\alpha} = \alpha$ in circular measure)

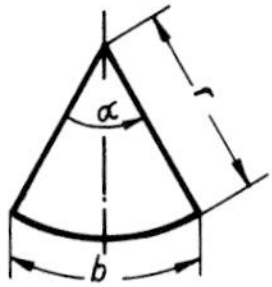

segment of a circle

b 40 $s = 2 r \sin\frac{\alpha}{2}$

b 41 $A = \frac{h}{6 s}(3h^2 + 4s^2) = \frac{r^2}{2}(\hat{\alpha} - \sin\alpha)$

b 42 $r = \frac{h}{2} + \frac{s^2}{8h}$

b 43 $h = r(1 - \cos\frac{\alpha}{2}) = \frac{s}{2} \tan\frac{\alpha}{4}$

b 44 $\hat{\alpha}$ see formula b 39

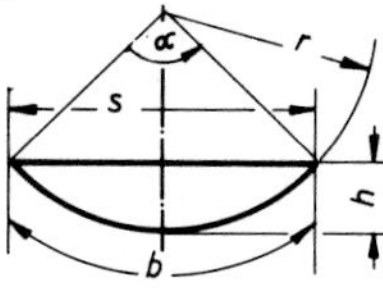

ellipse

b 45 $A = \frac{\pi}{4} D d = \pi a b$

b 46 $U \approx \pi \frac{D + d}{2}$

b 47 $= \pi(a+b)\left[1 + \frac{1}{4}\lambda^2 + \frac{1}{64}\lambda^4 + \frac{1}{256}\lambda^6 + \frac{25}{16384}\lambda^8 + \ldots\right]$, where $\lambda = \frac{a-b}{a+b}$

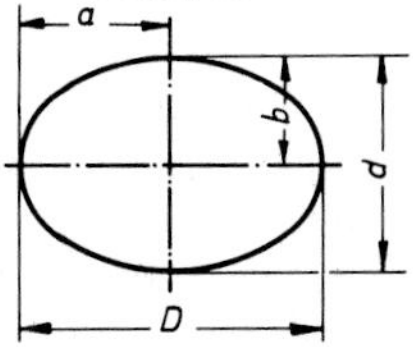

SOLID BODIES

C 1

cube

c 1 $V = a^3$

c 2 $A_o = 6\,a^2$

c 3 $d = \sqrt{3}\,a$

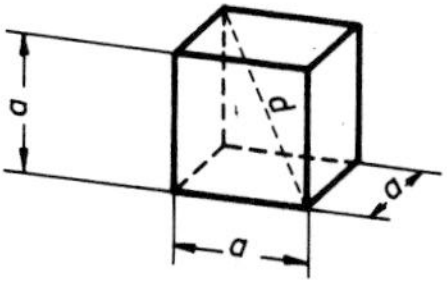

cuboid

c 4 $V = a\,b\,c$

c 5 $A_o = 2(a\,b + a\,c + b\,c)$

c 6 $d = \sqrt{a^2 + b^2 + c^2}$

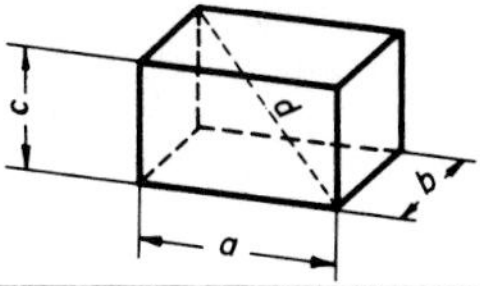

parallelepiped

c 7 $V = A_1\,h$

(Cavalieri principle)

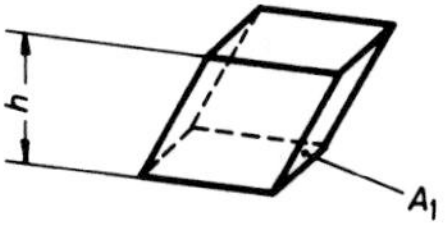

pyramid

c 8 $V = \dfrac{A_1\,h}{3}$

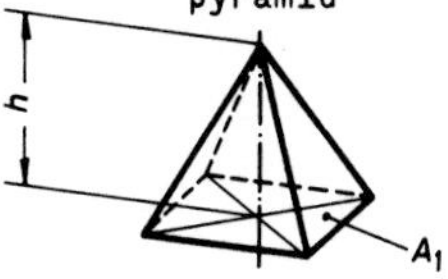

frustum of pyramid

c 9 $V = \dfrac{h}{3}\left(A_1 + A_2 + \sqrt{A_1\,A_2}\right)$

c10 $\approx h\,\dfrac{A_1 + A_2}{2}$

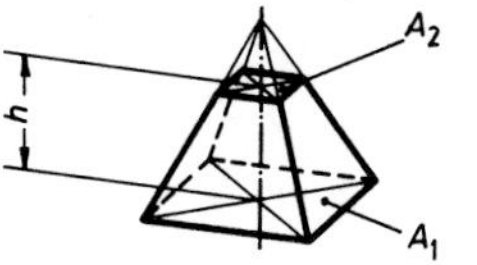

cylinder

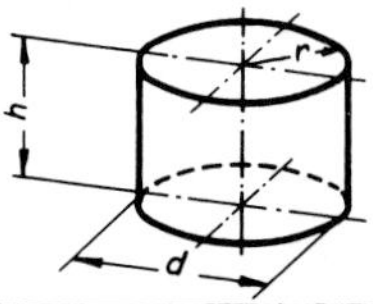

c 11 $$V = \frac{\pi}{4} d^2 h$$

c 12 $$A_m = 2 \pi r h$$

c 13 $$A_o = 2 \pi r (r + h)$$

hollow cylinder

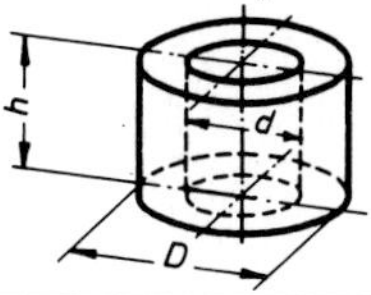

c 14 $$V = \frac{\pi}{4} h (D^2 - d^2)$$

cone

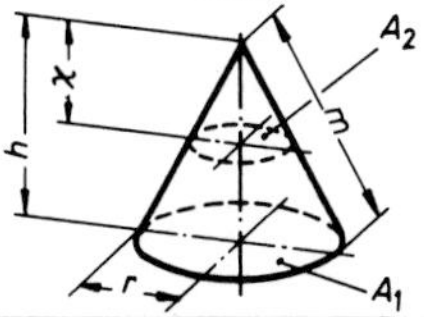

c 15 $$V = \frac{\pi}{3} r^2 h$$

c 16 $$A_m = \pi r m$$

c 17 $$A_o = \pi r (r + m)$$

c 18 $$m = \sqrt{h^2 + r^2}$$

c 19 $$A_2 : A_1 = x^2 : h^2$$

frustum of cone

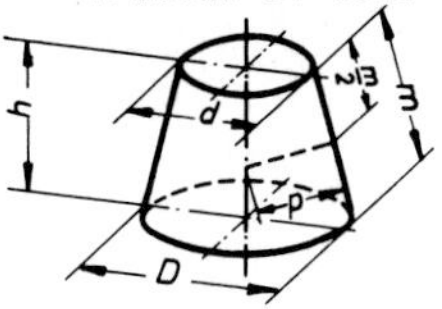

c 20 $$V = \frac{\pi}{12} h (D^2 + Dd + d^2)$$

c 21 $$A_m = \frac{\pi}{2} m (D + d) = 2 \pi p h$$

c 22 $$m = \sqrt{\left(\frac{D - d}{2}\right)^2 + h^2}$$

sphere

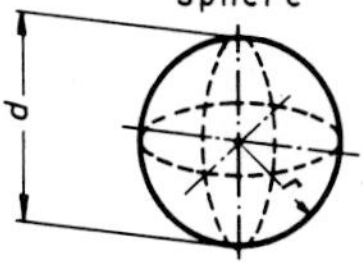

c 23 $$V = \frac{4}{3} \pi r^3 = \frac{1}{6} \pi d^3$$

c 24 $$\approx 4 \cdot 189 r^3$$

c 25 $$A_o = 4 \pi r^2 = \pi d^2$$

SOLID BODIES

zone of a sphere

c 26 $V = \frac{\pi}{6} h \left(3a^2 + 3b^2 + h^2\right)$

c 27 $A_m = 2 \pi r h$

c 28 $A_o = \pi \left(2 r h + a^2 + b^2\right)$

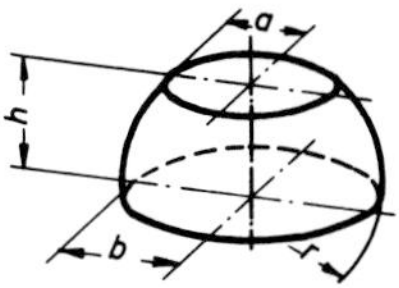

segment of a sphere

c 29 $V = \frac{\pi}{6} h \left(\frac{3}{4} s^2 + h^2\right)$

$= \pi h^2 \left(r - \frac{h}{3}\right)$

c 30 $A_m = 2 \pi r h$

c 31 $= \frac{\pi}{4} \left(s^2 + 4 h^2\right)$

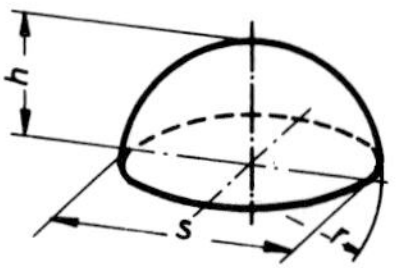

sector of a sphere

c 32 $V = \frac{2}{3} \pi r^2 h$

c 33 $A_o = \frac{\pi}{2} r \left(4 h + s\right)$

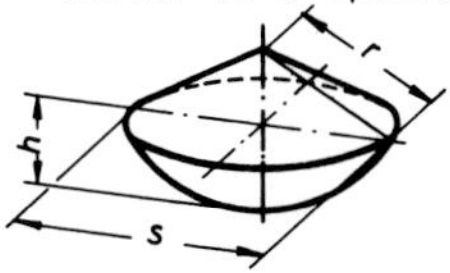

sphere with cylindrical boring

c 34 $V = \frac{\pi}{6} h^3$

c 35 $A_o = 2 \pi h \left(R + r\right)$

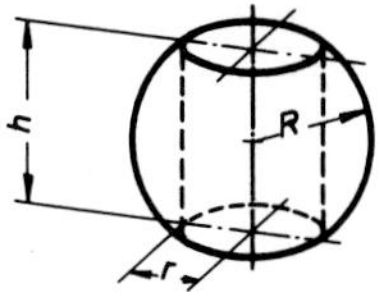

sphere with conical boring

c 36 $V = \frac{2}{3} \pi r^2 h$

c 37 $A_o = 2 \pi r \left(h + \sqrt{r^2 - \frac{h^2}{4}}\right)$

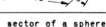

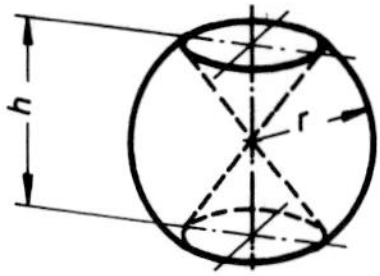

SOLID BODIES

torus

c 38 $V = \frac{\pi^2}{4} D d^2$

c 39 $A_o = \pi^2 D d$

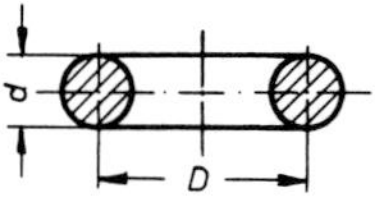

sliced cylinder

c 40 $V = \frac{\pi}{4} d^2 h$

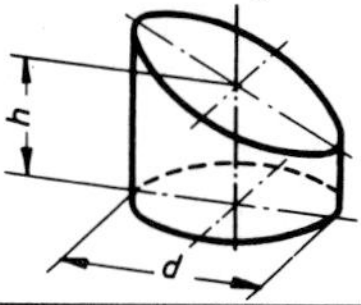

ungula

c 41 $V = \frac{2}{3} r^2 h$

c 42 $A_m = 2 r h$

c 43 $A_o = A_m + \frac{\pi}{2} r^2 + \frac{\pi}{2} r \sqrt{r^2 + h^2}$

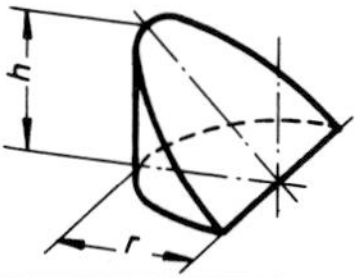

barrel

c 44 $V = \frac{\pi}{12} h (2 D^2 + d^2)$

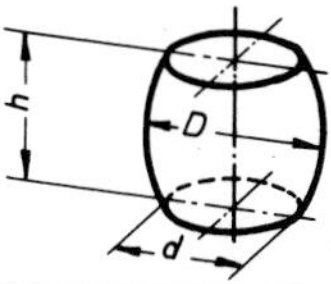

prismoid

c 45 $V = \frac{h}{6} (A_1 + A_2 + 4 A)$

This formula may be used for calculations involving solids shown in fig. C1...C3 and thus spheres and parts of spheres.

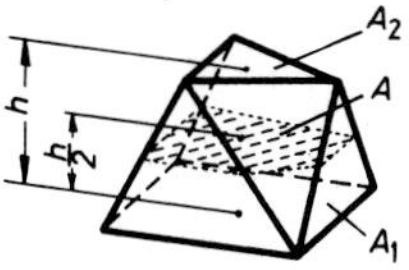

ARITHMETIC

Powers, Roots

D 1

Rules for powers and roots

	general	numerical examples
d 1	$p\,a^n \pm q\,a^n = (p \pm q)a^n$	$3a^4 + 4a^4 = 7a^4$
d 2	$a^m\,a^n = a^{m+n}$	$a^8\,a^4 = a^{12}$
d 3	$\dfrac{a^m}{a^n} = a^{m-n}$	$\dfrac{a^8}{a^2} = a^{8-2} = a^6$
d 4	$(a^m)^n = (a^n)^m = a^{mn}$	$(a^3)^2 = (a^2)^3 = a^{2\times 3} = a^6$
d 5	$a^{-n} = \dfrac{1}{a^n}$	$a^{-4} = \dfrac{1}{a^4}$
d 6	$\dfrac{a^n}{b^n} = \left(\dfrac{a}{b}\right)^n$	$\dfrac{a^3}{b^3} = \left(\dfrac{a}{b}\right)^3$
d 7	$p\sqrt[n]{a} \pm q\sqrt[n]{a} = (p \pm q)\sqrt[n]{a}$	$4\sqrt[3]{x} + 7\sqrt[3]{x} = 11\sqrt[3]{x}$
d 8	$\sqrt[n]{a\,b} = \sqrt[n]{a}\,\sqrt[n]{b}$	$\sqrt[4]{16\times 81} = \sqrt[4]{16}\times\sqrt[4]{81}$
d 9	$\dfrac{\sqrt[n]{a}}{\sqrt[n]{b}} = \sqrt[n]{\dfrac{a}{b}} = \left(\dfrac{a}{b}\right)^{\frac{1}{n}}$	$\dfrac{\sqrt{8}}{\sqrt{2}} = \sqrt{4} = 2$
d10	$\sqrt[nx]{a^{mx}} = \sqrt[n]{a^m}$	$\sqrt[6]{a^8} = \sqrt[3]{a^4}$
d11	$\sqrt[n]{a^m} = \left(\sqrt[n]{a}\right)^m = a^{\frac{m}{n}}$ +)	$\sqrt[4]{a^3} = \left(\sqrt[4]{a}\right)^3 = a^{\frac{3}{4}}$
d12	$\sqrt{-a} = i\sqrt{a}$	$\sqrt{-9} = i\sqrt{9} = i\,3$

+) Not applicable to special calculations, e.g. $\sqrt{(-2)^2} = \pm 2$ or $(\sqrt{-2})^2 = -2$

<u>Note:</u> Exponents of powers and roots have to be non-dimensional quantities!

ARITHMETIC
Logarithms
D 2

General

	system	log to the base of	terminology
d 13	$\log_a$	a	log to base a
d 14	$\log_{10} = \lg$	10	common log
d 15	$\log_e = \ln$	e	natural log
d 16	$\log_2 = \mathrm{ld}$	2	log to base 2

The symbols in $\log_a x = b$ are called: a base
x antilogarithm
b logarithm(log)

Rules for logarithmic calculations

d 17 $$\log_a (x\,y) = \log_a x + \log_a y$$

d 18 $$\log_a \frac{x}{y} = \log_a x - \log_a y$$

d 19 $$\log_a x^n = n \times \log_a x$$

d 20 $$\log_a \sqrt[n]{x} = \frac{1}{n} \log_a x$$

Exponential equation

d 21 $$a^x = b = e^{x \ln a}$$

d 22 whens: $$x = \frac{\log b}{\log a} \quad \Big| \quad a = \sqrt[x]{b}$$

Conversion of logarithms

d 23 $$\lg x = \lg e \times \ln x = 0{\cdot}434\,294 \times \ln x$$

d 24 $$\ln x = \frac{\lg x}{\lg e} = 2{\cdot}302\,585 \times \lg x$$

d 25 $$\mathrm{ld}\, x = 1{\cdot}442\,695 \times \ln x = 3{\cdot}321\,928 \times \lg x$$

Base of the natural logs $e = 2{\cdot}718\,281\,83\ldots$

Key to common logarithm of a number

d 26 $\lg 0{\cdot}01 = -2 = 8 - 10$
d 27 $\lg 0{\cdot}1 = -1 = 9 - 10$
d 28 $\lg 1 = 0$
d 29 $\lg 10 = 1$
d 30 $\lg 100 = 2$

etc.

Note: The antilogarithm always has to be a non-dimensional quantity.

Expansion of general algebraic expressions

d 31 $(a \pm b)^2 = a^2 \pm 2ab + b^2$

d 32 $(a \pm b)^3 = a^3 \pm 3a^2 b + 3ab^2 \pm b^3$

d 33 $(a + b)^n = a^n + \frac{n}{1} a^{n-1} b + \frac{n(n-1)}{1 \times 2} a^{n-2} b^2 + \frac{n(n-1)(n-2)}{1 \times 2 \times 3} a^{n-3} b^3 + \dots b^n$

d 34 $(a + b + c)^2 = a^2 + 2ab + 2ac + b^2 + 2bc + c^2$

d 35 $(a - b + c)^2 = a^2 - 2ab + 2ac + b^2 - 2bc + c^2$

d 36 $a^2 - b^2 = (a + b)(a - b)$

d 37 $a^3 + b^3 = (a + b)(a^2 - ab + b^2)$

d 38 $a^3 - b^3 = (a - b)(a^2 + ab + b^2)$

d 39 $a^n - b^n = (a - b)(a^{n-1} + a^{n-2} b + a^{n-3} b^2 + \dots + ab^{n-2} + b^{n-1})$

Quadratic equation (equation of the second degree)

d 40 Normal form $ax^2 + bx + c = 0$

d 41 Solutions $x_1; \; x_2 = \frac{-b \pm \sqrt{b^2 - 4ac}}{2a}$

d 42 Vieta's rule $x_1 + x_2 = -\frac{b}{a}; \quad x_1 x_2 = \frac{c}{a}$

Arithmetical calculation of a square root

example	explanation, ()-values corresponding to example
$\sqrt{21\,43{\cdot}69} = 46{\cdot}3$ 16 5 43 : 86 5 16 27 69 : 92 3 27 69 3 == ==	a) Divide number into pairs of figures starting at the decimal point. (If the number of figures is odd, the left hand digit counts as a pair). b) Calculate the highest square contained in the first pair (21) which is (16). The square root (4) is the first figure of the answer. Subtract (16) from (21) lea-

ving (5) and being down the next 2 figures to give (543).

c) Double the first figure of the answer ($2 \times 4 = 8$) and write it in the position shown. Divide it into the first 2 figures of the remainder ($54 \div 8 = 6$). Write (6) beside the (8) to give (86). Multiply (86) by (6) to give (516). Subtract (516) from (543) leaving new remainder (27).

d) Bring down next pair (69) and repeat.

Iterative calculation of an nth root

d 43 When $x = \sqrt[n]{A}$, then $x = \frac{1}{n}\left[(n-1)x_0 + \frac{A}{x_0^{\,n-1}}\right]$,

where x_0 is the initially estimated value of x. Repeatedly inserting the obtained x as a new value of x_0 gradually increases the accuracy of x.

Binomial theorem

d 44 $$(a+b)^n = \binom{n}{0}a^n + \binom{n}{1}a^{n-1}\ b + \binom{n}{2}a^{n-2}\ b^2 + \binom{n}{3}a^{n-3}\ b^3 + \dots$$

n must be a whole number.

d 45 $$\binom{n}{k} = \frac{n(n-1)(n-2)\ \dots\ (n-k+1)}{1\times2\times3\ \dots\ k}$$

n must be a whole number.

d 46 $$(a+b)^4 = 1a^4 + \frac{4}{1}a^{4-1}\ b + \frac{4\times3}{1\times2}a^{4-2}\ b^2 + \frac{4\times3\times2}{1\times2\times3}a^{4-3}\ b^3 + b^4$$

$$= a^4 + 4a^3\ b + 6a^2\ b^2 + 4a\ b^3 + b^4$$

Diagrammatic solution

Coefficient - Pascal triangle

d 47	$(a+b)^0$							1						
d 48	$(a+b)^1$						1		1					
d 49	$(a+b)^2$					1		2		1				
d 50	$(a+b)^3$				1		3		3		1			
d 51	$(a+b)^4$			1		4		6 ↓		4		1		
d 52	$(a+b)^5$		1		5 ↓		10		10		5		1	
d 53	$(a+b)^6$	1		6		15		20		15		6		1

Continue with each line starting and finishing with 1. The second and penultimate numbers should be the exponents, the others the sum of those to the right and left immediately above them.

Exponents

The sum of the exponents a and b in each separate term is equal to the binomial exponent n. As the power of a decreases the power of b increases.

Signs

$(a+b)$ is always positive

$(a-b)$ is initially positive and changes from term to term

d 54 ### Examples

$$(a+b)^5 = a^5 + 5a^4 b + 10a^3 b^2 + 10a^2 b^3 + 5ab^4 + b^5$$
$$(a-b)^5 = +a^5 - 5a^4 b + 10a^3 b^2 - 10a^2 b^3 + 5ab^4 - b^5$$

Permutations

An ordered selection or arrangement of r out of n things is called a "permutation" of the n things taken r at a time. The number of these permutations is denoted by:

d 55 $$P_r^n = n(n-1)(n-2)\dots(n-r+1), \quad n \geqq r$$

If $r = n$, this becomes

d 56 $$P_n^n = P_n = n(n-1)(n-2)\dots 1 = n!^{*)}$$

Example: The $n = 3$ things a, b, c can be permutated with each other (i.e. 3 at a time) in the followings 6 ways:

abc *bac* *cab*
acb *bca* *cba*. Here $r = n = 3$.

$$P_3 = 3! = 1 \times 2 \times 3 = 6.$$

Special case: The number of permutations of n things taken all together incorporating n_1 of one sort, n_2 of another sort and n_r of a rth sort is:

d 57 $$P_{n,r} = \frac{n!}{n_1! \times n_2! \times \dots n_r!}$$

Example: The $n = 3$ things a, a, b can be permutated the followings 3 ways:

aab *aba* *baa*. Here $n = 3$, $n_1 = 2$, $n_2 = 1$.

$$P_{3,2} = \frac{3!}{2! \times 1!} = \frac{1 \times 2 \times 3}{1 \times 2 \times 1} = 3.$$

Combinations

A selection of r out n things without regard to order is called "combination" of n things taken r at a time. The number of these combinations is denoted by

d 58 $$C_r^n = \frac{n!}{r!(n-r)!} = \binom{n}{r}^{**)}$$

Example: The $n = 3$ things a, b, c taken together give only the one combination *abc*. Here $n = 3$, $r = 3$.

Hence $$C_3^3 = \binom{3}{3} = \frac{3 \times 2 \times 1}{1 \times 2 \times 3} = 1.$$

The table on page D 6 compares combinations and permutations (with and without the things repeating).

*) $n!$ is pronounced "n factorial"
**) Symbol usual for binomial coefficient (see d 44)

Combinations and Permutations

(Explanations see D 5)

		d 62	d 61	d 60	d 59
		Number of combinations without repeating regardless of the positions of things	Number of combinations with repeating regardless of the positions of things	Number of permutations without repeating taking into account the positions of things	Number of permutations with repeating taking into account the positions of things
Formula		$C_r^n = \frac{n!}{r!(n-r)!} = \binom{n}{r}^{*)}$ $= \frac{n(n-1)\cdots(n-r+1)}{r!}$	${}^wC_r^n = \binom{n+r-1}{r}^{*)}$ $= \frac{n(n+1)\cdots(n+r-1)}{r!}$	$P_r^n = \frac{n!}{(n-r)!} = \binom{n}{r}r!^{*)}$ $= n(n-1)\cdots(n-r+1)$	${}^wP_r^n = n^r$
Explanation of symbols		C : Number of possible combinations n : Number of things given r : Number of things selected from n given things		P : Number of possible permutations	
Examples	Given	$n = 3$ things a, b, c	$r = 2$ things selected from the 3 given things		
	Possibilities	· ab ac · · bc · · ·	aa ab ac · bb bc · · cc	· ab ac ba · bc ca cb ·	aa ab ac ba bb bc ca cb cc
	Calculation of the number of possibilities	$C_2^3 = \frac{3!}{2!(3-2)!}$ $= \binom{3}{2} = \frac{3 \times 2}{1 \times 2} = 3$	${}^wC_2^3 = \binom{3+2-1}{2}$ $= \binom{4}{2}$ $= \frac{4 \times 3}{1 \times 2} = 6$	$P_2^3 = \binom{3}{2}2!$ $= \frac{3!}{(3-2)!} = \frac{3!}{1!}$ $= \frac{1 \times 2 \times 3}{1} = 6$	${}^wP_2^3 = 3^2$ $= 9$
	Note	ab and ba for example are the same combination		ab and ba for example are different permutations	

w : with repeating

*) calculation according to d 45

Determinants and linear equations

Second order determinants

d 63 $$\begin{aligned} a_{11}\,x + a_{12}\,y &= r_1 \\ a_{21}\,x + a_{22}\,y &= r_2 \end{aligned} \qquad D = \begin{vmatrix} a_{11} & a_{12} \\ a_{21} & a_{22} \end{vmatrix} = a_{11}\,a_{22} - a_{21}\,a_{12}$$

insert r column in place of

column x	column y
d 64 $D_1 = \begin{vmatrix} r_1 & a_{12} \\ r_2 & a_{22} \end{vmatrix} = r_1\,a_{22} - r_2\,a_{12}$	$D_2 = \begin{vmatrix} a_{11} & r_1 \\ a_{21} & r_2 \end{vmatrix} = r_2\,a_{11} - r_1\,a_{21}$
$x = \dfrac{D_1}{D}$	$y = \dfrac{D_2}{D}$

Third order determinants (Sarrus rule)

d 65 $$\begin{aligned} a_{11}\,x + a_{12}\,y + a_{13}\,z &= r_1 \\ a_{21}\,x + a_{22}\,y + a_{23}\,z &= r_2 \\ a_{31}\,x + a_{32}\,y + a_{33}\,z &= r_3 \end{aligned}$$

d 66 $$D = \left|\begin{matrix} a_{11} & a_{12} & a_{13} \\ a_{21} & a_{22} & a_{23} \\ a_{31} & a_{32} & a_{33} \end{matrix}\right|\begin{matrix} a_{11} & a_{12} \\ a_{21} & a_{22} \\ a_{31} & a_{32} \end{matrix}\right| \begin{aligned} = {} & a_{11}\,a_{22}\,a_{33} + a_{12}\,a_{23}\,a_{31} \\ & + a_{13}\,a_{21}\,a_{32} - a_{13}\,a_{22}\,a_{31} \\ & - a_{11}\,a_{23}\,a_{32} - a_{12}\,a_{21}\,a_{33} \end{aligned}$$

insert r column for x column:

d 67 $$D_1 = \left|\begin{matrix} r_1 & a_{12} & a_{13} \\ r_2 & a_{22} & a_{23} \\ r_3 & a_{32} & a_{33} \end{matrix}\right|\begin{matrix} r_1 & a_{12} \\ r_2 & a_{22} \\ r_3 & a_{32} \end{matrix}\right| \begin{aligned} = {} & r_1\,a_{22}\,a_{33} + a_{12}\,a_{23}\,r_3 \\ & + a_{13}\,r_2\,a_{32} - a_{13}\,a_{22}\,r_3 \\ & - r_1\,a_{23}\,a_{32} - a_{12}\,r_2\,a_{33} \end{aligned}$$

determine D_2 and D_3 similarly by replacing the y- and z-column by the r-column:

d 68 $$x = \frac{D_1}{D} \qquad y = \frac{D_2}{D} \qquad z = \frac{D_3}{D}$$

continued on D 8

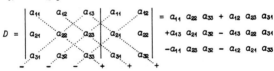

Determinants of more than the 2nd order:

(The Sarrus Rule, see D 7, may be used for determinants of higher order than the 3rd).

By adding or substracting suitable multiples of two rows or columns, endeavour to obtain zero values. Expand the determinant starting from the row or column containing most zeros. Alternate the signs of terms, starting with a_{11} as +.

Example:

d 69

$$\begin{array}{cccc} a_{11}^{+} & a_{12}^{-} & a_{13}^{+} & 0^{-} \\ a_{21} & a_{22} & a_{23} & a_{24}^{+} \\ a_{31} & a_{32} & a_{33} & a_{34}^{-} \\ a_{41} & a_{42} & a_{43} & 0^{+} \end{array}$$

Expand on 4th column:

d 70

$$a_{24}\begin{vmatrix} a_{11}^{+} & a_{12}^{-} & a_{13}^{+} \\ a_{31} & a_{32} & a_{33} \\ a_{41} & a_{42} & a_{43} \end{vmatrix} \qquad -a_{34}\begin{vmatrix} a_{11}^{+} & a_{12}^{-} & a_{13}^{+} \\ a_{21} & a_{22} & a_{23} \\ a_{41} & a_{42} & a_{43} \end{vmatrix}$$

Further expand as:

d 71

$$D = a_{24}\left(a_{11}\begin{vmatrix} a_{32} & a_{33} \\ a_{42} & a_{43} \end{vmatrix} - a_{12}\begin{vmatrix} a_{31} & a_{33} \\ a_{41} & a_{43} \end{vmatrix} + a_{13}\begin{vmatrix} a_{31} & a_{32} \\ a_{41} & a_{42} \end{vmatrix}\right) - a_{34}\left(\cdots\right)$$

To form the determinants D_1, D_2, ... (see D 5) substitute the r column for the first, second,... column of D, and evaluate in the same way as for D.

For determinant of the nth order, find $u_{1\ldots n}$ from the formulae:

d 72

$$u_1 = \frac{D_1}{D}, \quad u_2 = \frac{D_2}{D}, \quad \ldots \; u_n = \frac{D_n}{D}$$

Note: For determinants of the nth order continue until determinants of the 3rd order have been obtained.

Arithmetic series

The sequence 1, 4, 7, 10 etc is called an arithmetic series. (The difference between two consecutive terms is constant).

Formulae:

d 73 $$a_n = a_1 + (n-1)\,d$$

d 74 $$s_n = \frac{n}{2}(a_1 + a_n) = a_1\,n + \frac{n(n-1)\,d}{2}$$

Arithmetic mean:

Each term of an arithmetic series is the arithmetic mean a_m of its adjacent terms a_{m-1} and a_{m+1}.

Thus, the mth term is

d 75 $$a_m = \frac{a_{m-1} + a_{m+1}}{2} \qquad \text{for} \quad 1 < m < n$$

(e.g. in the above series $a_3 = \frac{4+10}{2} = 7$)

Geometric series

The sequence 1, 2, 4, 8 etc is called a geometric series. (The quotient of two consecutive terms is constant).

Formulae:

d 76 $$a_n = a_1\,q^{n-1}$$

d 77 $$s_n = a_1\,\frac{q^n - 1}{q - 1} = \frac{q \times a_n - a_1}{q - 1}$$

Geometric mean:

Each term of a geometric series is the geometric mean a_m of its adjacent terms a_{m-1} and a_{m+1}.

Thus, the mth term is

d 78 $$a_m = \sqrt{a_{m-1}\;a_{m+1}} \qquad \text{for} \quad 1 < m < n$$

(e.g. in the above series $a_3 = \sqrt{2 \times 8} = 4$)

For infinite geometric series ($n \to \infty$; $|q| < 1$) the following statements apply

d 79 $$a_n = \lim_{n \to \infty} a_n = 0\,; \qquad s_n = \lim_{n \to \infty} s_n = a_1 \frac{1}{1-q}$$

a_1 : initial term
a_n : final term
d : difference between two consecutive terms
n : number of terms
s_n : sum to n terms
q : quotient of two consecutive terms

Binomial series

d 80 $$f(x) = (1 \pm x)^{\alpha} = 1 \pm \binom{\alpha}{1}x + \binom{\alpha}{2}x^2 \pm \binom{\alpha}{3}x^3 + \dots$$

α may be either positive or negative, a whole number or a fraction.

Expansion of the binomial coefficient:

$$\binom{\alpha}{n} = \frac{\alpha(\alpha-1)(\alpha-2)(\alpha-3)\dots(\alpha-n+1)}{1\times2\times3\dots\times n}$$

Examples:

		for
d 81	$\frac{1}{1 \pm x} = (1 \pm x)^{-1} = 1 \mp x + x^2 \mp x^3 + \dots$	$\lvert x\rvert<1$
d 82	$\sqrt{1 \pm x} = (1 \pm x)^{\frac{1}{2}} = 1 \pm \frac{1}{2}x - \frac{1}{8}x^2 \pm \frac{1}{16}x^3 - \dots$	$\lvert x\rvert<1$
d 83	$\frac{1}{\sqrt{1 \pm x}} = (1 \pm x)^{-\frac{1}{2}} = 1 \mp \frac{1}{2}x + \frac{3}{8}x^2 \mp \frac{5}{16}x^3 + \dots$	$\lvert x\rvert<1$

Taylor series

d 84 $$f(x) = f(a) + \frac{f'(a)}{1!}(x-a) + \frac{f''(a)}{2!}(x-a)^2 + \dots$$

putting $a=0$ gives the MacLaurin series:

d 85 $$f(x) = f(0) + \frac{f'(0)}{1!}x + \frac{f''(0)}{2!}x^2 + \dots$$

Examples:

		for
d 86	$e^x = 1 + \frac{x}{1!} + \frac{x^2}{2!} + \frac{x^3}{3!} + \dots$	all x
d 87	$a^x = 1 + \frac{x \ln a}{1!} + \frac{(x \ln a)^2}{2!} + \frac{(x \ln a)^3}{3!} + \dots$	all x
d 88	$\ln x = 2\left[\frac{x-1}{x+1} + \frac{1}{3}\left(\frac{x-1}{x+1}\right)^3 + \frac{1}{5}\left(\frac{x-1}{x+1}\right)^5 + \dots\right]$	$x>0$
d 89	$\ln(1+x) = x - \frac{x^2}{2} + \frac{x^3}{3} - \frac{x^4}{4} + \frac{x^5}{5} - \dots$	$-1<x$, $x \leq +1$

d 90 $$\ln 2 = 1 - \frac{1}{2} + \frac{1}{3} - \frac{1}{4} + \frac{1}{5} - \dots$$

continued on D 11

Taylor series (continued)

Examples		for
d 91	$\sin x = x - \frac{x^3}{3!} + \frac{x^5}{5!} - \frac{x^7}{7!} + \dots$	all x
d 92	$\cos x = 1 - \frac{x^2}{2!} + \frac{x^4}{4!} - \frac{x^6}{6!} + \dots$	all x
d 93	$\tan x = x + \frac{1}{3}x^3 + \frac{2}{15}x^5 + \frac{17}{315}x^7 + \dots$	$\lvert x\rvert < \frac{\pi}{2}$
d 94	$\cot x = \frac{1}{x} - \frac{1}{3}x - \frac{1}{45}x^3 - \frac{2}{945}x^5 - \dots$	$0 < \lvert x\rvert$ $\lvert x\rvert < \pi$
d 95	$\operatorname{arcsin} x = x + \frac{1}{2}\frac{x^3}{3} + \frac{1\times3}{2\times4}\frac{x^5}{5} + \frac{1\times3\times5}{2\times4\times6}\frac{x^7}{7} + \dots$	$\lvert x\rvert \leqq 1$
d 96	$\operatorname{arccos} x = \frac{\pi}{2} - \operatorname{arcsin} x$	$\lvert x\rvert \leqq 1$
d 97	$\operatorname{arctan} x = x - \frac{x^3}{3} + \frac{x^5}{5} - \frac{x^7}{7} + \frac{x^9}{9} - \dots$	$\lvert x\rvert \leqq 1$
d 98	$\operatorname{arccot} x = \frac{\pi}{2} - \operatorname{arctan} x$	$\lvert x\rvert \leqq 1$
d 99	$\sinh x = x + \frac{x^3}{3!} + \frac{x^5}{5!} + \frac{x^7}{7!} + \frac{x^9}{9!} + \dots$	all x
d100	$\cosh x = 1 + \frac{x^2}{2!} + \frac{x^4}{4!} + \frac{x^6}{6!} + \frac{x^8}{8!} + \dots$	all x
d101	$\tanh x = x - \frac{1}{3}x^3 + \frac{2}{15}x^5 - \frac{17}{315}x^7 + \dots$	$\lvert x\rvert < \frac{\pi}{2}$
d102	$\coth x = \frac{1}{x} + \frac{1}{3}x - \frac{1}{45}x^3 + \frac{2}{945}x^5 - \dots$	$0 < \lvert x\rvert$ $\lvert x\rvert < \pi$
d103	$\operatorname{arsinh} x = x - \frac{1}{2}\frac{x^3}{3} + \frac{1\times3}{2\times4}\frac{x^5}{5} - \frac{1\times3\times5}{2\times4\times6}\frac{x^7}{7} + \dots$	$\lvert x\rvert < 1$
d104	$\operatorname{arcosh} x = \ln 2x - \frac{1}{2}\frac{1}{2x^2} - \frac{1\times3}{2\times4}\frac{1}{4x^4} - \frac{1\times3\times5}{2\times4\times6}\frac{1}{6x^6} + \dots$	$\lvert x\rvert > 1$
d105	$\operatorname{artanh} x = x + \frac{x^3}{3} + \frac{x^5}{5} + \frac{x^7}{7} + \frac{x^9}{9} + \dots$	$\lvert x\rvert < 1$
d106	$\operatorname{arcoth} x = \frac{1}{x} + \frac{1}{3x^3} + \frac{1}{5x^5} + \frac{1}{7x^7} + \dots$	$\lvert x\rvert > 1$

Fourier series

General: Each periodic function $f(x)$, which can be subdivided into a finite number of continuous intervals within it's period $-\pi \leqq x \leqq \pi$, may be expanded in this interval into convergent series of the following form ($x = \omega t$):

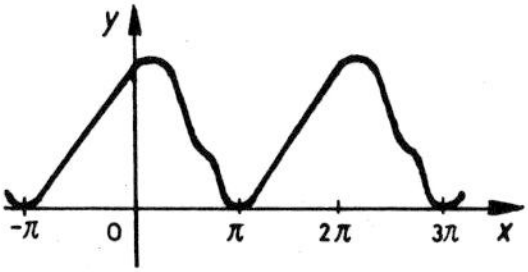

d107 $$f(x) = \frac{a_0}{2} + \sum_{n=1}^{\infty} \left[a_n \cos(nx) + b_n \sin(nx)\right]$$

The various coefficients can be calculated by:

d108 $$a_k = \frac{1}{\pi}\int_{-\pi}^{\pi} f(x)\cos(kx)\,dx \quad\Big|\quad b_k = \frac{1}{\pi}\int_{-\pi}^{\pi} f(x)\sin(kx)\,dx$$

with the index $k = 0, 1, 2, \ldots$

Simplified calculation of coeff. for symmetr. waveforms:

Even function: $f(x) = f(-x)$

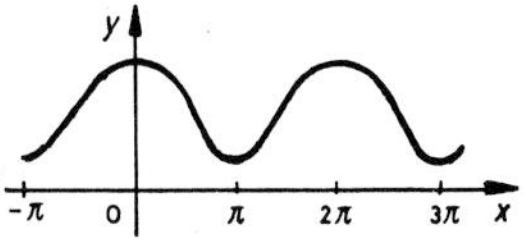

d109 $$a_k = \frac{2}{\pi}\int_{0}^{\pi} f(x)\cos(kx)\,dx$$

with the index $k = 0, 1, 2 \ldots$

d110 $b_k = 0$

Odd function: $f(x) = -f(-x)$

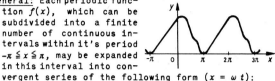

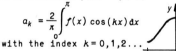

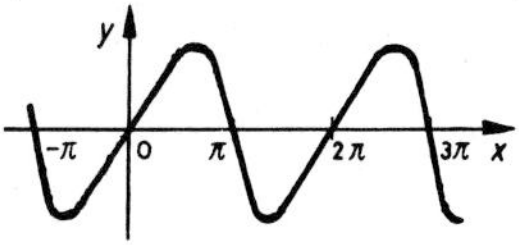

d111 $a_k = 0$

d112 $$b_k = \frac{2}{\pi}\int_{0}^{\pi} f(x)\sin(kx)\,dx$$

with the index $k = 0, 1, 2 \ldots$

	Even-harmonic functions	Odd-harmonic functions
d113	$f(x) = f(-x)$ and	$f(x) = -f(\ x)$ and
d114	$f(\frac{\pi}{2} + x) = -f(\frac{\pi}{2} - x)$ give:	$f(\frac{\pi}{2} + x) = -f(\frac{\pi}{2} - x)$ give:
d115	$a_k = \frac{4}{\pi}\int_{0}^{\pi/2} f(x)\cos(kx)\,dx$ for $k = 1, 3, 5, \ldots$	$b_k = \frac{4}{\pi}\int_{0}^{\pi/2} f(x)\sin(kx)\,dx$ for $k = 1, 3, 5, \ldots$
d116	$a_k = 0$ for $k = 0, 2, 4, \ldots$	$a_k = 0$ for $k = 0, 1, 2, \ldots$
d117	$b_k = 0$ for $k = 1, 2, 3, \ldots$	$b_k = 0$ for $k = 2, 4, 6, \ldots$

Table of Fourier expansions

d118 $y = a \quad \text{for} \quad 0 < x < \pi$

d119 $y = -a \quad \text{for} \quad \pi < x < 2\pi$

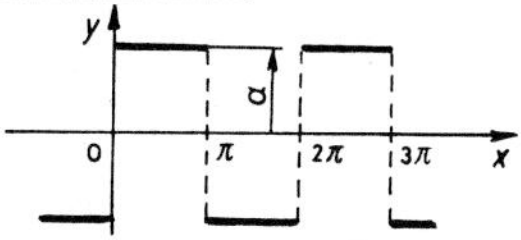

d120 $$y = \frac{4a}{\pi}\left[\sin x + \frac{\sin(3x)}{3} + \frac{\sin(5x)}{5} + \dots\right]$$

d121 $y = a \quad \text{for} \quad \alpha < x < \pi - \alpha$

d122 $y = -a \quad \text{for} \quad \pi + \alpha < x < 2\pi - \alpha$

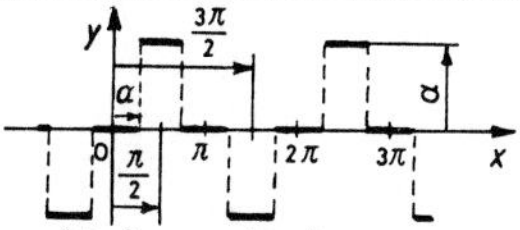

d123 $$y = \frac{4a}{\pi}\left[\cos\alpha \sin x + \frac{1}{3}\cos(3\alpha)\sin(3x) + \frac{1}{5}\cos(5\alpha)\sin(5x) + \dots\right]$$

d124 $y = a \quad \text{for} \quad \alpha < x < 2\pi - \alpha$

d125 $y = f(2\pi + x)$

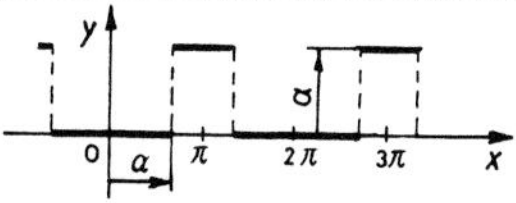

d126 $$y = \frac{2a}{\pi}\left[\frac{\pi-\alpha}{2} - \frac{\sin(\pi-\alpha)}{1}\cos x + \frac{\sin 2(\pi-\alpha)}{2}\cos(2x) - \frac{\sin 3(\pi-\alpha)}{3}\cos(3x) + \dots\right]$$

d127 $y = ax/b \quad \text{for} \quad 0 \leqq x \leqq b$

d128 $y = a \quad \text{for} \quad b \leqq x \leqq \pi - b$

d129 $y = a(\pi - x)/b \quad \text{for} \quad \pi - b \leqq x \leqq \pi$

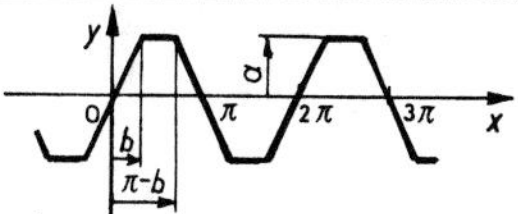

d130 $$y = \frac{4}{\pi}\frac{a}{b}\left[\frac{1}{1^2}\sin b \sin x + \frac{1}{3^2}\sin(3b)\sin(3x) + \frac{1}{5^2}\sin(5b)\sin(5x) + \dots\right]$$

d131 $y = \dfrac{ax}{2\pi} \quad \text{for} \quad 0 < x < 2\pi$

d132 $y = f(2\pi + x)$

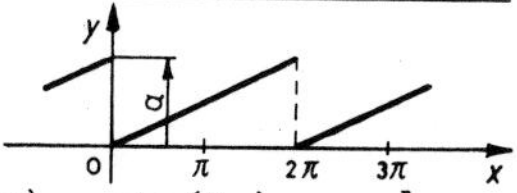

d133 $$y = \frac{a}{2} - \frac{a}{\pi}\left[\frac{\sin x}{1} + \frac{\sin(2x)}{2} + \frac{\sin(3x)}{3} + \dots\right]$$

continued on D 14

Continuation of D 13

d134 $y = 2ax/\pi$ for $0 \leqq x \leqq \pi/2$
d135 $y = 2a(\pi - x)/\pi$ for $\pi/2 \leqq x \leqq \pi$
d136 $y = -f(\pi + x)$

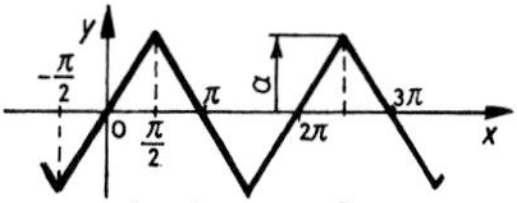

d137 $$y = \frac{8}{\pi^2} a \left[\sin x - \frac{\sin(3x)}{3^2} + \frac{\sin(5x)}{5^2} - \ldots \right]$$

d138 $y = ax/\pi$ for $0 \leqq x \leqq \pi$
d139 $y = a(2\pi - x)/\pi$ for $\pi \leqq x \leqq 2\pi$
d140 $y = f(2\pi + x)$

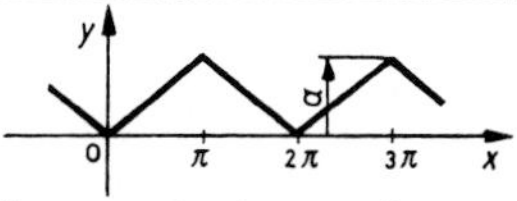

d141 $$y = \frac{a}{2} - \frac{4a}{\pi^2} \left[\frac{\cos x}{1^2} + \frac{\cos(3x)}{3^2} + \frac{\cos(5x)}{5^2} + \ldots \right]$$

d142 $y = a \sin x$ for $0 \leqq x \leqq \pi$
d143 $y = -a \sin x$ for $\pi \leqq x \leqq 2\pi$
d144 $y = f(\pi + x)$

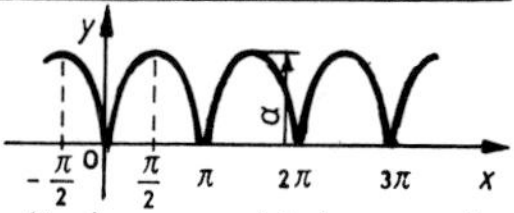

d145 $$y = \frac{2a}{\pi} - \frac{4a}{\pi} \left[\frac{\cos(2x)}{1 \times 3} + \frac{\cos(4x)}{3 \times 5} + \frac{\cos(6x)}{5 \times 7} + \ldots \right]$$

d146 $y = 0$ for $0 \leqq x \leqq \pi/2$
d147 $y = a \sin\left(x - \frac{\pi}{2}\right)$ for $\frac{\pi}{2} \leqq x \leqq \frac{3\pi}{2}$
d148 $y = f(2\pi + x)$

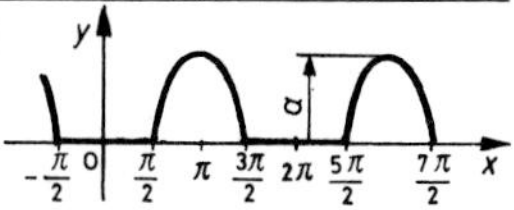

d149 $$y = \frac{2a}{\pi} \left[\frac{1}{2} - \frac{\pi}{4} \cos x + \frac{\cos(2x)}{2^2 - 1} - \frac{\cos(4x)}{4^2 - 1} + \frac{\cos(6x)}{6^2 - 1} - \ldots \right]$$

d150 $y = x^2$ for $-\pi \leqq x \leqq \pi$
d151 $y = f(-x) = f(2\pi + x)$

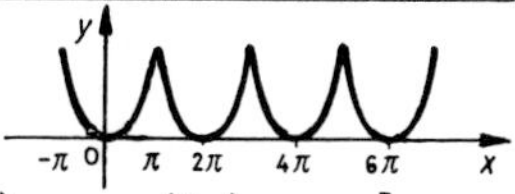

d152 $$y = \frac{\pi^2}{3} - 4 \left[\frac{\cos x}{1^2} - \frac{\cos(2x)}{2^2} + \frac{\cos(3x)}{3^2} - \ldots \right]$$

d153 $y = ax/\pi$ for $0 \leqq x \leqq \pi$
d154 $y = f(2\pi + x)$

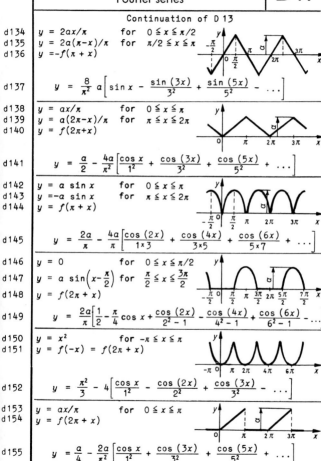

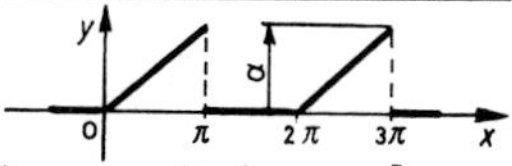

d155 $$y = \frac{a}{4} - \frac{2a}{\pi^2} \left[\frac{\cos x}{1^2} + \frac{\cos(3x)}{3^2} + \frac{\cos(5x)}{5^2} + \ldots \right] + \frac{a}{\pi} \left[\frac{\sin x}{1} - \frac{\sin(2x)}{2} + \frac{\sin(3x)}{3} - \ldots \right]$$

Complex numbers

General

$z = r\,e^{i\varphi} = a + i\,b$

a = real part of z

b = imaginary part of z

r = absolute value of z

= or modulus of z

φ = argument of z

a and b are real

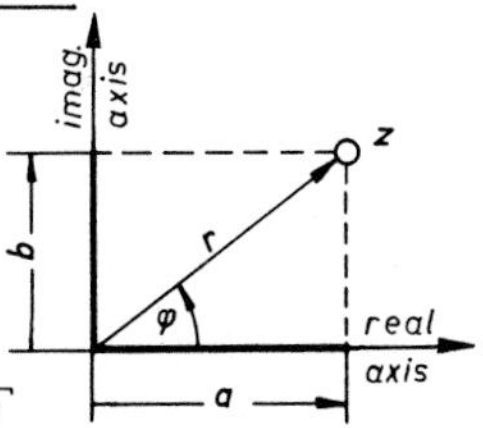

d156 $$i = \sqrt{-1}$$

d157	$i^1 = +i$	$i^{-1} = -i$
d158	$i^2 = -1$	$i^{-2} = -1$
d159	$i^3 = -i$	$i^{-3} = +i$
d160	$i^4 = +1$	$i^{-4} = +1$
d161	$i^5 = +i$	$i^{-5} = -i$

etc.

Note: In electrical engineering the letter j is used for i to avoid confusion.

In the Cartesian coordinate system:

d162 $$z = a + ib$$

d163 $$z_1 + z_2 = (a_1 + a_2) + i(b_1 + b_2)$$

d164 $$z_1 - z_2 = (a_1 - a_2) + i(b_1 - b_2)$$

d165 $$z_1 \times z_2 = (a_1 a_2 - b_1 b_2) + i(a_1 b_2 + a_2 b_1)$$

d166 $$\frac{z_1}{z_2} = \frac{a_1 a_2 + b_1 b_2}{a_2^2 + b_2^2} + i\frac{-a_1 b_2 + a_2 b_1}{a_2^2 + b_2^2}$$

d167 $$a^2 + b^2 = (a + ib)(a - ib)$$

d168 $$\sqrt{a \pm ib} = \sqrt{\frac{a + \sqrt{a^2 + b^2}}{2}} \pm i\sqrt{\frac{-a + \sqrt{a^2 + b^2}}{2}}$$

Where $a_1 = a_2$ and $b_1 = b_2$, then $z_1 = z_2$

continued on D 16

Complex numbers (continued)

In the polar coordinate system:

d169 $$z = r(\cos\varphi + i\sin\varphi) = a + ib$$

d170 $$r = +\sqrt{a^2 + b^2}$$

d171 $$\varphi = \arctan\frac{b}{a}$$

d172 $$\sin\varphi = \frac{b}{r} \quad \Big| \quad \cos\varphi = \frac{a}{r} \quad \Big| \quad \tan\varphi = \frac{b}{a}$$

d173 $$z_1 \times z_2 = r_1 \times r_2\left[\cos(\varphi_1 + \varphi_2) + i\sin(\varphi_1 + \varphi_2)\right]$$

d174 $$\frac{z_1}{z_2} = \frac{r_1}{r_2}\left[\cos(\varphi_1 - \varphi_2) + i\sin(\varphi_1 - \varphi_2)\right] \quad (z_2 \neq 0)$$

d175 $$z^n = r^n\left[\cos(n\varphi) + i\sin(n\varphi)\right] \quad (n > 0 \text{ integral})$$

d176 $$\sqrt[n]{z} = \left|\sqrt[n]{r}\right|\left(\cos\frac{\varphi + 2\pi k}{n} + i\sin\frac{\varphi + 2\pi k}{n}\right)$$

d177 $$\sqrt[n]{1} = \cos\frac{2\pi k}{n} + i\sin\frac{2\pi k}{n} \quad (n\text{th roots of unity})$$

in formula d 176 and d 177 $k = 0, 1, 2, \ldots, n-1$

d178 $$e^{i\widehat{\varphi}} = \cos\varphi + i\sin\varphi$$

d179 $$e^{-i\widehat{\varphi}} = \cos\varphi - i\sin\varphi = \frac{1}{\cos\varphi + i\sin\varphi}$$

d180 $$\left|e^{-i\widehat{\varphi}}\right| = \sqrt{\cos^2\varphi + \sin^2\varphi} = 1$$

d181 $$\cos\varphi = \frac{e^{i\widehat{\varphi}} + e^{-i\widehat{\varphi}}}{2} \quad \Big| \quad \sin\varphi = \frac{e^{i\widehat{\varphi}} - e^{-i\widehat{\varphi}}}{2i}$$

d182 $$\ln z = \ln r + i(\widehat{\varphi} + 2\pi k) \quad (k = 0, \pm 1, \pm 2, \ldots)$$

Where $r_1 = r_2$ and $\varphi_1 = \varphi_2 + 2\pi k$, then $z_1 = z_2$

Note: φ should be measured along the arc
k is any arbitrary whole number

Compound interest calculation

d 183 $$k_n = k_0 \; q^n$$

d 184 $$n = \frac{\lg \dfrac{k_n}{k_0}}{\lg q}$$

$$q = \sqrt[n]{\frac{k_n}{k_0}}$$

Annuity interest calculation

d 185 $$k_n = k_0 \; q^n - r \; q \, \frac{q^n - 1}{q - 1}$$

d 186 $$r = \frac{(k_0 q^n - k_n(q - 1)}{(q^n - 1)q}$$

d 187 $$n = \frac{\lg \dfrac{r \; q - k_n(q - 1)}{r \; q - k_0(q - 1)}}{\lg q}$$

Where $k_n = 0$, we get the „redemption formulae"

Deposit calculation
(savings bank formula)

d 188 $$k_n = k_0 q^n + r \; q \, \frac{q^n - 1}{q - 1}$$

d 189 $$r = \frac{(k_n - k_0 q^n)(q - 1)}{(q^n - 1)q}$$

d 190 $$n = \frac{\lg \dfrac{k_n(q - 1) + r \; q}{k_0(q - 1) + r \; q}}{\lg q}$$

Letters

k_0: initial capital
k_n: capital after n years
r : annual pensions (withdrawals)

n : number of years
q : $1 + p$
p : rate of interest (e.g. 0·06 at 6%)

d191 $$x = \frac{b\ c}{a}$$

d192 $$a : b = c : x$$

x : 4th proportional

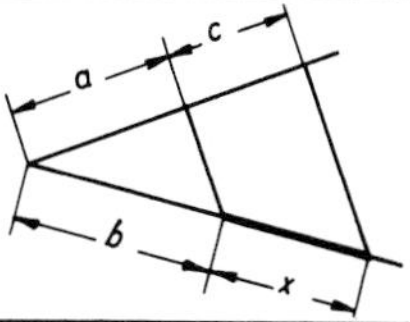

d193 $$x = \frac{b^2}{a}$$

d194 $$a : b = b : x$$

x : 3rd proportional

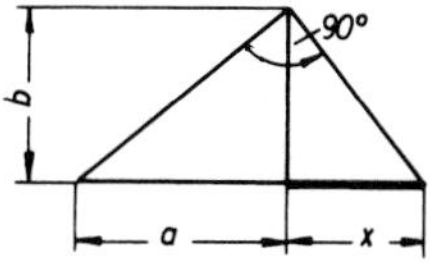

d195 $$x = \sqrt{a\ b}$$

d196 $$a : x = x : b$$

x : mean proportional

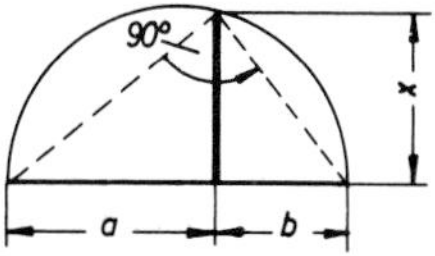

d197 $$x^2 = a^2 + b^2$$

d198 or $$x = \sqrt{a^2 + b^2}$$

x : hypothenuse of a right-angled triangle

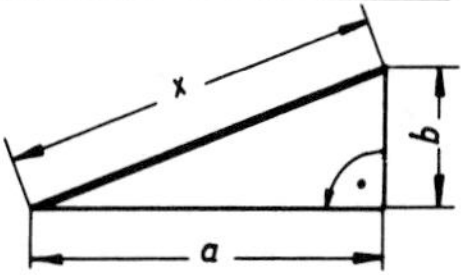

d199 $$x = \frac{a}{2}\sqrt{3}$$

x : height of an equilateral triangle

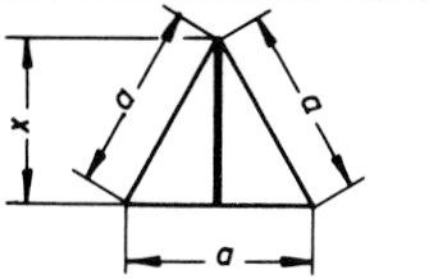

d200 $$x = \frac{a}{2}(\sqrt{5} - 1)$$

d201 $$\approx a\ 0{\cdot}618$$

d202 $$a : x = x : (a-x)$$

x : larger section of a repeatedly subdivided line (golden section)

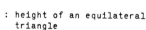

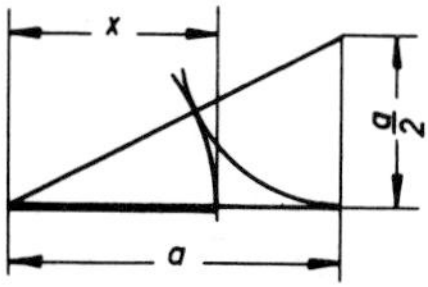

FUNCTIONS OF A CIRCLE

Basic terms

E 1

Circular and angular measure of a plane angle

Circular measure

Circular measure is the ratio of the distance d measured along the arc to the radius r.

It is given the unit "radian" which has no dimensions.

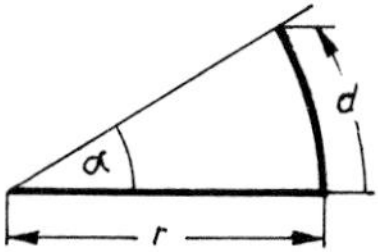

e 1 $$\alpha = \frac{d}{r} \text{ (rad)}$$

Unit: rad

Angular measure

Angular measure is obtained by dividing the angle subtended at the centre of a circle into 360 equal divisions known as "degrees".

Unit: °

e 2 A degree is divided into 60 minutes (unit: '),
e 3 a minute is divided into 60 seconds (unit: ").

Relation between circular and angular measure

By considering a circle, it may be seen that

e 4 $$360^\circ = 2\pi \text{ radians}$$

e 5 or $$1 \text{ rad} = 57{\cdot}2958^\circ$$

e 6

degrees	0°	30°	45°	60°	75°	90°	180°	270°	360°
radians	0	$\frac{\pi}{6}$	$\frac{\pi}{4}$	$\frac{\pi}{3}$	$\frac{5}{12}\pi$	$\frac{\pi}{2}$	π	$\frac{3}{2}\pi$	2π
	0	0·52	0·79	1·05	1·31	1·57	3·14	4·71	6·28

Right angled triangle

e 7 $\sin \alpha = \dfrac{\text{opposite}}{\text{hypotenuse}} = \dfrac{a}{c}$

e 8 $\cos \alpha = \dfrac{\text{adjacent}}{\text{hypotenuse}} = \dfrac{b}{c}$

e 9 $\tan \alpha = \dfrac{\text{opposite}}{\text{adjacent}} = \dfrac{a}{b}$ | $\cot \alpha = \dfrac{\text{adjacent}}{\text{opposite}} = \dfrac{b}{a}$

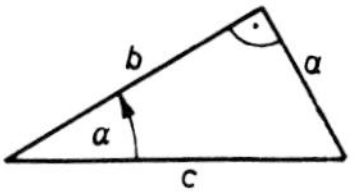

Functions of the more important angles

e 10

angle α	0°	30°	45°	60°	75°	90°	180°	270°	360°
$\sin \alpha$	0	0·500	0·707	0·866	0·966	1	0	−1	0
$\cos \alpha$	1	0·866	0·707	0·500	0·259	0	−1	0	1
$\tan \alpha$	0	0·577	1·000	1·732	3·732	∞	0	∞	0
$\cot \alpha$	∞	1·732	1·000	0·577	0·268	0	∞	0	∞

Relations between sine and cosine functions

Basic equations

e 11 Sine function $y = A \sin(k\alpha - \varphi)$

e 12 Cosine function $y = A \cos(k\alpha - \varphi)$

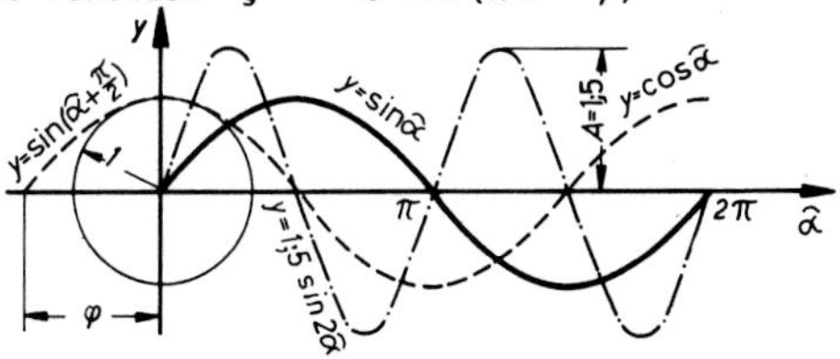

——— sine curve	with an amplitude of	$A = 1$ and $k = 1$
—·— sine curve		$A = 1 \cdot 5$ and $k = 2$
- - - - cosine curve		$A = 1$ and $k = 1$

or sine curve with a phase shift of $\varphi = -\dfrac{\pi}{2}$

FUNCTIONS OF A CIRCLE

Quadrants

E 3

e 15	$\sin(90^\circ - \alpha)$	$= +\cos\alpha$	$\sin(90^\circ + \alpha)$	$= +\cos\alpha$
e 16	$\cos(\;''\;)$	$= +\sin\alpha$	$\cos(\;''\;)$	$= -\sin\alpha$
e 17	$\tan(\;''\;)$	$= +\cot\alpha$	$\tan(\;''\;)$	$= -\cot\alpha$
e 18	$\cot(\;''\;)$	$= +\tan\alpha$	$\cot(\;''\;)$	$= -\tan\alpha$
e 19	$\sin(180^\circ - \alpha)$	$= +\sin\alpha$	$\sin(180^\circ + \alpha)$	$= -\sin\alpha$
e 20	$\cos(\;''\;)$	$= -\cos\alpha$	$\cos(\;''\;)$	$= -\cos\alpha$
e 21	$\tan(\;''\;)$	$= -\tan\alpha$	$\tan(\;''\;)$	$= +\tan\alpha$
e 22	$\cot(\;''\;)$	$= -\cot\alpha$	$\cot(\;''\;)$	$= +\cot\alpha$
e 23	$\sin(270^\circ - \alpha)$	$= -\cos\alpha$	$\sin(270^\circ + \alpha)$	$= -\cos\alpha$
e 24	$\cos(\;''\;)$	$= -\sin\alpha$	$\cos(\;''\;)$	$= +\sin\alpha$
e 25	$\tan(\;''\;)$	$= +\cot\alpha$	$\tan(\;''\;)$	$= -\cot\alpha$
e 26	$\cot(\;''\;)$	$= +\tan\alpha$	$\cot(\;''\;)$	$= -\tan\alpha$
e 27	$\sin(360^\circ - \alpha)$	$= -\sin\alpha$	$\sin(360^\circ + \alpha)$	$= +\sin\alpha$
e 28	$\cos(\;''\;)$	$= +\cos\alpha$	$\cos(\;''\;)$	$= +\cos\alpha$
e 29	$\tan(\;''\;)$	$= -\tan\alpha$	$\tan(\;''\;)$	$= +\tan\alpha$
e 30	$\cot(\;''\;)$	$= -\cot\alpha$	$\cot(\;''\;)$	$= +\cot\alpha$
e 31	$\sin(-\alpha)$	$= -\sin\alpha$	$\sin(\alpha \pm n \times 360^\circ)$	$= +\sin\alpha$
e 32	$\cos(\;''\;)$	$= +\cos\alpha$	$\cos(\;''\;)$	$= +\cos\alpha$
e 33	$\tan(\;''\;)$	$= -\tan\alpha$	$\tan(\alpha \pm n \times 180^\circ)$	$= +\tan\alpha$
e 34	$\cot(\;''\;)$	$= -\cot\alpha$	$\cot(\;''\;)$	$= +\cot\alpha$

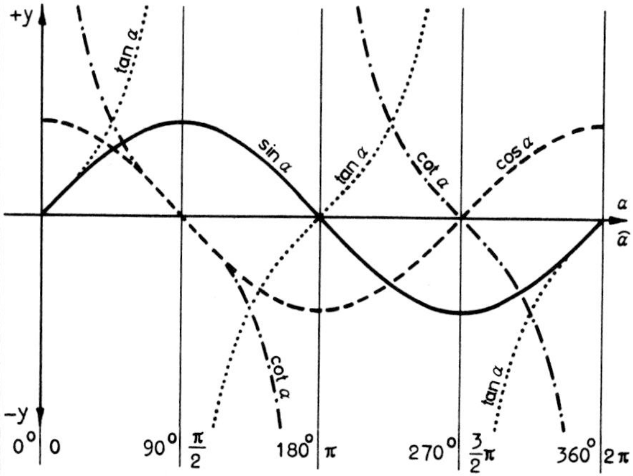

FUNCTIONS OF A CIRCLE

Trigonometric conversions

Basic identities

e 35 $\sin^2\alpha + \cos^2\alpha = 1$ | $\tan\alpha \; \cot\alpha = 1$

e 36 $1 + \tan^2\alpha = \dfrac{1}{\cos^2\alpha}$ | $1 + \cot^2\alpha = \dfrac{1}{\sin^2\alpha}$

Sum and difference of angles

e 37 $\sin(\alpha \pm \beta) = \sin\alpha \; \cos\beta \pm \cos\alpha \; \sin\beta$

e 38 $\cos(\alpha \pm \beta) = \cos\alpha \; \cos\beta \mp \sin\alpha \; \sin\beta$

e 39 $\tan(\alpha \pm \beta) = \dfrac{\tan\alpha \pm \tan\beta}{1 \mp \tan\alpha \; \tan\beta}$; $\cot(\alpha \pm \beta) = \dfrac{\cot\alpha \; \cot\beta \mp 1}{\pm\cot\alpha + \cot\beta}$

Sum and difference of functions of angles

e 40 $\sin\alpha + \sin\beta = 2\sin\dfrac{\alpha+\beta}{2} \cos\dfrac{\alpha-\beta}{2}$

e 41 $\sin\alpha - \sin\beta = 2\cos\dfrac{\alpha+\beta}{2} \sin\dfrac{\alpha-\beta}{2}$

e 42 $\cos\alpha + \cos\beta = 2\cos\dfrac{\alpha+\beta}{2} \cos\dfrac{\alpha-\beta}{2}$

e 43 $\cos\alpha - \cos\beta = -2\sin\dfrac{\alpha+\beta}{2} \sin\dfrac{\alpha-\beta}{2}$

e 44 $\tan\alpha \pm \tan\beta = \dfrac{\sin(\alpha \pm \beta)}{\cos\alpha \; \cos\beta}$

e 45 $\cot\alpha \pm \cot\beta = \dfrac{\sin(\beta \pm \alpha)}{\sin\alpha \; \sin\beta}$

e 46 $\sin\alpha \; \cos\beta = \dfrac{1}{2}\sin(\alpha+\beta) + \dfrac{1}{2}\sin(\alpha-\beta)$

e 47 $\cos\alpha \; \cos\beta = \dfrac{1}{2}\cos(\alpha+\beta) + \dfrac{1}{2}\cos(\alpha-\beta)$

e 48 $\sin\alpha \; \sin\beta = \dfrac{1}{2}\cos(\alpha-\beta) - \dfrac{1}{2}\cos(\alpha+\beta)$

e 49 $\tan\alpha \; \tan\beta = \dfrac{\tan\alpha + \tan\beta}{\cot\alpha + \cot\beta} = -\dfrac{\tan\alpha - \tan\beta}{\cot\alpha - \cot\beta}$

e 50 $\cot\alpha \; \cot\beta = \dfrac{\cot\alpha + \cot\beta}{\tan\alpha + \tan\beta} = -\dfrac{\cot\alpha - \cot\beta}{\tan\alpha - \tan\beta}$

e 51 $\cot\alpha \; \tan\beta = \dfrac{\cot\alpha + \tan\beta}{\tan\alpha + \cot\beta} = -\dfrac{\cot\alpha - \tan\beta}{\tan\alpha - \cot\beta}$

Sum of 2 harmonic oscillations of the same frequency

e 52 $a\sin(\omega t + \varphi_1) + b\cos(\omega t + \varphi_2) = \sqrt{c^2 + d^2}\,\sin(\omega t + \varphi)$

with $c = a\sin\varphi_1 + b\cos\varphi_2$; $d = a\cos\varphi_1 - b\sin\varphi_2$

$\varphi = \arctan\dfrac{c}{d}$ and $\varphi = \arcsin\dfrac{c}{\sqrt{c^2+d^2}}$ } both conditions must be satisfied

FUNCTIONS OF A CIRCLE

Trigonometric conversions

Ratios between simple, double and half angles

	$\sin\alpha =$	$\cos\alpha =$	$\tan\alpha =$	$\cot\alpha =$
e 53	$\cos(90^\circ-\alpha)$	$\sin(90^\circ-\alpha)$	$\cot(90^\circ-\alpha)$	$\tan(90^\circ-\alpha)$
e 54	$\sqrt{1-\cos^2\alpha}$	$\sqrt{1-\sin^2\alpha}$	$\dfrac{1}{\cot\alpha}$	$\dfrac{1}{\tan\alpha}$
e 55	$2\sin\dfrac{\alpha}{2}\cos\dfrac{\alpha}{2}$	$\cos^2\dfrac{\alpha}{2}-\sin^2\dfrac{\alpha}{2}$	$\dfrac{\sin\alpha}{\cos\alpha}$	$\dfrac{\cos\alpha}{\sin\alpha}$
e 56	$\dfrac{\tan\alpha}{\sqrt{1+\tan^2\alpha}}$	$\dfrac{\cot\alpha}{\sqrt{1+\cot^2\alpha}}$	$\dfrac{\sin\alpha}{\sqrt{1-\sin^2\alpha}}$	$\dfrac{\cos\alpha}{\sqrt{1-\cos^2\alpha}}$
e 57	$\sqrt{\cos^2\alpha-\cos 2\alpha}$	$1-2\sin^2\dfrac{\alpha}{2}$	$\sqrt{\dfrac{1}{\cos^2\alpha}-1}$	$\sqrt{\dfrac{1}{\sin^2\alpha}-1}$
e 58	$\sqrt{\dfrac{1-\cos 2\alpha}{2}}$	$\sqrt{\dfrac{1+\cos 2\alpha}{2}}$	$\dfrac{\sqrt{1-\cos^2\alpha}}{\cos\alpha}$	$\dfrac{\sqrt{1-\sin^2\alpha}}{\sin\alpha}$
e 59	$\dfrac{1}{\sqrt{1+\cot^2\alpha}}$	$\dfrac{1}{\sqrt{1+\tan^2\alpha}}$		
e 60	$\dfrac{2\tan\dfrac{\alpha}{2}}{1+\tan^2\dfrac{\alpha}{2}}$	$\dfrac{1-\tan^2\dfrac{\alpha}{2}}{1+\tan^2\dfrac{\alpha}{2}}$	$\dfrac{2\tan\dfrac{\alpha}{2}}{1-\tan^2\dfrac{\alpha}{2}}$	$\dfrac{\cot^2\dfrac{\alpha}{2}-1}{2\cot\dfrac{\alpha}{2}}$

	$\sin 2\alpha =$	$\cos 2\alpha =$	$\tan 2\alpha =$	$\cot 2\alpha =$
e 61	$2\sin\alpha\cos\alpha$	$\cos^2\alpha-\sin^2\alpha$	$\dfrac{2\tan\alpha}{1-\tan^2\alpha}$	$\dfrac{\cot^2\alpha-1}{2\cot\alpha}$
e 62		$2\cos^2\alpha-1$	$\dfrac{2}{\cot\alpha-\tan\alpha}$	$\dfrac{1}{2}\cot\alpha-\dfrac{1}{2}\tan\alpha$
e 63		$1-2\sin^2\alpha$		

	$\sin\dfrac{\alpha}{2} =$	$\cos\dfrac{\alpha}{2} =$	$\tan\dfrac{\alpha}{2} =$	$\cot\dfrac{\alpha}{2} =$
e 64			$\dfrac{\sin\alpha}{1+\cos\alpha}$	$\dfrac{\sin\alpha}{1-\cos\alpha}$
e 65	$\sqrt{\dfrac{1-\cos\alpha}{2}}$	$\sqrt{\dfrac{1+\cos\alpha}{2}}$	$\dfrac{1-\cos\alpha}{\sin\alpha}$	$\dfrac{1+\cos\alpha}{\sin\alpha}$
e 66			$\sqrt{\dfrac{1-\cos\alpha}{1+\cos\alpha}}$	$\sqrt{\dfrac{1+\cos\alpha}{1-\cos\alpha}}$

FUNCTIONS OF A CIRCLE

Acute angle triangle

E 6

Oblique angle triangle

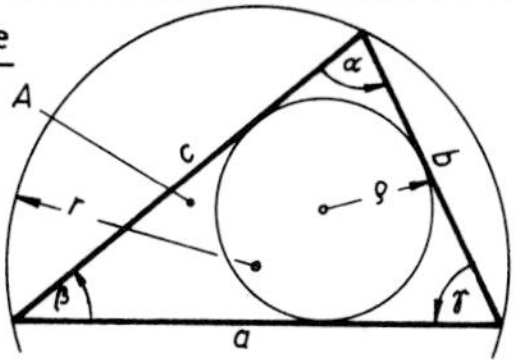

Sine Rule

e 67 $\sin\alpha : \sin\beta : \sin\gamma = a : b : c$

e 68 $a = \dfrac{b}{\sin\beta}\sin\alpha = \dfrac{c}{\sin\gamma}\sin\alpha$

e 69 $b = \dfrac{a}{\sin\alpha}\sin\beta = \dfrac{c}{\sin\gamma}\sin\beta$

e 70 $c = \dfrac{a}{\sin\alpha}\sin\gamma = \dfrac{b}{\sin\beta}\sin\gamma$

Cosine Rule

e 71 $a^2 = b^2 + c^2 - 2\,b\,c\cos\alpha$

e 72 $b^2 = c^2 + a^2 - 2\,a\,c\cos\beta$

e 73 $c^2 = a^2 + b^2 - 2\,a\,b\cos\gamma$

(for obtuse angles the cosine is negative)

Tangent Rule

e 74 $\dfrac{a+b}{a-b} = \dfrac{\tan\frac{\alpha+\beta}{2}}{\tan\frac{\alpha-\beta}{2}}$ | $\dfrac{a+c}{a-c} = \dfrac{\tan\frac{\alpha+\gamma}{2}}{\tan\frac{\alpha-\gamma}{2}}$ | $\dfrac{b+c}{b-c} = \dfrac{\tan\frac{\beta+\gamma}{2}}{\tan\frac{\beta-\gamma}{2}}$

Half-angle Rule

e 75 $\tan\dfrac{\alpha}{2} = \dfrac{\varrho}{s-a}$ | $\tan\dfrac{\beta}{2} = \dfrac{\varrho}{s-b}$ | $\tan\dfrac{\gamma}{2} = \dfrac{\varrho}{s-c}$

Area, radius of incircle and circumcircle

e 76 $A = \dfrac{1}{2}\,b\,c\sin\alpha = \dfrac{1}{2}\,a\,c\sin\beta = \dfrac{1}{2}\,a\,b\sin\gamma$

e 77 $A = \sqrt{s(s-a)(s-b)(s-c)} = \varrho\, s$

e 78 $\varrho = \sqrt{\dfrac{(s-a)(s-b)(s-c)}{s}}$

e 79 $r = \dfrac{1}{2}\,\dfrac{a}{\sin\alpha} = \dfrac{1}{2}\,\dfrac{b}{\sin\beta} = \dfrac{1}{2}\,\dfrac{c}{\sin\gamma}$

e 80 $s = \dfrac{a+b+c}{2}$

Inverse circular functions

Definitions

		function $y =$		
	$\arcsin x$	$\arccos x$	$\arctan x$	$\text{arccot}\, x$
e 81 identical with	$x = \sin y$	$x = \cos y$	$x = \tan y$	$x = \cot y$
e 82 defined within	$-1 \leqq x \leqq +1$	$-1 \leqq x \leqq +1$	$-\infty < x < +\infty$	$-\infty < x < +\infty$
e 83 principal value	$-\frac{\pi}{2} \leqq y \leqq +\frac{\pi}{2}$	$\pi \geqq y \geqq 0$	$-\frac{\pi}{2} < y < +\frac{\pi}{2}$	$\pi > y > 0$

Basic properties

e 84 $\arccos x = \frac{\pi}{2} - \arcsin x$ | $\text{arccot}\, x = \frac{\pi}{2} - \arctan x$

Ratios between inverse circular functions

For $x \geqq 0$:

	$\arcsin x =$	$\arccos x =$	$\arctan x =$	$\text{arccot}\, x =$
e 85	$\arccos\sqrt{1-x^2}$	$\arcsin\sqrt{1-x^2}$	$\arcsin\frac{x}{\sqrt{1+x^2}}$	$\arcsin\frac{1}{\sqrt{1+x^2}}$
e 86	$\arctan\frac{x}{\sqrt{1-x^2}}$	$\arctan\frac{\sqrt{1-x^2}}{x}$	$\arccos\frac{1}{\sqrt{1+x^2}}$	$\arccos\frac{x}{\sqrt{1+x^2}}$
e 87	$\text{arccot}\frac{\sqrt{1-x^2}}{x}$	$\text{arccot}\frac{x}{\sqrt{1-x^2}}$	$\text{arccot}\frac{1}{x}$	$\arctan\frac{1}{x}$

For $x \leqq 0$:

e 88 $\arcsin(-x) = -\arcsin x$ | $\arccos(-x) = \pi - \arccos x$

e 89 $\arctan(-x) = -\arctan x$ | $\text{arccot}(-x) = \pi - \text{arccot}\, x$

Addition theorems

e 90 $$\arcsin a \pm \arcsin b = \arcsin\left(a\sqrt{1-b^2} \pm b\sqrt{1-a^2}\right)$$

e 91 $$\arccos a \pm \arccos b = \arccos\left(a\,b \mp \sqrt{1-a^2}\,\sqrt{1-b^2}\right)$$

e 92 $$\arctan a \pm \arctan b = \arctan\frac{a \pm b}{1 \mp a\,b}$$

e 93 $$\text{arccot}\, a \pm \text{arccot}\, b = \text{arccot}\frac{a\,b \mp 1}{b \pm a}$$

Straight line

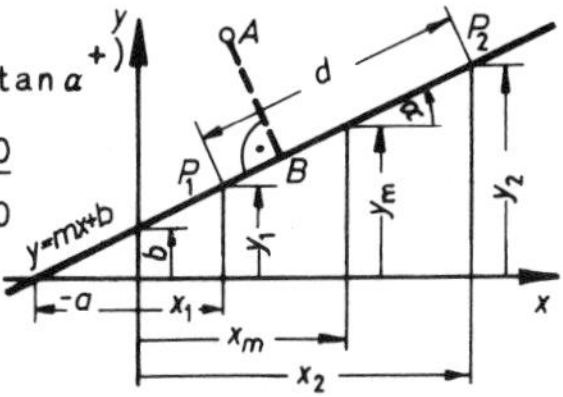

f 1 Equation $y = mx + b$

f 2 Gradient $m = \dfrac{y_2 - y_1}{x_2 - x_1} = \tan\alpha$ +)

Interc. form for $a \neq 0$; $b \neq 0$

f 3 $$\frac{x}{a} + \frac{y}{b} - 1 = 0$$

Gradient m_l of perpendicular AB

f 4 $$m_l = \frac{-1}{m}$$

Line joining two points $P_1(x_1, y_1)$ and $P_2(x_2, y_2)$

f 5 $$\frac{y - y_1}{x - x_1} = \frac{y_2 - y_1}{x_2 - x_1}$$

Line through one point $P_1(x_1, y_1)$ and gradient m

f 6 $$y - y_1 = m(x - x_1)$$

f 7 Distance between two points $d = \sqrt{(x_2 - x_1)^2 + (y_2 - y_1)^2}$ +)

Mid point of a line joining two points

f 8 $$x_m = \frac{x_1 + x_2}{2} \qquad y_m = \frac{y_1 + y_2}{2}$$

Point of intersection of two straight lines (see diagram triangle)

f 9 $$x_3 = \frac{b_2 - b_1}{m_1 - m_2} \qquad y_3 = m_1 x_3 + b_1 = m_2 x_3 + b_2$$

f10 Angle of intersection φ of two straight lines $\tan\varphi = \dfrac{m_2 - m_1}{1 + m_2 \cdot m_1}$ +) (see diagram-triangle)

Triangle

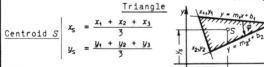

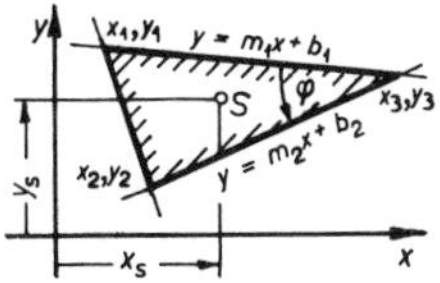

Centroid S

f11 $$x_s = \frac{x_1 + x_2 + x_3}{3}$$

f12 $$y_s = \frac{y_1 + y_2 + y_3}{3}$$

Area

f13 $$A = \frac{(x_1 y_2 - x_2 y_1) + (x_2 y_3 - x_3 y_2) + (x_3 y_1 - x_1 y_3)}{2}$$

+) Where x and y have same dimension and are represented in equal scales (see also h 1).

ANALYTICAL GEOMETRY — F 2

Circle, Parabola

Circle

Circle equation

centre: at the origin	centre: elsewhere
14 $x^2 + y^2 = r^2$	$(x-x_0)^2 + (y-y_0)^2 = r^2$

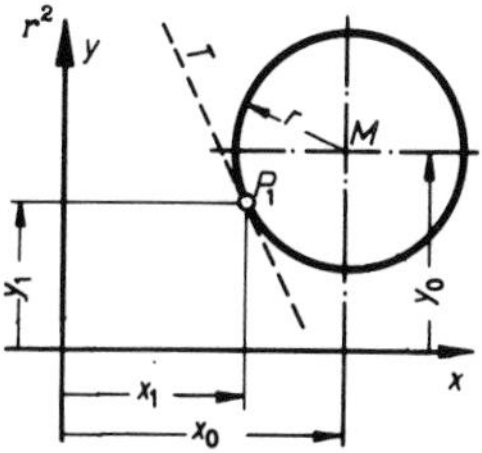

Basic equation

15 $x^2 + y^2 + ax + by + c = 0$

Radius of circle

16 $r = \sqrt{x_0^2 + y_0^2 - c}$

Coordinates of the centre M

17 $x_0 = -\frac{a}{2}$ | $y_0 = -\frac{b}{2}$

Tangent T at point $P_1(x_1, y_1)$

18 $y = \frac{r^2 - (x-x_0)(x_1-x_0)}{y_1 - y_0} + y_0$

Parabola

Parabola equation (by converting to this equation the vertex and parameter p may be ascertained)

	vertex: at the origin	vertex: elsewhere	parabola open at
19	$x^2 = 2py$	$(x-x_0)^2 = 2p(y-y_0)$	top
20	$x^2 = -2py$	$(x-x_0)^2 = -2p(y-y_0)$	bottom

F: focus
L: directrix
S: tangent at the vertex

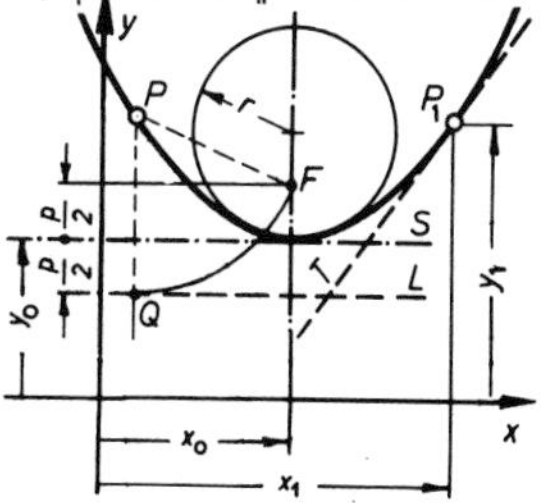

Basic equation

21 $y = ax^2 + bx + c$

22 **Vertex radius** $r = p$

23 **Basic property** $\overline{PF} = \overline{PQ}$

Tangent T at point $P_1(x_1, y_1)$

24 $y = \frac{2(y_1 - y_0)(x - x_1)}{x_1 - x_0} + y_1$

Hyperbola

Hyperbolic equation

point of intersection of asymptotes	
at the origin	elsewhere
$\dfrac{x^2}{a^2} - \dfrac{y^2}{b^2} - 1 = 0$	$\dfrac{(x - x_0)^2}{a^2} - \dfrac{(y - y_0)^2}{b^2} - 1 = 0$

25

Basic equation

$$ax^2 + by^2 + cx + dy + e = 0$$

26

Basic property

$$\overline{F_2 P} - \overline{F_1 P} = 2a$$

27

Eccentricity

$$e = \sqrt{a^2 + b^2} \quad +)$$

28

Gradient of asymptotes

$$\tan\alpha = m = \pm\frac{b}{a} \quad +)$$

29

Vertex radius $\quad p = \dfrac{b^2}{a}$

Tangent T at $P_1(x_1, y_1)$

$$y = \frac{b^2}{a^2} \frac{(x_1 - x_0)(x - x_1)}{y_1 - y_0} + y_1$$

30

Rectangular hyperbola

Explanation in a rectangular hyperbola $\quad a = b \quad$ thus

Gradient of asymptotes

$$\tan\alpha^{+)} = m = \pm 1 \qquad (\alpha = 45°)$$

31

Equation (for asymptotes parallel to x and y axes):

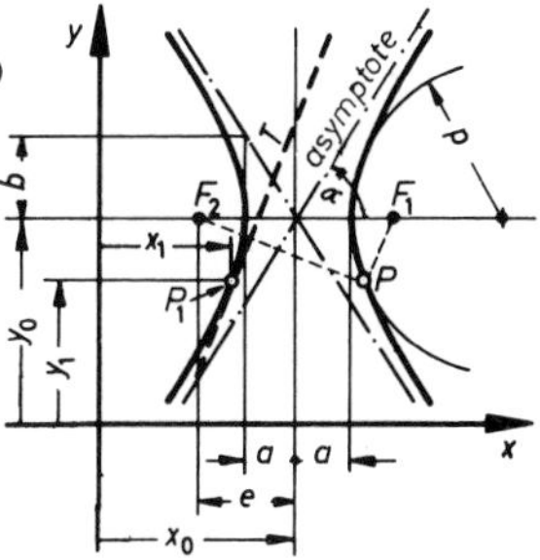

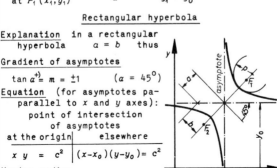

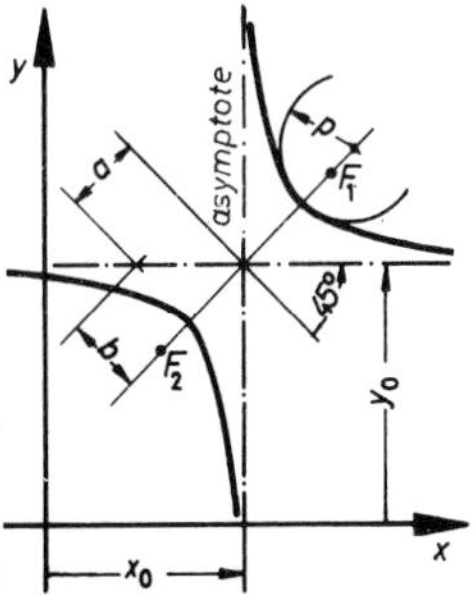

point of intersection of asymptotes	
at the origin	elsewhere
$xy = c^2$	$(x - x_0)(y - y_0) = c^2$

32

Vertex radius

$$p = a \qquad \text{(parameter)}$$

33

+) Conditions according to note on page F 1

Ellipse

Ellipse equation

	point of intersection of axes	
	at the origin	elsewhere
f 34	$\frac{x^2}{a^2} + \frac{y^2}{b^2} - 1 = 0$	$\frac{(x - x_0)^2}{a^2} + \frac{(y - y_0)^2}{b^2} - 1 = 0$

Vertex radii

f 35 $r_N = \frac{b^2}{a}$ | $r_H = \frac{a^2}{b}$

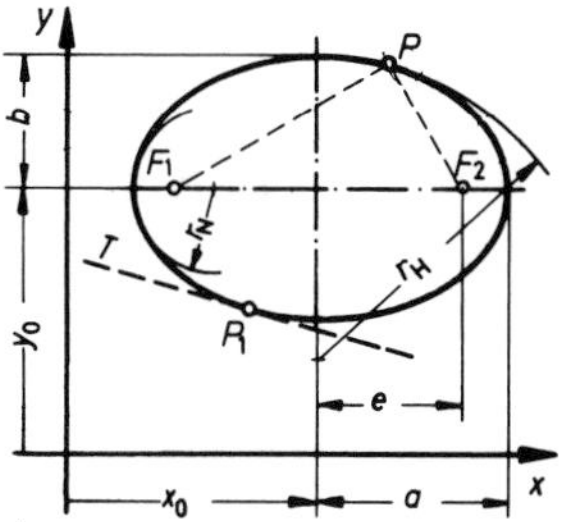

Eccentricity

f 36 $e = \sqrt{a^2 - b^2}$

Basic property

f 37 $\overline{F_1 P} + \overline{F_2 P} = 2a$

Tangent T at $P_1(x_1; y_1)$

f 38 $y = -\frac{b^2}{a^2} \frac{(x_1 - x_0)(x - x_1)}{y_1 - y_0} + y_1$

Note: F_1 and F_2 are focal points

Exponential curve

Basic equation

f 39 $y = a^x$

Here a is a positive constant, and x is a number.

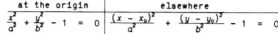

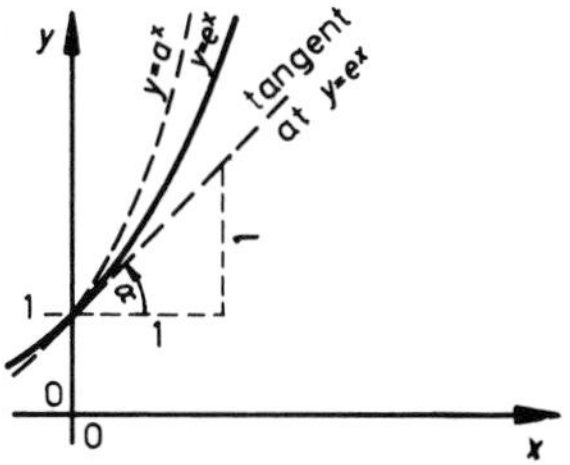

Note

All exponential curves pass through the point $x = 0$; $y = 1$.

The derivative of the curve passing through this point with a gradient of 45° ($\tan\alpha$[+)] = 1) is equal to the curve itself. The constant a now becomes e (Euler number) and is the base of the natural log.

$$e = 2 \cdot 718\,281\,828\,459\ldots$$

+) Conditions according to note on page F 1

HYPERBOLIC FUNCTIONS — G 1

Hyperbolic functions

Hyperbolic functions

Definition

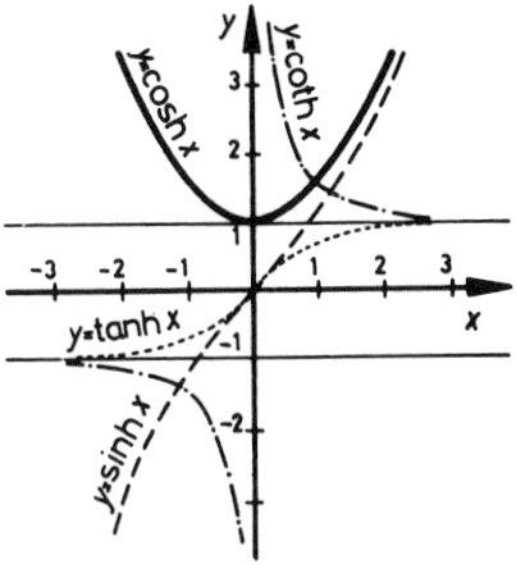

g 1 $\sinh x = \dfrac{e^x - e^{-x}}{2}$

g 2 $\cosh x = \dfrac{e^x + e^{-x}}{2}$

g 3 $\tanh x = \dfrac{e^x - e^{-x}}{e^x + e^{-x}} = \dfrac{e^{2x} - 1}{e^{2x} + 1}$

g 4 $\coth x = \dfrac{e^x + e^{-x}}{e^x - e^{-x}} = \dfrac{e^{2x} + 1}{e^{2x} - 1}$

Basic properties

g 5 $\cosh^2 x - \sinh^2 x = 1$

g 6 $\tanh x \times \coth x = 1$

g 7 $\tanh x = \dfrac{\sinh x}{\cosh x}$ | $1 - \tanh^2 x = \dfrac{1}{\cosh^2 x}$ | $1 - \coth^2 x = \dfrac{-1}{\sinh^2 x}$

Ratios between hyperbolic functions

where x is positive

	$\sinh x =$	$\cosh x =$	$\tanh x =$	$\coth x =$
g 8	$\sqrt{\cosh^2 x - 1}$	$\sqrt{\sinh^2 x + 1}$	$\dfrac{\sinh x}{\sqrt{\sinh^2 x + 1}}$	$\dfrac{\sqrt{\sinh^2 x + 1}}{\sinh x}$
g 9	$\dfrac{\tanh x}{\sqrt{1 - \tanh^2 x}}$	$\dfrac{1}{\sqrt{1 - \tanh^2 x}}$	$\dfrac{\sqrt{\cosh^2 x - 1}}{\cosh x}$	$\dfrac{\cosh x}{\sqrt{\cosh^2 x - 1}}$
g 10	$\dfrac{1}{\sqrt{\coth^2 x - 1}}$	$\dfrac{\coth x}{\sqrt{\coth^2 x - 1}}$	$\dfrac{1}{\coth x}$	$\dfrac{1}{\tanh x}$

g 11 For negative argument: $\sinh(-x) = -\sinh x$ | $\cosh(-x) = +\cosh x$

g 12 $\tanh(-x) = -\tanh x$ | $\coth(-x) = -\coth x$

Addition theorems

g 13 $\sinh(a \pm b) = \sinh a \cosh b \pm \cosh a \sinh b$

g 14 $\cosh(a \pm b) = \cosh a \cosh b \pm \sinh a \sinh b$

g 15 $\tanh(a \pm b) = \dfrac{\tanh a \pm \tanh b}{1 \pm \tanh a \tanh b}$

g 16 $\coth(a \pm b) = \dfrac{\coth a \coth b \pm 1}{\coth a \pm \coth b}$

Inverse hyperbolic functions

Definition

		function $y =$		
	arsinh x	arcosh x	artanh x	arcoth x
g 17 identical with	$x = \sinh y$	$x = \cosh y$	$x = \tanh y$	$x = \coth y$
g 18 logarithmic equivalents	$= \ln(x+\sqrt{x^2+1})$	$= \ln(x+\sqrt{x^2-1})$	$= \frac{1}{2}\ln\frac{1+x}{1-x}$	$= \frac{1}{2}\ln\frac{x+1}{x-1}$
g 19 defined within	$-\infty < x < +\infty$	$1 \leqq x < +\infty$	$\|x\| < 1$	$\|x\| > 1$
g 20 primary value	$-\infty < y < +\infty$	$0 \leqq y < +\infty$	$-\infty < y < +\infty$	$-\infty < y < +\infty$

Ratios between inverse hyperbolic functions

where x is positive:

	arsinh x =	arcosh x =	artanh x =	arcoth x =
g 21	$\operatorname{arcosh}\sqrt{1+x^2}$	$\operatorname{arsinh}\sqrt{x^2-1}$	$\operatorname{arsinh}\frac{x}{\sqrt{1-x^2}}$	$\operatorname{arsinh}\frac{1}{\sqrt{x^2-1}}$
g 22	$\operatorname{artanh}\frac{x}{\sqrt{1+x^2}}$	$\operatorname{artanh}\frac{\sqrt{x^2-1}}{x}$	$\operatorname{arcosh}\frac{1}{\sqrt{1-x^2}}$	$\operatorname{arcosh}\frac{x}{\sqrt{x^2-1}}$
g 23	$\operatorname{arcoth}\frac{\sqrt{1+x^2}}{x}$	$\operatorname{arcoth}\frac{x}{\sqrt{x^2-1}}$	$\operatorname{arcoth}\frac{1}{x}$	$\operatorname{artanh}\frac{1}{x}$

For negative argument:

g 24 $\operatorname{arsinh}(-x) = -\operatorname{arsinh} x$

g 25 $\operatorname{artanh}(-x) = -\operatorname{artanh} x$ | $\operatorname{arcoth}(-x) = -\operatorname{arcoth} x$

Addition theorems

g 26 $$\operatorname{arsinh} a \pm \operatorname{arsinh} b = \operatorname{arsinh}\left(a\sqrt{b^2+1} \pm b\sqrt{a^2+1}\right)$$

g 27 $$\operatorname{arcosh} a \pm \operatorname{arcosh} b = \operatorname{arcosh}\left[ab \pm \sqrt{(a^2-1)(b^2-1)}\right]$$

g 28 $$\operatorname{artanh} a \pm \operatorname{artanh} b = \operatorname{artanh}\frac{a \pm b}{1 \pm ab}$$

g 29 $$\operatorname{arcoth} a \pm \operatorname{arcoth} b = \operatorname{arcoth}\frac{ab \pm 1}{a \pm b}$$

Differential coefficients (or derivatives)

Gradient of a curve

The gradient of a curve $y = f(x)$ varies from point to point. By the gradient of a curve at point P we mean the gradient of the tangent at the point. If x and y have equal dimensions - which is not the case in most technical diagrams - and are presented at equal scales, the gradient may be expressed by the tangent of angle α between the tangent at point P and the horizontal axis:

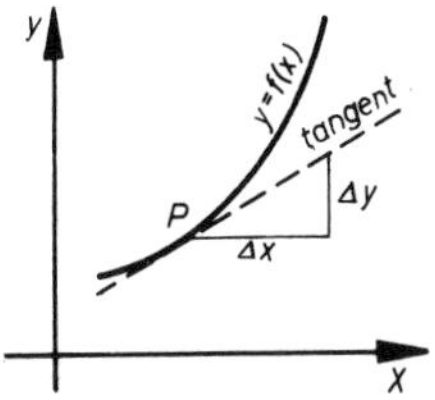

$$m = \tan\alpha$$

Always applicable is:

h 1 gradient $m = \dfrac{\Delta y}{\Delta x}$

Difference coefficient

The difference coefficient or mean gradient of the function $y = f(x)$ between PP_1 is:

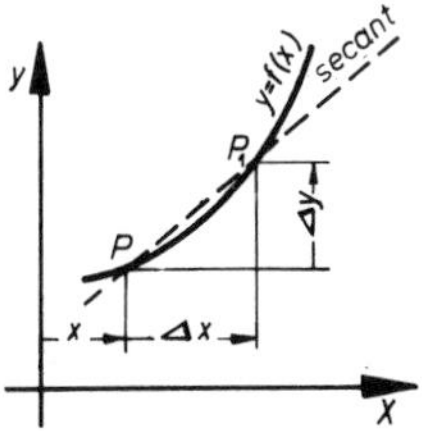

h 2 $$\frac{\Delta y}{\Delta x} = \frac{f(x+\Delta x)-f(x)}{\Delta x}$$

Differential coefficient

Where Δx is infinitely small, i.e. where Δx approches zero, the slope at P becomes the limiting value of the slope of one of the secants. This slope is the "derivative" or "differential coefficient" of the function at P.

h 3 $$y' = \frac{dy}{dx} = f'(x)$$

$$y' = \lim_{\Delta x \to 0} \frac{\Delta y}{\Delta x} = \lim_{\Delta x \to 0} \frac{f(x+\Delta x)-f(x)}{\Delta x} = \frac{dy}{dx} = f'(x)$$

DIFFERENTIAL CALCULUS
Meaning of derivative

H 2

Geometric meaning of derivative

Gradient of a curve

If, for each point x of a curve, we plot its corresponding gradient as an ordinate y', we obtain the first gradient curve $y' = f'(x)$ or the first derivative of the original curve $y = f(x)$. If we now take the derivative of the first gradient $y' = f'(x)$, we obtain $y'' = f''(x)$ or the second derivative of the original curve $y = f(x)$ etc.

Example: $y = Ax^3 + Bx^2 + Cx + D$

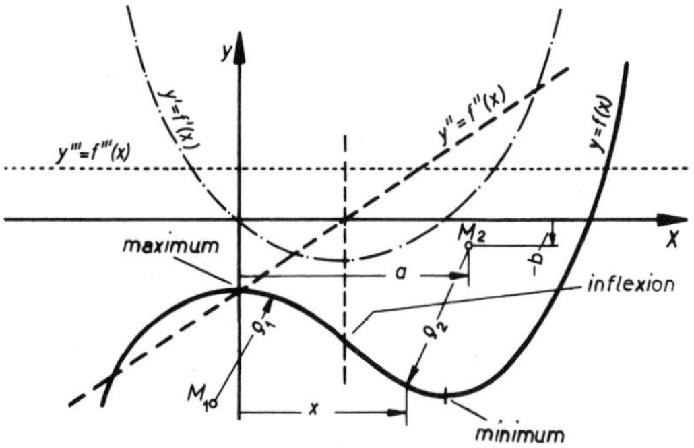

Radius of curvature ρ at any point x

h 4 $$\rho = \frac{\sqrt{(1 + y'^2)^3}}{y''}$$

M is below the curve where ρ is −
M is above the curve where ρ is +

Centre coordinates for radius ρ

h 5 $$a = x - \frac{1 + y'^2}{y''} y'$$

h 6 $$b = y + \frac{1 + y'^2}{y''}$$

continued on H 3

Determination of minima, maxima and inflexions

Minima and maxima

The value $x = a$ obtained for $y' = 0$ is inserted in y''.

7 For $y''(a) > 0$ there is a minimum at $x = a$,
8 for $y''(a) < 0$ there is a maximum at $x = a$,
9 for $y''(a) = 0$ see h 19.

Inflexion

The value $x = a$ obtained for $y'' = 0$ is inserted in y'''.

10 For $y'''(a) \neq 0$ there is an inflexion at $x = a$.

Shape of the curve $y = f(x)$

Rise and fall

11 $y'(x) > 0$ $y(x)$ increases as x increases
12 $y'(x) < 0$ $y(x)$ decreases as x increases
13 $y'(x) = 0$ $y(x)$ is tangentially parallel the x-axis at x

Curve

14 $y''(x) < 0$ $y(x)$ is convex (viewed from above)
15 $y''(x) > 0$ $y(x)$ is concave (viewed from above)

16 $y''(x) = 0$	with	a change of sign	flexion
	without	$y'(x)$ at x has a	bottom point

Exceptional case

Where at a point $x = a$

17 $y'(a) = y''(a) = y'''(a) = \ldots\ y^{(n-1)}(a) = 0$, but
18 $y^{n}(a) \neq 0$, one of the 4 conditions is present:

n = even number		n = uneven number	
19 $y^{(n)}(a) > 0$	$y^{(n)}(a) < 0$	$y^{(n)}(a) > 0$	$y^{(n)}(a) < 0$

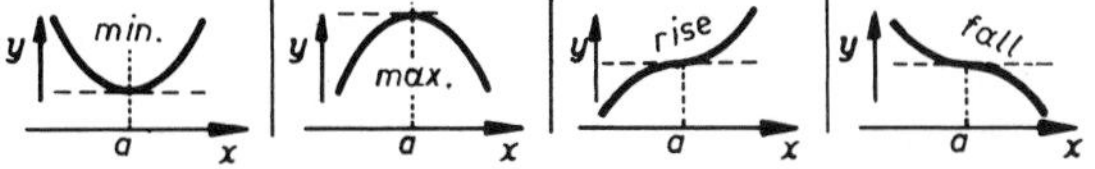

Derivatives

Basic rules

	function	derivative
h 21	$y = c\,x^n + C$	$y' = c\,n\,x^{n-1}$
h 22	$y = u(x) \pm v(x)$	$y' = u'(x) \pm v'(x)$
h 23	$y = u(x)\,v(x)$	$y' = u'v + u\,v'$
h 24	$y = \dfrac{u(x)}{v(x)}$	$y' = \dfrac{u'v - u\,v'}{v^2}$
h 25	$y = \sqrt{x}$	$y' = \dfrac{1}{2\sqrt{x}}$
h 26	$y = u(x)^{v(x)}$	$y' = u^v\left(\dfrac{u'v}{u} + v' \ln u\right)$

Derivative of a function of a function (chain rule)

h 27	$y = f\left[u(x)\right]$	$y' = f'(u)\;u'(x) = \dfrac{dy}{dx} = \dfrac{dy}{du}\,\dfrac{du}{dx}$

Parametric form of derivative

h 28	$y = f(x) \quad \begin{cases} x = f(t) \\ y = f(t) \end{cases}$	$y' = \dfrac{dy}{dt}\,\dfrac{dt}{dx} = \dfrac{\dot{y}}{\dot{x}}$
h 29		$y'' = \dfrac{d^2y}{dx^2} = \dfrac{\dot{x}\ddot{y} - \dot{y}\ddot{x}}{\dot{x}^3}$

Derivative of inverse functions

The equation $y = f(x)$ solved for x gives the inverse function $x = \varphi(y)$.

h 30	$x = \varphi(y)$	$f'(x) = \dfrac{1}{\varphi'(x)}$

Example

h 31	$y = f(x) = \arccos x$	$f'(x) = \dfrac{1}{-\sin y} = -\dfrac{1}{\sqrt{1-x^2}}$
h 32	gives $x = \varphi(y) = \cos y$	

Derivatives

Exponential functions

	function	derivative
h 33	$y = e^x$	$y' = e^x = y'' = \dots$
h 34	$y = e^{-x}$	$y' = -e^{-x}$
h 35	$y = e^{ax}$	$y' = a\, e^{ax}$
h 36	$y = x\ e^x$	$y' = e^x\ (1 + x)$
h 37	$y = \sqrt{e^x}$	$y' = \frac{\sqrt{e^x}}{2}$
h 38	$y = a^x$	$y' = a^x \ln a$
h 39	$y = a^{nx}$	$y' = n\, a^{nx} \ln a$
h 40	$y = a^{x^2}$	$y' = a^{x^2}\ 2x \ln a$

Trigonometrical functions

h 41	$y = \sin x$	$y' = \cos x$
h 42	$y = \cos x$	$y' = -\sin x$
h 43	$y = \tan x$	$y' = \frac{1}{\cos^2 x} = 1 + \tan^2 x = \sec^2 x$
h 44	$y = \cot x$	$y' = \frac{-1}{\sin^2 x} = -(1 + \cot^2 x) = -\operatorname{cosec}^2 x$
h 45	$y = a \sin(kx)$	$y' = a\ k \cos(kx)$
h 46	$y = a \cos(kx)$	$y' = -a\ k \sin(kx)$
h 47	$y = \sin^n x$	$y' = n \sin^{n-1} x \cos x$
h 48	$y = \cos^n x$	$y' = -n \cos^{n-1} x \sin x$
h 49	$y = \tan^n x$	$y' = n \tan^{n-1} x\ (1 + \tan^2 x)$
h 50	$y = \cot^n x$	$y' = -n \cot^{n-1} x\ (1 + \cot^2 x)$
h 51	$y = \frac{1}{\sin x}$	$y' = \frac{-\cos x}{\sin^2 x}$
h 52	$y = \frac{1}{\cos x}$	$y' = \frac{\sin x}{\cos^2 x}$

Derivatives

Logarithmic functions

	function	derivative
h 53	$y = \ln x$	$y' = \frac{1}{x}$
h 54	$y = {}^{a}\log x$	$y' = \frac{1}{x \ln a}$
h 55	$y = \ln(1 \pm x)$	$y' = \frac{\pm 1}{1 \pm x}$
h 56	$y = \ln x^n$	$y' = \frac{n}{x}$
h 57	$y = \ln \sqrt{x}$	$y' = \frac{1}{2x}$

Hyperbolic functions

	function	derivative
h 58	$y = \sinh x$	$y' = \cosh x$
h 59	$y = \cosh x$	$y' = \sinh x$
h 60	$y = \tanh x$	$y' = \frac{1}{\cosh^2 x}$
h 61	$y = \coth x$	$y' = \frac{-1}{\sinh^2 x}$

Inverse trigonometrical functions

	function	derivative
h 62	$y = \arcsin x$	$y' = \frac{1}{\sqrt{1 - x^2}}$
h 63	$y = \arccos x$	$y' = -\frac{1}{\sqrt{1 - x^2}}$
h 64	$y = \arctan x$	$y' = \frac{1}{1 + x^2}$
h 65	$y = \text{arccot}\, x$	$y' = -\frac{1}{1 + x^2}$
h 66	$y = \text{arsinh}\, x$	$y' = \frac{1}{\sqrt{x^2 + 1}}$
h 67	$y = \text{arcosh}\, x$	$y' = \frac{1}{\sqrt{x^2 - 1}}$
h 68	$y = \text{artanh}\, x$	$y' = \frac{1}{1 - x^2}$
h 69	$y = \text{arcoth}\, x$	$y' = \frac{1}{1 - x^2}$

Integration

Integration reverse of differentiation

By integral calculus we mean the problem of finding a function $y = f(x)$, the derivative of $F(x)$ being equal to $f(x)$. Thus

j 1 $$F'(x) = \frac{dF(x)}{dx} = f(x)$$

hence, by integration

the indefinite integral

j 2 $$\int f(x)\, dx = F(x) + C$$

Here C is unknown constant which disappears on differentiation, since the derivative of a constant equals zero.

Geometric meaning of the indefinite integral

As this figure shows, there are an infinite number of $y = F(x)$ curves with a gradient $y' = f(x)$. All $y = F(x)$ curves are the same shape, but intersect the y-axis at varying distances. The constant C, however, establishes a fixed curve. If the curve is to pass through the point x_0/y_0, then

j 3 $$C = y_0 - F(x_0)$$

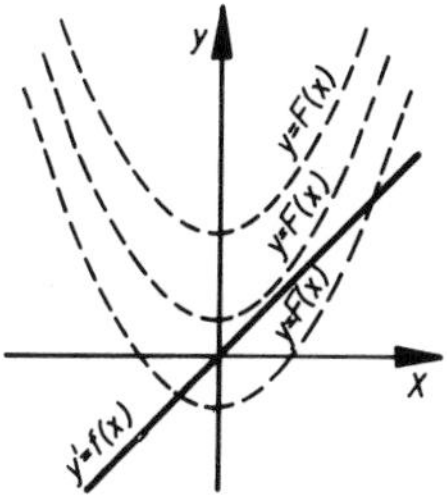

The definite integral

The definite integral is represented by

j 4 $$\int_a^b f(x)\, dx = F(x)\Big|_a^b = F(b) - F(a)$$

Here integration takes place between the limits a and b, the second substitution resultant is subtracted from the first causing the constant C to disappear.

INTEGRAL CALCULUS

Integration rules

J 2

Integration

Basic formulae

j 5 $$\int x^n \, dx = \frac{x^{n+1}}{n+1} + C, \quad \text{here} \quad n \neq -1$$

j 6 $$\int \frac{dx}{x} = \ln x + C$$

j 7 $$\int \left[u(x) \pm v(x)\right] dx = \int u(x) \, dx \pm \int v(x) \, dx$$

j 8 $$\int \frac{u'(x)}{u(x)} \, dx = \ln u(x) + C$$

j 9 $$\int u(x) \, u'(x) \, dx = \frac{1}{2}\left[u(x)\right]^2 + C$$

Integration by parts

j10 $$\int u(x) \, v'(x) \, dx = u(x) \, v(x) - \int u'(x) \, v(x) \, dx$$

Integration by substitution

j11 $$\int f(x) \, dx = \int f\left[\varphi(z)\right] \varphi'(z) \, dz$$

here $x = \varphi(z)$ and $dx = \varphi'(z) \, dz$

Example:

j12 $F(x) = \int \sqrt{3x-5} \, dx.$

Where $3x - 5 = z$, the derivative is $z = \frac{dz}{dx} = 3.$

Thus $dx = \frac{dz}{3}$ expressed in terms of z, the integral becomes $F(x) = \frac{1}{3}\int \sqrt{z} \, dz = \frac{2}{9} z\sqrt{z} + C.$ Insert value of z in above expression: $F(x) = \frac{2}{9}(3x-5)\sqrt{3x-5} + C$

Integrals

(omitting integral constant C)

J 13 $\int e^x \, dx = e^x$ $\qquad\qquad$ $\int \ln x \, dx = x \ln x - x$

J 14 $\int a^x \, dx = \dfrac{a^x}{\ln a}$ $\qquad\qquad$ $\int \dfrac{dx}{x - a} = \ln (x - a)$

J 15 $\int \dfrac{dx}{(x-a)(x-b)} = \dfrac{1}{a - b} \ln \dfrac{x - a}{x - b}$ $\qquad (a \neq b)$

J 16 $\int \dfrac{dx}{(x - a)^n} = -\dfrac{1}{(n - 1)(x - a)^{n-1}}$ $\qquad (n \neq 1)$

J 17 $\int \dfrac{dx}{x^2 - a^2} = -\dfrac{1}{a} \operatorname{arcoth} \dfrac{x}{a} = \dfrac{1}{2a} \ln \dfrac{x-a}{x+a}$ $\qquad (x > a)$

J 18 $\int \dfrac{dx}{a^2 - x^2} = \dfrac{1}{a} \operatorname{artanh} \dfrac{x}{a} = \dfrac{1}{2a} \ln \dfrac{a+x}{a-x}$ $\qquad (x < a)$

J 19 $\int \dfrac{dx}{x^2 + a^2} = \dfrac{1}{a} \arctan \dfrac{x}{a}$ $\qquad\qquad$ $\int \dfrac{x \, dx}{x^2 + a^2} = \dfrac{1}{2} \ln(x^2 + a^2)$

J 20 $\int \dfrac{dx}{(x^2 + a^2)^2} = \dfrac{x}{2a^2 (x^2 + a^2)} + \dfrac{1}{2a^3} \arctan \dfrac{x}{a}$

J 21 $\int \dfrac{dx}{(x^2 + a^2)^n} = \dfrac{x}{2a^2 (n-1)(x^2 + a^2)^{n-1}} + \dfrac{2n-3}{2a^2 (n-1)} \int \dfrac{dx}{(x^2 + a^2)^{n-1}}$ $\qquad (n \neq 1)$

J 22 $\int \sqrt{x} \, dx = \dfrac{2}{3} \sqrt{x^3}$ $\qquad\qquad$ $\int \dfrac{dx}{\sqrt{x}} = 2 \sqrt{x}$

J 23 $\int \dfrac{dx}{\sqrt{a^2 - x^2}} = \arcsin \dfrac{x}{a}$ $\qquad\qquad$ $\int \dfrac{dx}{\sqrt{ax + b}} = \dfrac{2}{a} \sqrt{ax + b}$

J 24 $\int \dfrac{dx}{\sqrt{x^2 - a^2}} = \operatorname{arcosh} \dfrac{x}{a} = \ln \left(x + \sqrt{x^2 - a^2}\right)$

J 25 $\int \dfrac{dx}{\sqrt{x^2 + a^2}} = \operatorname{arsinh} \dfrac{x}{a} = \ln \left(x + \sqrt{x^2 + a^2}\right)$

J 26 $\int \sqrt{a^2 - x^2} \, dx = \dfrac{x}{2} \sqrt{a^2 - x^2} + \dfrac{a^2}{2} \arcsin \dfrac{x}{a}$

J 27 $\int \sqrt{x^2 - a^2} \, dx = \dfrac{x}{2} \sqrt{x^2 - a^2} - \dfrac{a^2}{2} \operatorname{arcosh} \dfrac{x}{a}$

J 28 $\int \sqrt{x^2 + a^2} \, dx = \dfrac{x}{2} \sqrt{x^2 + a^2} + \dfrac{a^2}{2} \operatorname{arsinh} \dfrac{x}{a}$

Integrals

(omitting integral constant C)

J 29 $\int \sin x \, dx = -\cos x$

J 30 $\int \sin^2 x \, dx = \frac{x}{2} - \frac{1}{4}\sin(2x)$

J 31 $\int \sin^3 x \, dx = -\frac{3}{4}\cos x + \frac{1}{12}\cos(3x)$

J 32 $\int \sin^n x \, dx = -\frac{1}{n}\cos x \quad \sin^{n-1} x + \frac{n-1}{n}\int \sin^{n-2} x \, dx$

J 33 $\int \sin(ax)\,dx = -\frac{1}{a}\cos(ax)$

34 $\int \cos x \, dx = \sin x$

35 $\int \cos^2 x \, dx = \frac{x}{2} + \frac{1}{4}\sin(2x)$

36 $\int \cos^3 x \, dx = \frac{3}{4}\sin x + \frac{1}{12}\sin(3x)$

37 $\int \cos^n x \, dx = \frac{1}{n}\sin x \quad \cos^{n-1} x + \frac{n-1}{n}\int \cos^{n-2} x \, dx$

38 $\int \cos(ax)\,dx = \frac{1}{a}\sin(ax)$

39 $\int \tan x \, dx = -\ln \cos x$ ‖ $\int \tan(ax)\,dx = -\frac{1}{a}\ln \cos(ax)$

40 $\int \tan^2 x \, dx = \tan x - x$

41 $\int \tan^n x \, dx = \frac{\tan^{n-1} x}{n-1} - \int \tan^{n-2} x \, dx \qquad (n \neq 1)$

42 $\int \cot x \, dx = \ln \sin x$ ‖ $\int \cot(ax)\,dx = \frac{1}{a}\ln \sin(ax)$

43 $\int \cot^2 x \, dx = -x - \cot x$

44 $\int \cot^n x \, dx = -\frac{\cot^{n-1} x}{n-1} - \int \cot^{n-2} x \, dx \qquad (n \neq 1)$

45 $\int \frac{dx}{\sin x} = \ln \tan \frac{x}{2}$ ‖ $\int \frac{dx}{\sin^2 x} = -\cot x$

46 $\int \frac{dx}{\sin^n x} = -\frac{1}{n-1}\frac{\cos x}{\sin^{n-1} x} + \frac{n-2}{n-1}\int \frac{dx}{\sin^{n-2} x} \qquad (n \neq 1)$

47 $\int \frac{dx}{\cos x} = \ln \tan\left(\frac{x}{2} + \frac{\pi}{4}\right)$ ‖ $\int \frac{dx}{\cos^2 x} = \tan x$

48 $\int \frac{dx}{\cos^n x} = \frac{1}{n-1}\frac{\sin x}{\cos^{n-1} x} + \frac{n-2}{n-1}\int \frac{dx}{\cos^{n-2} x} \qquad (n \neq 1)$

Integrals

(omitting integral constant C)

49 $\int \frac{dx}{1+\sin x} = \tan\left(\frac{x}{2} - \frac{\pi}{4}\right)$ | $\int \frac{dx}{1-\sin x} = -\cot\left(\frac{x}{2} - \frac{\pi}{4}\right)$

50 $\int \frac{dx}{1+\cos x} = \tan \frac{x}{2}$ | $\int \frac{dx}{1-\cos x} = -\cot \frac{x}{2}$

51 $\int \sin(ax)\sin(bx)\,dx = -\frac{\sin(ax+bx)}{2(a+b)} + \frac{\sin(ax-bx)}{2(a-b)} \quad (|a| \neq |b|)$

52 $\int \sin(ax)\cos(bx)\,dx = -\frac{\cos(ax+bx)}{2(a+b)} - \frac{\cos(ax-bx)}{2(a-b)} \quad (|a| \neq |b|)$

53 $\int \cos(ax)\cos(bx)\,dx = \frac{\sin(ax+bx)}{2(a+b)} + \frac{\sin(ax-bx)}{2(a-b)} \quad (|a| \neq |b|)$

54 $\int x^n \sin(ax)\,dx = -\frac{x^n}{a}\cos(ax) + \frac{n}{a}\int x^{n-1}\cos(ax)\,dx$

55 $\int x^n \cos(ax)\,dx = \frac{x^n}{a}\sin(ax) - \frac{n}{a}\int x^{n-1}\sin(ax)\,dx$

---

56 $\int \arcsin x\,dx = x \arcsin x + \sqrt{1-x^2}$

57 $\int \arccos x\,dx = x \arccos x - \sqrt{1-x^2}$

58 $\int \arctan x\,dx = x \arctan x - \frac{1}{2}\ln(1+x^2)$

59 $\int \operatorname{arccot} x\,dx = x \operatorname{arccot} x + \frac{1}{2}\ln(1+x^2)$

---

60 $\int \sinh x\,dx = \cosh x$

61 $\int \sinh^2 x\,dx = \frac{1}{4}\sinh(2x) - \frac{x}{2}$

62 $\int \sinh^n x\,dx = \frac{1}{n}\cosh x \sinh^{n-1} x - \frac{n-1}{n}\int \sinh^{n-2} x\,dx$

63 $\int \sinh(ax)\,dx = \frac{1}{a}\cosh(ax)$

---

64 $\int \cosh x\,dx = \sinh x$

65 $\int \cosh^2 x\,dx = \frac{1}{4}\sinh(2x) + \frac{x}{2}$

66 $\int \cosh^n x\,dx = \frac{1}{n}\sinh x \cosh^{n-1} x + \frac{n-1}{n}\int \cosh^{n-2} x\,dx$

67 $\int \cosh(ax)\,dx = \frac{1}{a}\sinh(ax)$

Integrals

(omitting integral constant C)

68 $\int \tanh x \, dx = \ln \cosh x$

69 $\int \tanh^2 x \, dx = x - \tanh x$

70 $\int \tanh^n x \, dx = -\frac{1}{n-1} \tanh^{n-1} x + \int \tanh^{n-2} x \, dx \quad (n \neq 1)$

71 $\int \tanh(ax) \, dx = \frac{1}{a} \ln \cosh(ax)$

---

72 $\int \coth x \, dx = \ln \sinh x$

73 $\int \coth^2 x \, dx = x - \coth x$

74 $\int \coth^n x \, dx = -\frac{1}{n-1} \coth^{n-1} x + \int \coth^{n-2} x \, dx \quad (n \neq 1)$

75 $\int \coth(ax) \, dx = \frac{1}{a} \ln \sinh(ax)$

---

76 $\int \frac{dx}{\sinh x} = \ln \tanh \frac{x}{2}$

77 $\int \frac{dx}{\sinh^2 x} = -\coth x$

78 $\int \frac{dx}{\cosh x} = 2 \arctan e^x$

79 $\int \frac{dx}{\cosh^2 x} = \tanh x$

---

80 $\int \operatorname{arsinh} x \, dx = x \operatorname{arsinh} x - \sqrt{x^2 + 1}$

81 $\int \operatorname{arcosh} x \, dx = x \operatorname{arcosh} x - \sqrt{x^2 - 1}$

82 $\int \operatorname{artanh} x \, dx = x \operatorname{artanh} x + \frac{1}{2} \ln(1 - x^2)$

83 $\int \operatorname{arcoth} x \, dx = x \operatorname{arcoth} x + \frac{1}{2} \ln(x^2 - 1)$

---

84 $\int \sin^m x \cos^n x \, dx = \frac{1}{m+n} \sin^{m+1} x \cos^{n-1} x + \frac{n-1}{m+n} \int \sin^m x \cos^{n-2} x \, dx$

Where n is an uneven number the residual integral is:

85 $\int \sin^m x \cos x \, dx = \frac{\sin^{m+1} x}{m+1}$

INTEGRAL CALCULUS

Application of integration

J 7

Arc differential $\quad ds = \sqrt{dx^2 + dy^2} = \sqrt{1 + \left(\frac{dy}{dx}\right)^2}\, dx$

	arc length	surface area where the curve rotates around the x-axis
j 86	$s = \int_a^b \sqrt{1 + y'^2}\, dx$	$A_m = 2\pi \int_a^b y \sqrt{1 + y'^2}\, dx$

static moment of a curve

	x-axis	y-axis
j 87	$M_x = \int_a^b y\sqrt{1+y'^2}\,dx$	$M_y = \int_a^b x\sqrt{1+y'^2}\,dx$

coordinates of centre of gravity

j 88	$x_S = \frac{M_y}{s}$	$y_S = \frac{M_x}{s}$

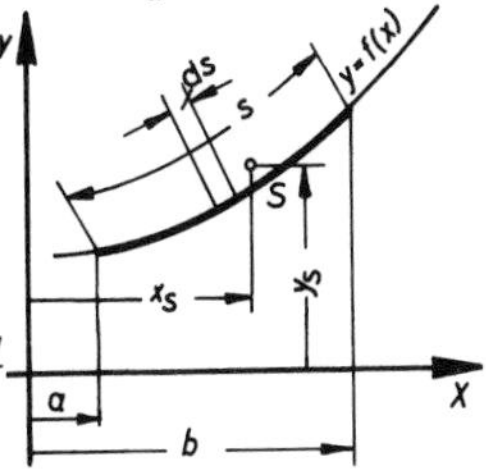

	area	rotating body where area A rotates around the x-axis	volume of a body, the cross section A_1 of which is a function of x
j 89	$A = \int_a^b y\, dx$	$V = \pi \int_a^b y^2\, dx$	$V = \int_a^b A_1(x)\, dx$

static moment of a curve in relation to the

	x-axis	y-axis
j 90	$H_x = \int_a^b \frac{y^2}{2}\,dx$	$H_y = \int_a^b xy\, dx$

coordinates of centre of gravity

j 91	$x_S = \frac{H_y}{A}$	$y_S = \frac{H_x}{A}$

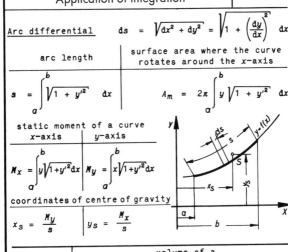

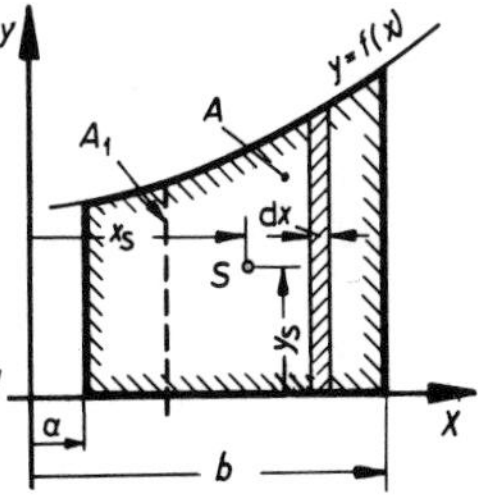

Static moment of a body
(in relation to the y-z plane)

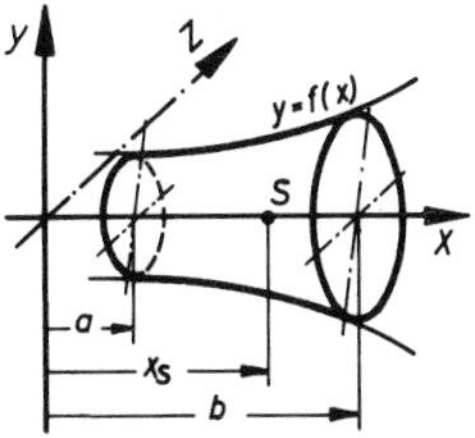

92 $$M_{yz} = \pi \int_a^b x \; y^2 \; dx$$

Coordinates of centre of gravity

93 $$x_S = \frac{M_{yz}}{V}$$

Pappus theorems

Surface area of a revolving body

A_m = arc length s times the distance covered by the centre of gravity

94 $= 2\pi s y_S$ (see also formulae j 86 and j 88)

Volume of a revolving body

V = area A times the distance covered by the centre of gravity

95 $= 2\pi A y_S$ (see also formulae j 89 and j 91)

Numerical integration

Division of area into an even number n of strips of equal

width $h = \dfrac{b_1 - b_0}{n}$.

96 Then, according to the

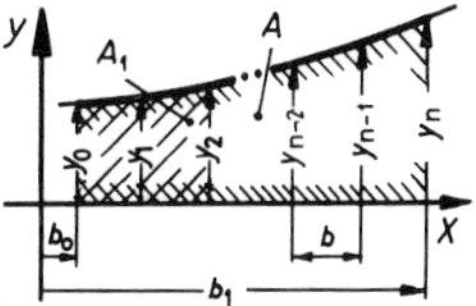

Trapezium rule

97 $$A = \frac{b}{2}\left(y_0 + 2y_1 + 2y_2 + \ldots + y_n\right)$$

Simpson's rule for three ordinates:

98 $$A_1 = \frac{b}{3}\left(y_0 + 4y_1 + y_2\right)$$

Simpson's rule for more than three ordinates:

99 $$A = \frac{b}{3}\left[y_0 + y_n + 2\left(y_2 + y_4 + \ldots + y_{n-2}\right) + 4\left(y_1 + y_3 + \ldots + y_{n-1}\right)\right]$$

Moment of inertia
(Second moment of area)

General

By moment of inertia in relation to an axis x or a point O, we mean the sum of the products of line-, area-, volume- or mass-elements and the squares of their distances from the x-axis or point O:

Moment of inertia	Second moment of area
100 $I = \int x^2\,dm \quad \mathrm{kg\,m^2}$	$I = \int x^2\,dA \quad \mathrm{m^4}$

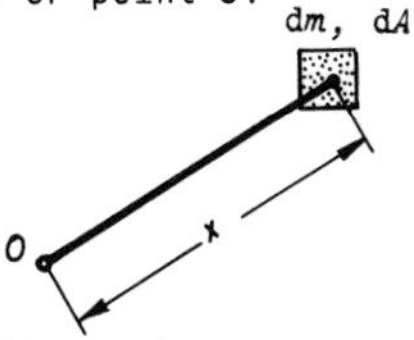

Steiner's theorem (Parallel axis theorem)

For every mass moment of inertia, both axial and polar, the following equation will apply:

101 $I_{xx} = I_{yy} + m\,l_s^2 \quad \mathrm{kg\,m^2}$

Similar equations will apply for line, area and volume moments of inertia:

$I_{xx} = I_{yy} + A\,l_s^2 \quad \mathrm{m^4}$

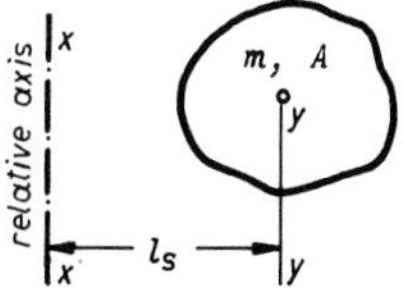

Moment of inertia of plane curves

in relation to the x-axis	in relation to the y-axis
102 $I_{0x} = \int_a^b y^2 \sqrt{1+y'^2}\,dx$	$I_{0y} = \int_a^b x^2 \sqrt{1+y'^2}\,dx$

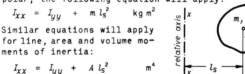

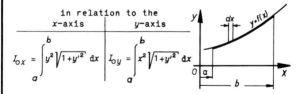

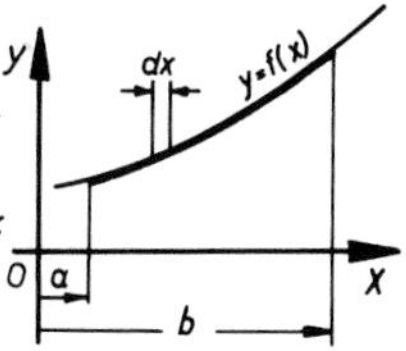

I_{xx} : moment of inertia about a general axis xx
I : moment of inertia about the centre of gravity
m, A : total length, area, volume, or mass
l_s : distance of centre of gravity from axis or point

Moments of inertia and centrif. moments of plane surfaces

By axial second moment of area of a plane surface in relation to an axis x or y within the plane we mean the sum of the products of the area-elements dA and the squares of their distances from axis x or y, respectively:

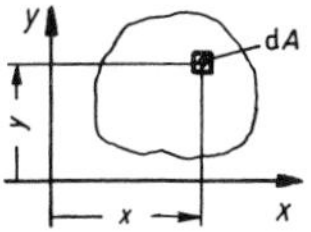

j 103 $I_x = \int y^2 \, dA \; ; \quad I_y = \int x^2 \, dA$

A given function $y = f(x)$ yields:

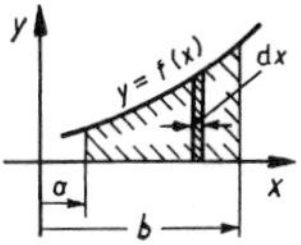

x-axis	y-axis
$I_x = \int_a^b \frac{y^3}{3} \, dx$	$I_y = \int_a^b x^2 y \, dx$

j 104

By polar second moment of area of a plane surface in relation to a point O within the plane we mean the sum of the products of the area-elements dA and the squares of their distances r from point O.

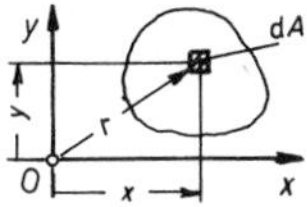

j 105 $I_p = \int r^2 \, dA$

Where the relative axes of I_x and I_y are perpendicular to each other, the polar second moment of area in relation to the pole (intersect. O of axes x and y) is:

j 106 $I_p = \int r^2 \, dA = \int (y^2 + x^2) \, dA = I_x + I_y$

By centrifugal moment (product of inertia) of a plane surface in relation to 2 axes within the plane we mean the sum of the products of the area-elements dA and the products of their distances x and y from the two axes:

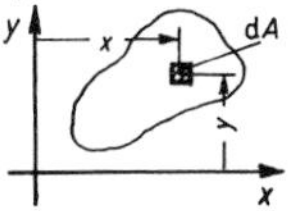

j 107 $I_{xy} = \int x \, y \, dA \gtreqless 0$

One of the relative axes being an axis of symmetry of the plane surface results in $I_{xy} = 0$.

Conversion to an inclined axis x': Where moments I_x, I_y, and I_{xy} in relation to two perpendicular axes x and y are known, the second moment of area I_α in relation to an axis inclined x' by an angle α with respect to the x-axis can be calculated by:

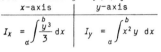

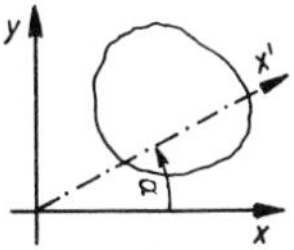

j 108 $I_\alpha = I_x \cos^2\alpha + I_y \sin^2\alpha - I_{xy} \sin 2\alpha$

Examples in conjunction to second moments of area on page J 10

on page J 10

Rectangle

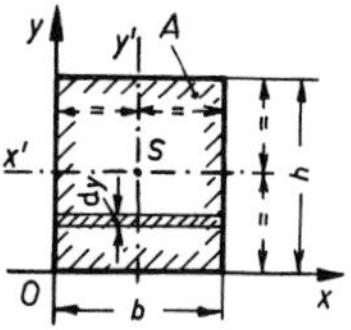

j 109 $I_x = \int_0^h y^2 b \, dy = b\left[\frac{y^3}{3}\right]_0^h = \frac{b\,h^3}{3}$

j 110 $I_{x'} = I_x - A\left(\frac{h}{2}\right)^2 = \frac{b\,h^3}{12}$

j 111 $I_y = \frac{b^3\,h}{3}$; $I_{y'} = \frac{b^3\,h}{12}$

j 112 $I_{po} = I_x + I_y = \frac{b\,h^3}{3} + \frac{b^3\,h}{3} = \frac{b\,h}{3}(b^2+h^2)$; $I_{ps} = \frac{b\,h}{12}(b^2+h^2)$

j 113 $I_{xy} = I_{x'y'} + \frac{b}{2}\times\frac{h}{2}A$. As x' and/or y' are axes of symmetry, $I_{x'y'}$ is zero. Hence:

j 114 $I_{xy} = \frac{b}{2}\times\frac{h}{2}(b\,h) = \left(\frac{b\,h}{2}\right)^2$

Circle

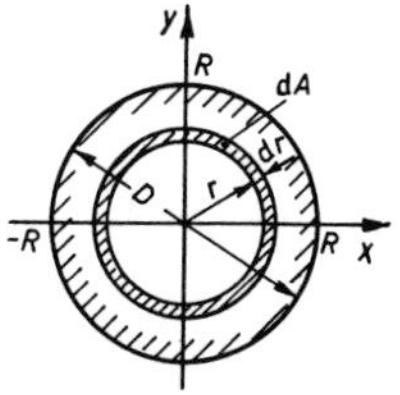

j 115 $I_p = \int_0^R r^2 \, dA = \int_0^R r^2 \, 2\pi r \, dr$

j 116 $= 2\pi\left[\frac{r^4}{4}\right]_0^R = \frac{\pi R^4}{2}$

j 117 $I_x = I_y = \frac{I_p}{2} = \frac{\pi R^4}{4} = \frac{\pi D^4}{64}$

j 118 $I_{xy} = 0$, as x and y are axes of symmetry.

Semicircle

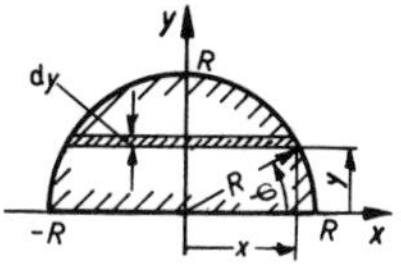

j 119 $I_x = \int_0^R y^2 \, dA = \int_0^R y^2 \, 2\,x \, dy$

j 120 $= 2\int_0^R y^2 \sqrt{R^2 - y^2} \, dy = \frac{\pi R^4}{8} = I_y$

j 121 $I_p = 2\,\frac{\pi R^4}{8} = \frac{\pi R^4}{4}$; $I_{xy} = 0$, as y is axis of symmetry.

Regular polygon

j 122 $I_x = I_y = \frac{I_p}{2} = \frac{n\,a\,r}{2\times 48}(12r^2+a^2) = \frac{n\,a\,R}{2\times 24}(6R^2-a^2)$; $I_{xy} = 0$

r : radius of inscribed circle
R : radius of circumscribed circle
a : length of side
n : number of sides

Second moment of volume of a solid

'Moment of inertia' of a cuboid

Where $\left(\frac{b\,h^3}{12} + \frac{b^3\,h}{12}\right)$ is the polar moment of inertia of a rectangle (see J 11), the equation for the z-axis:

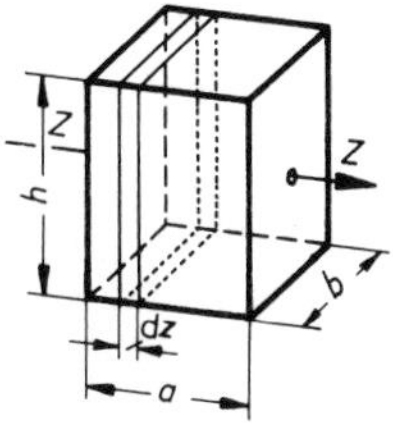

j 123 $$I_{v,zz} = \int_0^a \left(\frac{bh^3}{12} + \frac{b^3\,h}{12}\right) dz = \frac{abh}{12}(b^2 + h^2)$$

'Moment of inertia' of a circular cylinder

for the axis zz:

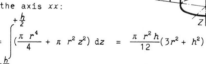

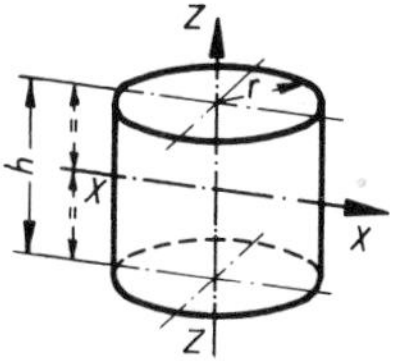

j 124 $$I_{v,zz} = \int_{-\frac{h}{2}}^{+\frac{h}{2}} \frac{\pi\ r^4}{2}\ dz = \frac{\pi\ r^4\ h}{2}$$

for the axis xx:

j 125 $$I_{v,xx} = \int_{-\frac{h}{2}}^{+\frac{h}{2}} \left(\frac{\pi\ r^4}{4} + \pi\ r^2\,z^2\right) dz = \frac{\pi\ r^2\,h}{12}(3r^2 + h^2)$$

Dynamic moment of inertia (mass moment of inertia)

The mass moment of inertia I about a particular axis is the product of the second moment of volume I_v about that axis and the density ϱ.

j 126 $$I = I_v \times \varrho \qquad \text{kg m}^2,\ \text{kgf m s}^2,\ \text{lb ft}^2$$

j 127 where $$\varrho = \frac{m}{V} \qquad \text{kg m}^{-3},\ \text{kg dm}^{-3},\ \text{lb ft}^{-3}$$

e.g. for a cylinder about the axis ZZ:

j 128 $$I_{zz} = I_{v,zz} \times \frac{m}{V} = \frac{\pi\,r^4\,h}{2} \times \frac{m}{r^2\,\pi\,h} = \frac{m\,r^2}{2}$$

For other mass moments of inertia see M 3

STATICS
General terms

General

Statics deals with the theory of equilibrium and with the determination of external forces acting on stationary solid bodies (e.g. support reactions). The contents of page K1...K14 are applicable only to forces acting in one plane.

The most important quantities of statics and their units

Length l

Is a base quantity, see preface.
Units: m; cm; km

Force F (see explanation on M 1)

Being a vector a force is defined by its magnitude, direction, and point of application (P).

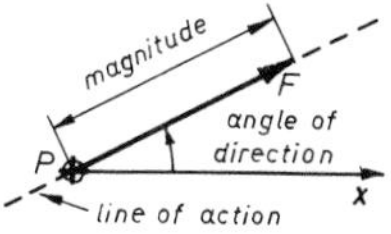

Gravitational force G

definition: force of earth's attract.
point of applicat: centre of grav. S
line of action: vertical line intersecting centre of gravity
direction: downwards (towards earth's centre)
magnitude: determined by spring balance.

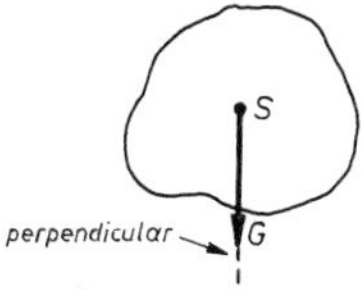

Support reaction F_A

force applied to body by support A.

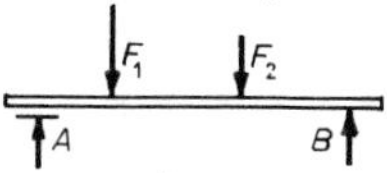

Resultant force F_R

force representing the total action of several external forces.

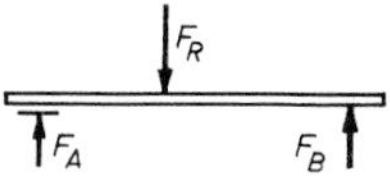

Moment M of a force F about a point O

The perpendicular distance from point O to the line of action of force F is called lever arm l. F_1 and F_2 form a pair of forces. The moment may be represented by a vector.

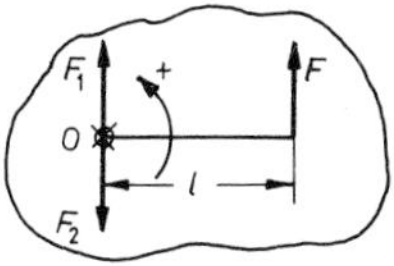

k 1 $F_1 = F$; $F_2 = -F_1$

k 2 moment $M = \pm F\,l$

Moment theorem: The moment of the resultant force is equal to the sum of the moments of the individual forces.

STATICS

Forces composition

K 2

Graphical composition of forces

diagram of forces	force polygon

Two forces

k 4 

A number of forces and a common point of application

k 5 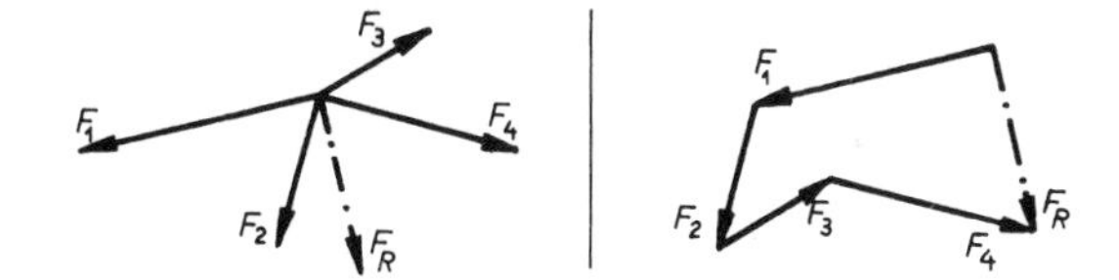

Parallel forces

k 6 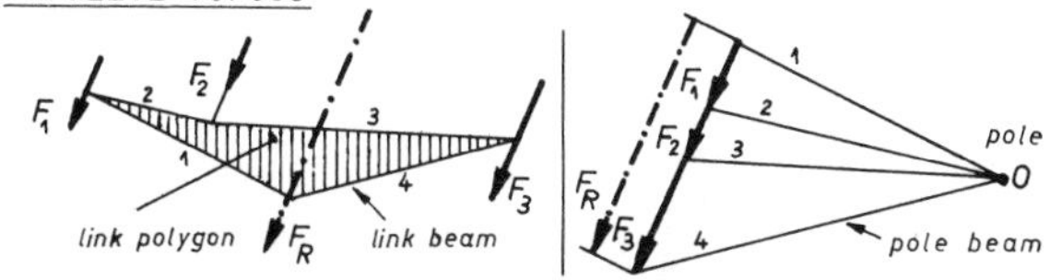

A number of forces and any random point of application

k 7 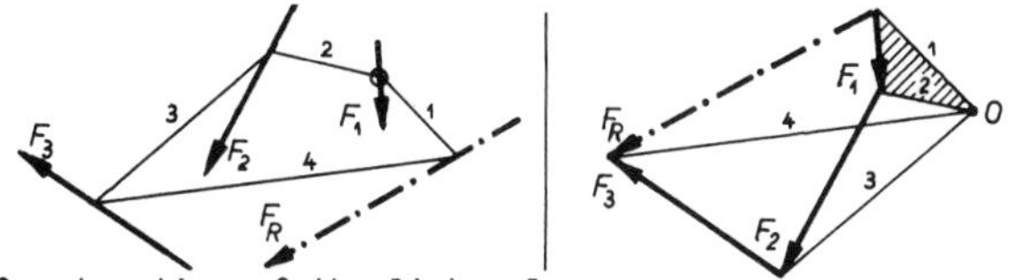

Construction of the link polygon

Draw force polygon and determine pole O so as to avoid any link rays running parallel. Draw pole rays. Construct link polygon such that link rays run parallel to corresponding pole rays. Thereby each point of intersection in the link polygon corresponds to a triangle in the force polygon (e.g. triangle F_1-1-2 of force polygon corresponds to point of intersection F_1-1-2 of link polygon).

STATICS

Forces composition

K 3

Mathematical composition of forces

Resolution of a force

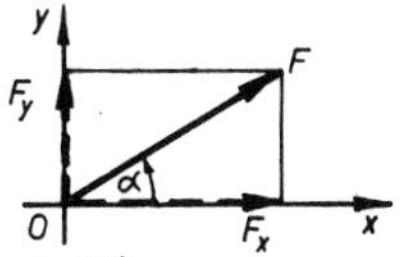

k 8 $F_x = F \times \cos\alpha$ | $F_y = F \times \sin\alpha$

k 9 $F = +\sqrt{F_x^2 + F_y^2}$ | $\tan\alpha = \dfrac{F_y}{F_x}$

(for signs of trigonometrical functions of α see tables k 16 to k 19)

Moment M_0 of a force about a point O

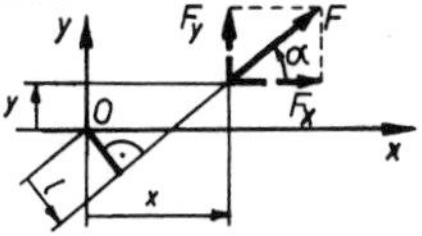

k 10 $M_0 = +F \times l = F_y\,x - F_x\,y$

(for signs of trigonom. funct. of α see table k16 through k19)

Resultant force F_R of any random given forces

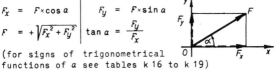

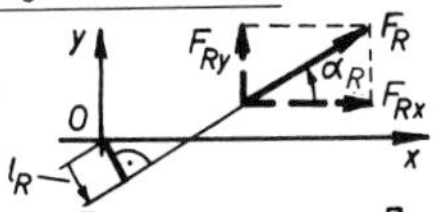

k 11 components $F_{Rx} = \Sigma F_x$ | $F_{Ry} = \Sigma F_y$

k 12 magnitude $F_R = +\sqrt{F_{Rx}^2 + F_{Ry}^2}$

k 13 angle of direction α_{F_R} $\tan\alpha_R = \dfrac{F_{Ry}}{F_{Rx}}$; $\sin\alpha_R = \dfrac{F_{Ry}}{F_R}$; $\cos\alpha_R = \dfrac{F_{Rx}}{F_R}$

k 14 distance $l_R = \dfrac{|\Sigma M_0|}{|F_R|}$ (moment theorem)

k 15 sign of $F_R \times l_R$ = sign of ΣM_0

Signs of trigonometr. functions of x, y; F_x, F_y; F_{Rx}, F_{Ry}

	quadrant	α, α_R	$\cos\alpha$	$\sin\alpha$	$\tan\alpha$	x, F_x, F_{Rx}	y, F_y, F_{Ry}
k 16	I	0... 90°	+	+	+	+	+
k 17	II	90...180°	−	+	−	−	+
k 18	III	180...270°	−	−	+	−	−
k 19	IV	270...360°	+	−	−	+	−

F_x, F_y : components of F parallel to x-axis and y-axis
F_{Rx}, F_{Ry} : components of F_R parallel to x-axis and y-axis
x, y : coordinates of F
α, α_R : angles of F and F_R
l, l_R : distances of F and F_R from reference point

Conditions of equilibrium

A body is said to be in equilibrium when both the resultant force and the sum of the moments of all external forces about any random point are equal to zero.

	forces	graphical	mathematical
k 20	with common point of application	closed force polygon	$\Sigma F_x = 0;\ \Sigma F_y = 0;$
k 21	parallel to vertical axis	force polyg. and link polygon closed	$\Sigma F_y = 0;\ \Sigma M = 0$
k 22	arbitrary		$\Sigma F_x = 0;\ \Sigma F_y = 0;\ \Sigma M = 0$

Simply supported beam with point loads W_1 and W_2

Find reactions R_A and R_B:

Graph. solution:

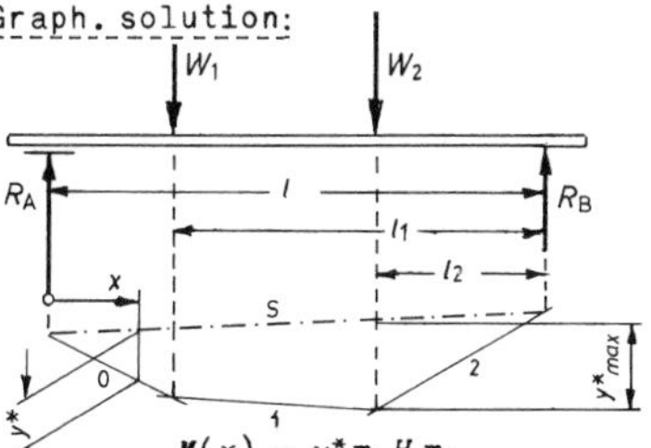

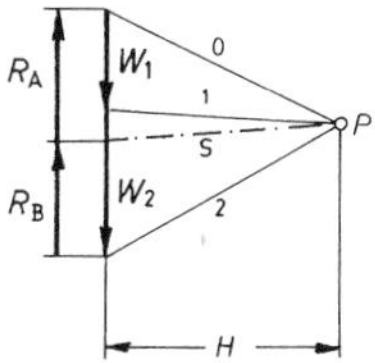

$M(x) = y^* m_L H m_F$ kNm, Ncm, Nmm

k 23 m_L : scale of lengths = true length/diagram length

k 24 m_F : scale of forces = force/diagram length

H : pole spacing | y^*: vertical distance between closing line s and link polygon.

k 25 Mathem. solution: $R_A = W_1\ l_1/l + W_2\ l_2/l$; $R_B = (W_1 + W_2) - R_A$

Distributed loads are divided into small sections and considered as corresponding point forces acting through the centres of mass of the sections.

Wall mounted crane (3 forces): Find reactions F_A, F_B.

problem

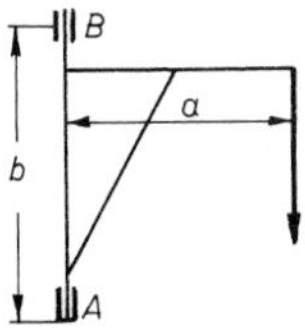

solution

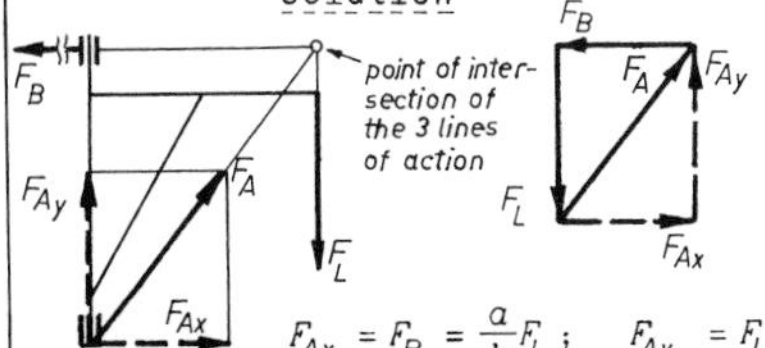

$F_{Ax} = F_B = \frac{a}{b} F_L$; $F_{Ay} = F_L$

Mathematical determination of member loads

(Ritter method - Method of Sections)

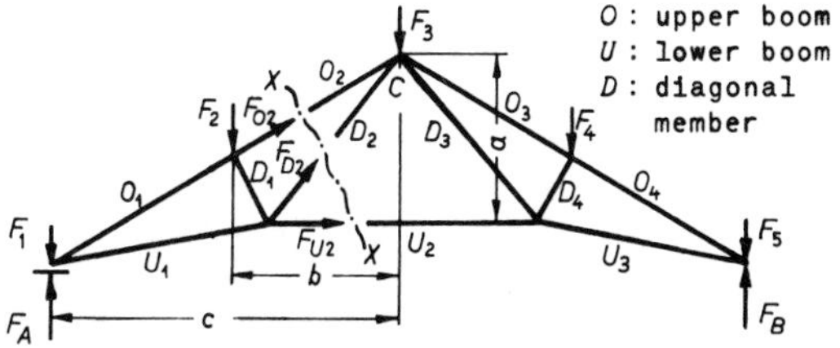

k 26

Determine the support reactions from K 4 (girder on two supports). Draw a line $X...X$ through the framework to the bar in question, but intersecting no more than 3 bars. Take tensile forces as positive, so that compressive forces are negative.

Establish the equation of moments $\Sigma M = 0$ with moments of external and internal forces taken about the point of intersection of two unknown forces.

Rule for moment signs

Where turning moment is counter-clockwise the sign is positive

Where turning moment is clockwise the sign is negative

Example (from the above girder)

problem: to find force F_{U2} in bar U_2.

solution:

Draw a line $X...X$ through $O_2 - D_2 - U_2$. Since the lines O_2 and D_2 meet at C, this is the relative point of intersection selected so that the moment O_2 and D_2 may equal zero.

Proceed as follows: $\Sigma M_C = 0$

$$+ a\,F_{U2} + b\,F_2 - c(F_A - F_1) = 0$$

$$F_{U2} = \frac{-b\,F_2 + c(F_A - F_1)}{a}$$

STATICS
Lattice girder

K 6

Graphical determination of forces in members

(Cremona method – Bow diagram)

k 27

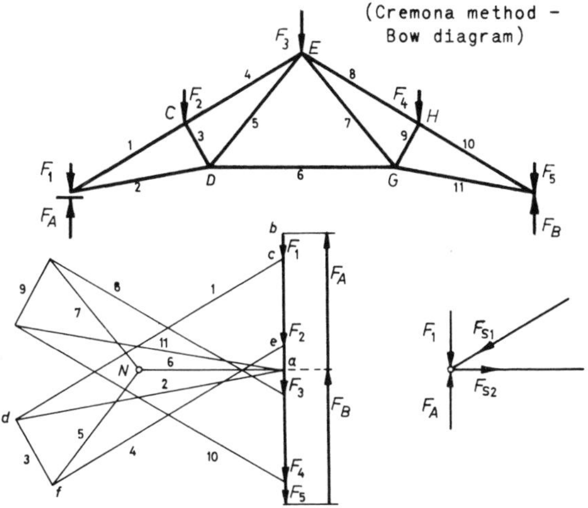

Basic principles

Each bar is confined by two adjacent joints. The external forces only act through the joints.

Procedure

Establish a scale of forces and determine the support reactions. Since each force polygon must not contain more than two unknown forces, start at joint *A*. Establish identical order of forces (cw or ccw) for all joints (e.g. $F_A - F_1 - F_{S1} - F_{S2}$).

Joint *A*: Force polygon *a* – *b* – *c* – *d* – *a*. Keep a record of forces being tensile or compressional.

Joint *C*: Force polygon *d* – *c* – *e* – *f* – *d*.

u.s.w.

Check

Forces acting through a single joint in the framework form a polygon in the Bow diagram.

Forces acting through a single point in the diagram of forces form a triangle in the framework.

Arc of circle

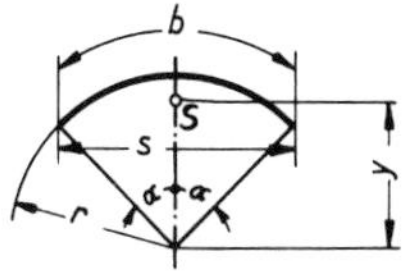

k 28 $y = \dfrac{r \sin \alpha \, 180^\circ}{\pi \, \alpha} = \dfrac{r \, s}{b}$

k 29 $y = 0{\cdot}6366 \, r$ bei $2\alpha = 180^\circ$

k 30 $y = 0{\cdot}9003 \, r$ bei $2\alpha = 90^\circ$

k 31 $y = 0{\cdot}9549 \, r$ bei $2\alpha = 60^\circ$

Triangle

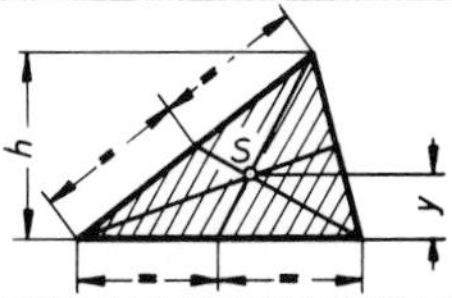

k 31 $y = \dfrac{1}{3} h$

S is the point of intersection of the medians

Sector of a circle

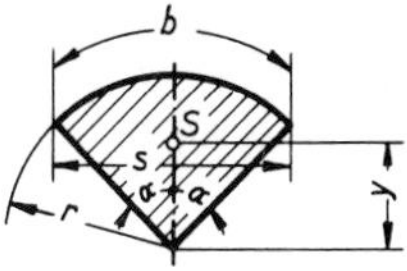

k 32 $y = \dfrac{2r \sin \alpha \, 180^\circ}{3 \, \pi \, \alpha} = \dfrac{2r \, s}{3b}$

k 33 $y = 0{\cdot}4244 \, r$ bei $2\alpha = 180^\circ$

k 34 $y = 0{\cdot}6002 \, r$ bei $2\alpha = 90^\circ$

k 35 $y = 0{\cdot}6366 \, r$ bei $2\alpha = 60^\circ$

Trapezium

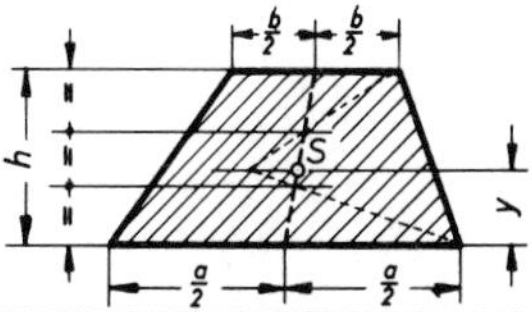

k 36 $y = \dfrac{h}{3} \times \dfrac{a + 2b}{a + b}$

Sector of an annulus

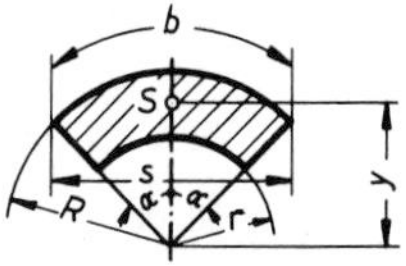

k 37 $y = \dfrac{2}{3} \times \dfrac{R^3 - r^3}{R^2 - r^2} \times \dfrac{\sin \alpha}{\operatorname{arc} \alpha}$

k 38 $ = \dfrac{2}{3} \times \dfrac{R^3 - r^3}{R^2 - r^2} \times \dfrac{s}{b}$

Segment of a circle

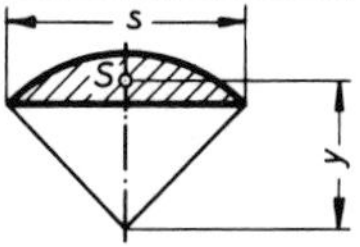

k 39 $y = \dfrac{s^3}{12 \, A}$

for area A see B 3

For determination of centre of gravity S, see also J 7

Determination of centre of gravity of any random surface area

Graphical solution

Subdivide the total area A into partial areas A_1, $A_2 \ldots A_n$ the centres of gravity of which are known. The size of each partial area is represented as a force applied to the centre of area of each partial area. Use the force polygon (see K 2) to determine the mean forces A_{R_x} and A_{R_y} operating in any two directions (preferably at right angles). The point of intersection of the lines of application will indicate the position of the centre of area A.

40

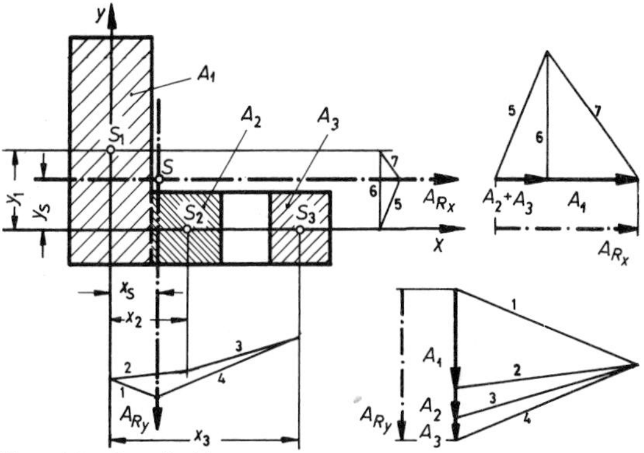

Mathematical solution

Subdivide the above total area A into partial areas A_1, $A_2 \ldots A_n$; we now get

	distance	in general	in the above example
41	$x_S=$	$\dfrac{\sum_{i=1}^{n} A_i \; x_i}{A}$	$\dfrac{A_1 \; x_1 + A_2 \; x_2 + A_3 \; x_3}{A}$
42	$y_S=$	$\dfrac{\sum_{i=1}^{n} A_i \; y_i}{A}$	$\dfrac{A_1 \; y_1 + A_2 \; y_2 + A_3 \; y_3}{A}$

Note: In the above example the distances x_1, y_2 and y_3 each equal zero.

Force acting parallel to a sliding plane

	static friction	limiting frict.	sliding friction
43	$\upsilon = 0$	$\upsilon = 0$	$\rightarrow \upsilon > 0$

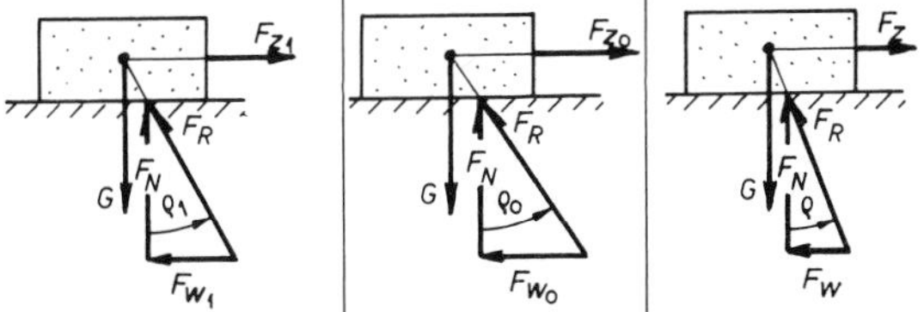

	static friction	limiting frict.	sliding friction
44	$F_{W_1} = -F_{Z_1} = G \tan \varrho_1$	$F_{W_0} = -F_{Z_0} = G \tan \varrho_0$	$F_W = -F_Z = G \tan \varrho$
45	$F_N = -G$	$F_N = -G$	$F_N = -G$
46		$\mu_0 = \tan \varrho_0 > \mu$	$\mu = \tan \varrho < \mu_0$
47	$0 < \varrho_1 (\text{variable}) < \varrho_0$	$\varrho_0 = \text{const.} > \varrho$	$\varrho = \text{const.} < \varrho_0$

A force F_{Z1} gradually increasing from zero is compensated by an increasing F_{W1} without causing the body to move, until F_{Z1} reaches the value

48 $$F_{Z0} = G\mu_0 .$$

As soon as this happens, the body starts sliding, whereby F_Z drops to $G\mu$. Any excessive force will now accelerate the body.

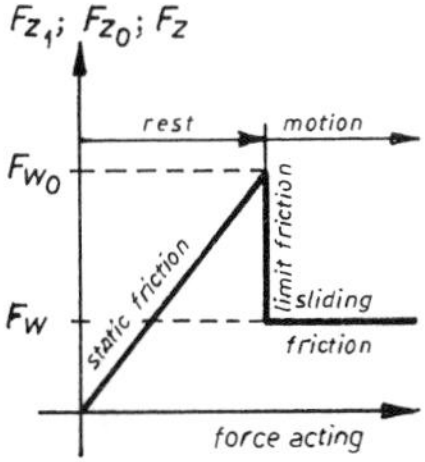

Force applied obliquely

The force F needed to set in motion a body, weight G:

49 $$F = G \frac{\mu_0}{\sin \alpha - \mu_0 \cos \alpha} = G \frac{\sin \varrho_0}{\sin(\alpha - \varrho_0)}$$

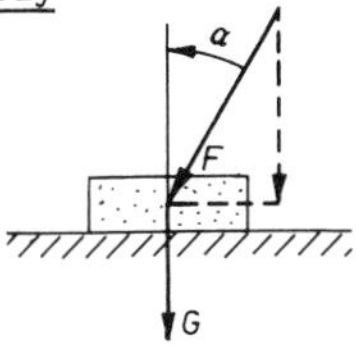

The force needed to maintain the motion is ascertained by replacing μ_0 by μ. No motion possible when result of F is negative.

F_{W_1}, F_{W_0}, F_W : frict. coeff.	$-, \mu_0, \mu$: frict. force	see
F_{Z_1}, F_{Z_0}, F_Z : angle of frict.	$\varrho_1, \varrho_0, \varrho$: tract. force	Z 7

Inclined plane

General

The angle α at which a body will move easily down an inclined plane is the angle of friction ϱ.

$$\tan\alpha = \tan\varrho = \mu$$

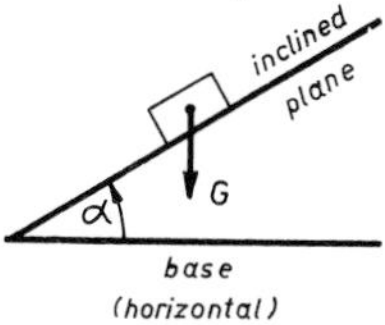

K 50 Application in the experimental determination of the angle of friction or the friction coefficient:

$$\mu = \tan\varrho$$

K 51 Condition of automatic locking: $\alpha < \varrho$

Friction properties

motion	constant velocity maintained by tractive force F parallel to inclined plane	constant velocity maintained by tractive force F parallel to base
upwards $0 < \alpha < \alpha^*$	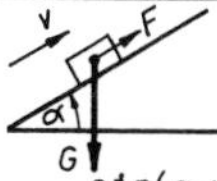 $F = G\,\dfrac{\sin(\alpha+\varrho)}{\cos\varrho}$	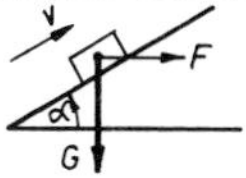 $F = G\,\tan(\alpha+\varrho)$
downwards $0 < \alpha < \varrho$	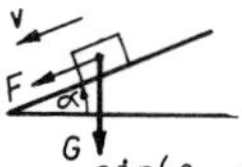 $F = G\,\dfrac{\sin(\varrho-\alpha)}{\cos\varrho}$	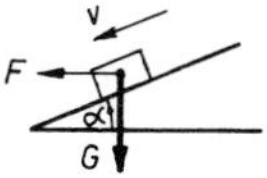 $F = G\,\tan(\varrho-\alpha)$
downwards $\varrho < \alpha < \alpha^*$	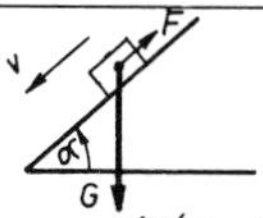 $F = G\,\dfrac{\sin(\alpha-\varrho)}{\cos\varrho}$	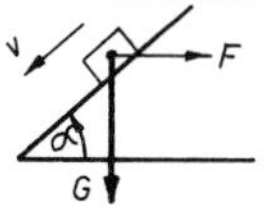 $F = G\,\tan(\alpha-\varrho)$

K 52 (upwards), K 53 (downwards, $0 < \alpha < \varrho$), 54 (downwards, $\varrho < \alpha < \alpha^*$)

Note: For static friction replace μ by μ_0 and ϱ by ϱ_0.

α^*: Tilting angle of body

Wedges

		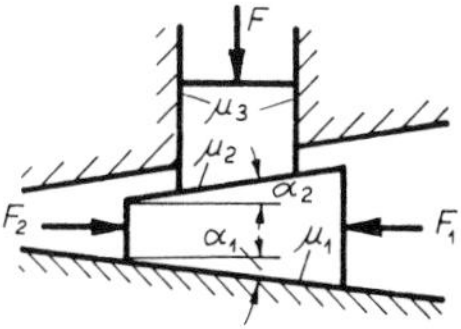	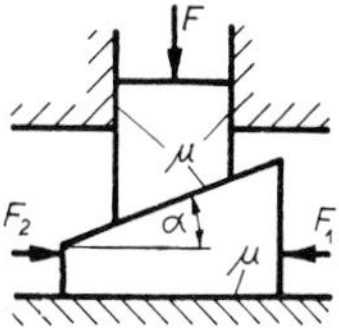
k 55	driving in	$F_1 = F\ \dfrac{\tan(\alpha_1+\varrho_1)+\tan(\alpha_2+\varrho_2)}{1-\tan\varrho_3\ \tan(\alpha_2+\varrho_2)}$	$F_1 = F\ \tan(\alpha+2\varrho)$
k 56	loosening	$F_2 = F\ \dfrac{\tan(\alpha_1-\varrho_1)+\tan(\alpha_2-\varrho_2)}{1+\tan\varrho_3\ \tan(\alpha_2-\varrho_2)}$	$F_2 = F\ \tan(\alpha-2\varrho)$
k 57	autom. locking	$\alpha_1 + \alpha_2 \leqq \varrho_{0_1} + \varrho_{0_2}$	$\alpha \leqq 2\varrho_0$

Screws

			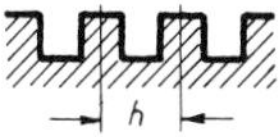	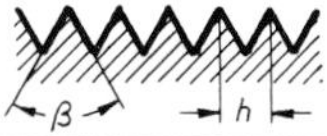
k 58	turning moment when	raising	$M_1 = F\ r\ \tan(\alpha+\varrho)$	$M_1 = F\ r\ \tan(\alpha+\varrho')$
k 59		lowering	$M_2 = F\ r\ \tan(\alpha-\varrho)$	$M_2 = F\ r\ \tan(\alpha-\varrho')$
k 60	conditions of automatic locking when lowering		$\alpha < \varrho$	$\alpha < \varrho'$
k 61	efficiency of screw when	raised	$\eta = \dfrac{\tan\alpha}{\tan(\alpha+\varrho)}$	$\eta = \dfrac{\tan\alpha}{\tan(\alpha+\varrho')}$
k 62		lowered	$\eta = \dfrac{\tan(\alpha-\varrho)}{\tan\alpha}$	$\eta = \dfrac{\tan(\alpha-\varrho')}{\tan\alpha}$

M_1 : raising moment — N m, [kgf m]

M_2 : lowering moment — N m, [kgf m]

α : lead of thread $\left(\tan\alpha = \dfrac{h}{2\pi r}\right)$

k 63 ϱ : angle of friction $(\tan\varrho = \mu)$

k 64 ϱ' : angle of friction $\left(\tan\varrho' = \dfrac{\mu}{\cos\beta/2}\right)$

k 65 r : mean radius of thread — m, mm

Bearing friction

radial bearing	longitudinal bearing
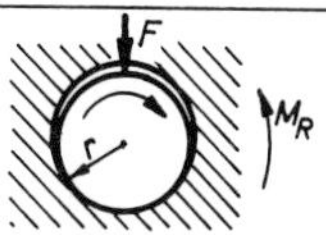	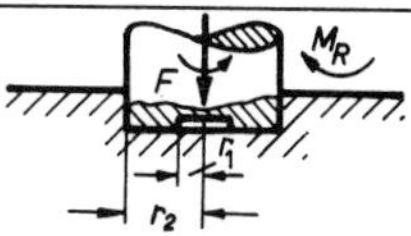
K 66 $M_R = \mu_q \, r \, F$	$M_R = \mu_L \dfrac{r_1 + r_2}{2} F$

M_R: moment of friction

μ_q: coefficient of friction of a radial bearing (not constant values)

μ_l: coefficient of friction of a longitudinal bearing (not constant values)

Note: μ_q and μ_l are determined experimentally as a function of bearing condition, bearing clearance,

K 67 and lubrication. For run in condition: $\mu_o \approx \mu_l \approx \mu_q$.

K 68 Always use $r_1 > 0$ to allow for lubrication.

Rolling resistance

Rolling of a cylinder

K 69 $F = \dfrac{f}{r} F_N \approx \dfrac{f}{r} G$

K 70 Rolling condit.: $F_W < \mu_o F_N$

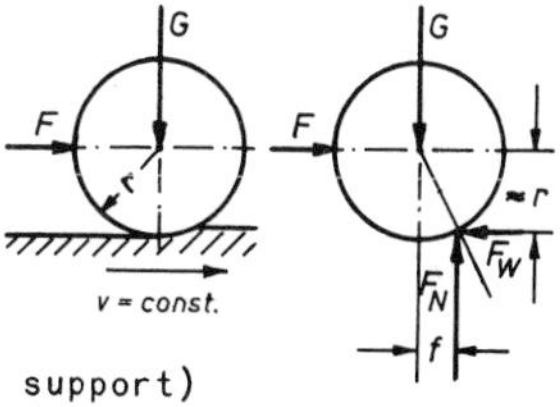

F_W: force of rolling resist.

f: lever arm of rolling resistance - value on Z 7 (caused by deformation of cylinder and support)

μ_o: coeff. of static frict. between cylinder and support

Displacement of a plate supported by cylinders

K 71 $F = \dfrac{(f_1 + f_2)\, G_1 + n\, f_2\, G_2}{2\, r}$

K 72 where $f_1 = f_2 = f$ and $n\, G_2 < G_1$:

K 73 $F = \dfrac{f}{r} G_1$

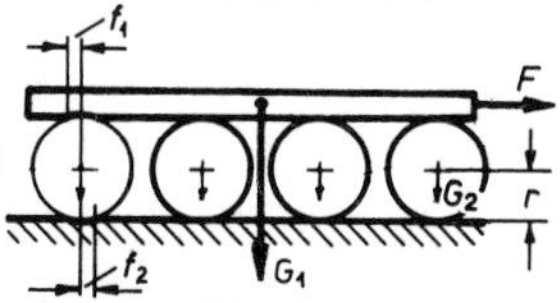

G_1, G_2: weight of plate and one cylinder

f_1, f_2: lever arms of force of rolling resistance

F: tractive force

r: radius of cylinder | n: number of cylinders

Rope friction

	tract. force and frict. force for raising load	lowering load
k 75	$F_1 = e^{\mu\hat{\alpha}}\,G$	$F_2 = e^{-\mu\hat{\alpha}}\,G$
k 76	$F_R = (e^{\mu\hat{\alpha}} - 1)G$	$F_R = (1 - e^{-\mu\hat{\alpha}})G$

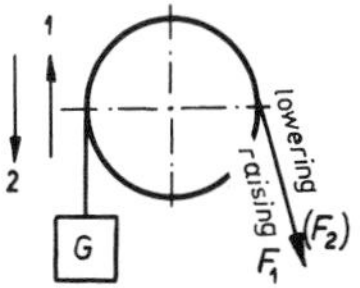

Formulae apply where cylinder is stationary and rope is moving at constant velocity, or where rope is stationary and cylinder is rotating at constant angular velocity.

k 77 Condit. of equilibrium: $F_2 < F < F_1$ | $G\,e^{-\mu\hat{\alpha}} < F < G\,e^{\mu\hat{\alpha}}$

(F : force without friction)

Belt drive

k 78 $$F_U = \frac{M_a}{r}$$

k 79 $$F_U = F_R$$

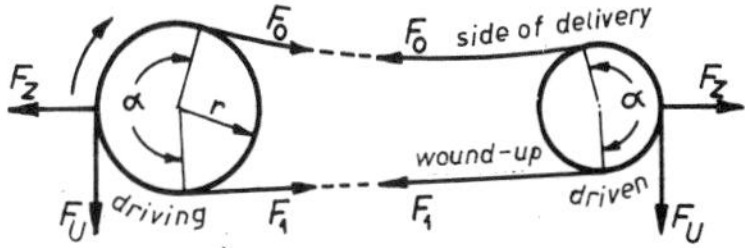

	forces	in motion	at rest
k 80	F_o	$F_o = \dfrac{F_U}{e^{\mu\hat{\alpha}} - 1}$	$F_o = F_1 = \dfrac{F_z}{2}\,\dfrac{(e^{\mu\hat{\alpha}} + 1)}{e^{\mu\hat{\alpha}} - 1)}$
k 81	F_1	$F_1 = F_U\,\dfrac{e^{\mu\hat{\alpha}}}{e^{\mu\hat{\alpha}} - 1}$	
k 82	F_z	$F_z = F_U\,\dfrac{e^{\mu\hat{\alpha}} + 1}{e^{\mu\hat{\alpha}} - 1}$	

F_U : tangential force of driving wheel
F_R : frictional force of rope
M_a: driving torque
$\hat{\alpha}$: angle of contact (radians). Always introduce lowest value into formula
μ : coeff. of sliding frict. (value of exper. obtained for
k 83 leather belt on steel drum: $\mu = 0{,}22 + 0{,}012\,v$ s/m
v : belt velocity
e = 2.718 281 83... (base of natural logs)

Rope operated machines

The following figures deal solely with rope rigidity, disregarding bearing friction.

	unknown quantity	fixed sheave	free sheave	pulley block ordinary	pulley block differential
		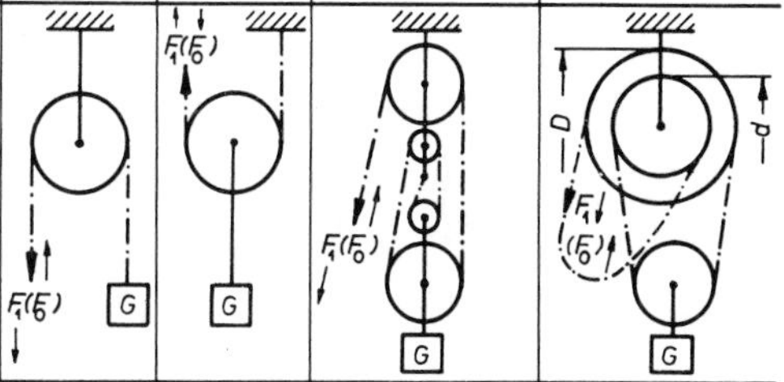			
k 84	$F_1 =$	$\varepsilon\, G$	$\dfrac{\varepsilon}{1+\varepsilon}\, G$	$\dfrac{\varepsilon^n(\varepsilon - 1)}{\varepsilon^n - 1}\, G$	$\dfrac{\varepsilon^2 - \dfrac{d}{D}}{\varepsilon + 1}\, G$
k 85	$F_0 =$	$\dfrac{1}{\varepsilon}\, G$	$\dfrac{1}{1+\varepsilon}\, G$	$\dfrac{\dfrac{1}{\varepsilon^n}\left(\dfrac{1}{\varepsilon} - 1\right)}{\dfrac{1}{\varepsilon^n} - 1}\, G$	$\dfrac{\varepsilon}{1+\varepsilon}\left(\dfrac{1}{\varepsilon^2} - \dfrac{d}{D}\right) G$
k 86	$F =$	G	$\dfrac{1}{2}\, G$	$\dfrac{1}{n}\, G$	$\dfrac{1}{2}\left(1 - \dfrac{d}{D}\right) G$
k 87	$s =$	h	$2\, h$	$n\, h$	$\dfrac{2}{1 - \dfrac{d}{D}}\, h$

k 88 mechanical advantage $= \dfrac{\text{force}}{\text{effort}} = \dfrac{F}{G} = \dfrac{h}{s}$

F_1 : force required to raise load, disregarding bearing friction
F_0 : force required to lower load, disregarding bearing friction
F : force, disregarding both rope rigidity and bearing friction

k 89 $\varepsilon = \dfrac{1}{\eta}$: loss factor for rope rigidity (for wire ropes and chains $\approx 1{\cdot}05$)

η : efficiency
n : number of sheaves
h : path of load
s : path of force

KINEMATICS
General terms
L1

General

Kinematics deals with motions of bodies as a function of time.

The most important quantities of kinematics and their units

Length l, see K 1
Units: m; km

Rotational angle φ
Unit: rad

Time t
Is a base quantity, see preface.
Units: s; min; h

Frequency f
The frequency of a harmonic or sinusoidal oscillation is the ratio of the number of periods (full cycles) and the corresponding time.

1 1 $$f = \frac{\text{number of oscillations}}{\text{corresponding time}}$$

Units: Hz (Hertz) = 1/s = cycle/s; 1/min

Period T
The period T is the time required for one full cycle. It is the reciprocal of the frequency f.

1 2 $$T = \frac{1}{f}$$

Units: s; min; h

Rotational speed n
Where an oscillation is tightly coupled with the rotation of a shaft, and one revolution of the shaft corresponds exactly to one full cycle of the oscillation, the rotational speed n of the shaft is equal to the frequency f of the oscillation.

1 3 $$n = f\,; \quad f = \frac{n \text{ min}}{60 \text{ s}}$$

Units: revolutions/second (1/s)
rev/min, r.p.m. (1/min)

continued on L 2

continued from L 1

Velocity v

The velocity v is the first derivative of the distance s with respect to the time t:

L 4 $$v = \frac{ds}{dt} = \dot{s}$$

Where the velocity is constant, the following relation applies:

L 5 $$v = \frac{s}{t}$$

Units: m/s; km/h

Angular velocity ω, angular frequency ω

The angular velocity ω is the first derivative of the angle turned through φ, with respect to the time t:

L 6 $$\omega = \frac{d\varphi}{dt} = \dot{\varphi}$$

Hence, for constant angular velocity:

L 7 $$\omega = \frac{\varphi}{t}$$

Where $f = n$ (see 13), the angular velocity ω is equal to the angular frequency ω.

L 8 $$\omega = 2\pi f = 2\pi n = \dot{\varphi}$$

Units: 1/s; rad/s; 1°/s

Acceleration a

The acceleration a is the first derivative of the velocity v with respect to the time t:

L 9 $$a = \frac{dv}{dt} = \dot{v} = \frac{d^2 s}{dt^2} = \ddot{s}$$

Units: m/s^2; km/h^2

Angular acceleration α

The angular acceleration α is the first derivative of the angular velocity ω with respect to the time t:

L 10 $$\alpha = \frac{d\omega}{dt} = \dot{\omega} = \frac{d^2\varphi}{dt^2} = \ddot{\varphi}$$

Units: $1/s^2$; rad/s^2; $1^\circ/s^2$

Distance, velocity, and acceleration of mass point in motion

Distance-time curve

An s–t curve is recorded for the motion. The first derivative of this curve is the instantaneous velocity v:

11 $$v = \frac{ds}{dt} = \dot{s}$$

It is the slope dt the tangent to the s–t curve.

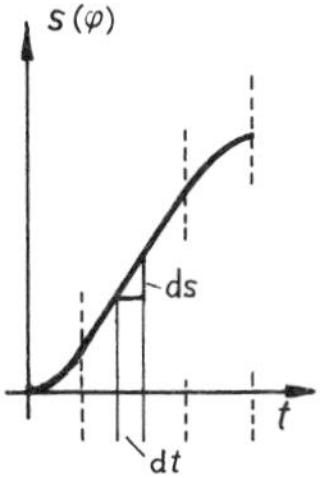

Velocity-time curve

The velocity-time history is expressed as a v–t curve. The first derivative of this curve is the instantaneous acceleration a. Hence the acceleration is the second derivative of the distance-time curve.

12 $$a = \frac{dv}{dt} = \dot{v} = \ddot{s}$$

It is the slope dt the tangent to the v–t curve.

The shaded area represents the distance travelled $s(t)$.

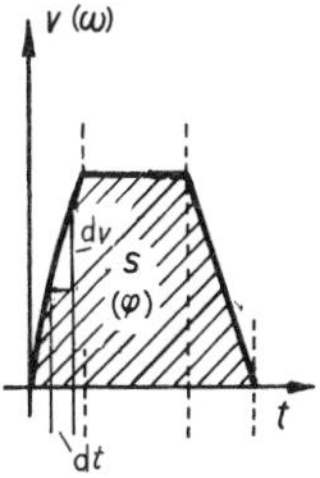

Acceleration-time curve

The acceleration-time history is shown as an a–t curve, which enables peak accelerations to be determined.

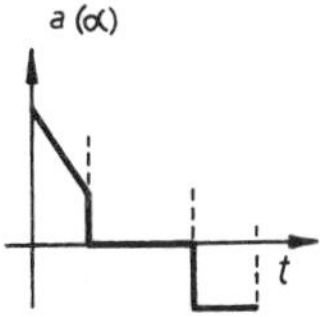

13 $a > 0$: acceleration (increasing velocity)

14 $a < 0$: retardation (decreasing velocity)

Note to diagrams

The letters in brackets apply to rotations (for explanation see L 5 and L 6).

KINEMATICS

The most important kinds of motion

Linear motion

Paths are straight lines. All points of a body cover congruent paths.

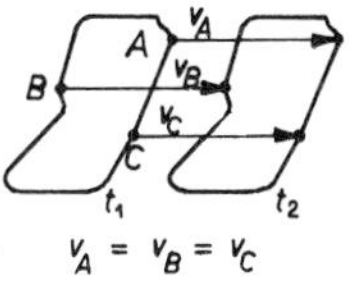

Special linear motions

	uniform	uniform accelerated
	motion	
L 15	$v = v_o = \text{const.}$	$a = a_o = \text{constant}$

Rotational motion

Paths are circles about the axis. Angle turned through $\widehat{\varphi}$, angular velocity ω, and angular acceleration α are identical for all points of the body.

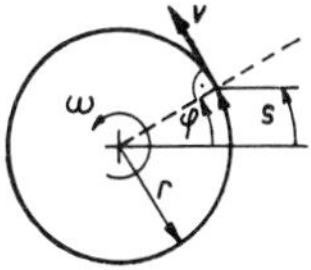

Special rotational motions

	uniform	uniform accelerated
	motion	
L 16	$\omega = \omega_o = \text{const.}$	$\alpha = \alpha_o = \text{constant}$

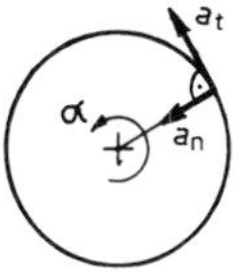

Distance s, velocity v, and tangential acceleration a_t are proportional to the radius:

L 17 $s = r\widehat{\varphi}\ ; \quad v = r\omega\ ; \quad a = r\alpha = a_t$

L 18 centripetal acceleration $a_n = \omega^2 r = \dfrac{v^2}{r}$

Harmonic oscillation

Paths are straight lines or circles. The body moves back and forth about a position of rest. The maximum deflection from this position is called "amplitude".

Instantaneous position, velocity, and acceleration are harmonic functions of time.

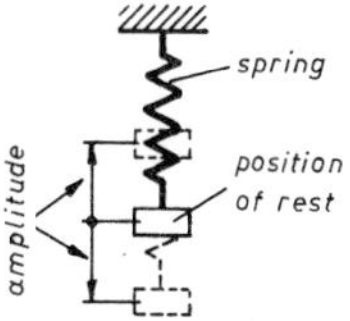

KINEMATICS

Linear motion

L 5

Uniform and uniform accelerated linear motion

	unknown parameter	uniform $a = 0$ $\upsilon = \text{const.}$	uniform { accelerated ($a > 0$), retarded ($a < 0$) } a = constant, $\upsilon_0 = 0$	$\upsilon_0 > 0$	EU
		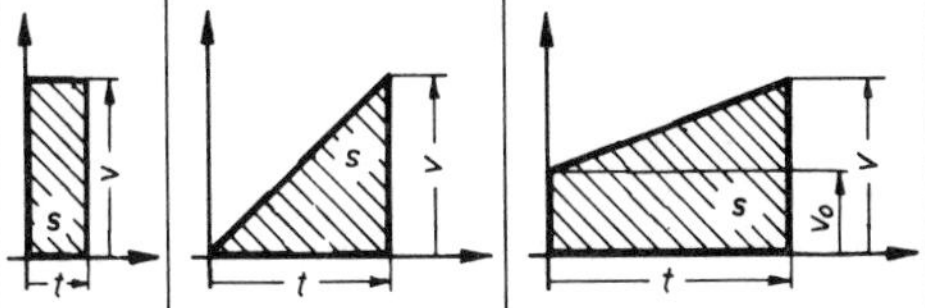			
L 19	$s =$	$\upsilon\, t$	$\dfrac{\upsilon\, t}{2} = \dfrac{a\, t^2}{2} = \dfrac{\upsilon^2}{2\, a}$	$\dfrac{t}{2}(\upsilon_0 + \upsilon) = \upsilon_0 t + \dfrac{1}{2} a\, t^2$	m cm km
L 20	$\upsilon =$	$\dfrac{s}{t}$	$\sqrt{2as} = \dfrac{2\, s}{t} = a\, t$	$\upsilon_0 + a\, t = \sqrt{\upsilon_0^2 + 2as}$	m/s cm/s km/h
L 21	$\upsilon_0 =$	const.	0	$\upsilon - a\, t = \sqrt{\upsilon^2 - 2as}$	
L 22	$a =$	0	$\dfrac{\upsilon}{t} = \dfrac{2s}{t^2} = \dfrac{\upsilon^2}{2s}$	$\dfrac{\upsilon - \upsilon_0}{t} = \dfrac{\upsilon^2 - \upsilon_0^2}{2s}$	m/s^2 cm/h^2 km/h^2
L 23	$t =$	$\dfrac{s}{\upsilon}$	$\sqrt{\dfrac{2\, s}{a}} = \dfrac{\upsilon}{a} = \dfrac{2s}{\upsilon}$	$\dfrac{\upsilon - \upsilon_0}{a} = \dfrac{2s}{\upsilon_0 + \upsilon}$	s min h

Note

The shaded areas represent the distance s covered during the time period t.

KINEMATICS

Rotational motion

Uniform and uniform accelerated rotation about a fixed axis

	unknown parameter	uniform $\alpha = 0$ ω=const.	uniform accelerated ($\alpha > 0$) retarded ($\alpha < 0$) α = constant $\omega_0 = 0$	$\omega_0 > 0$	EU
		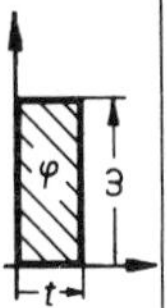	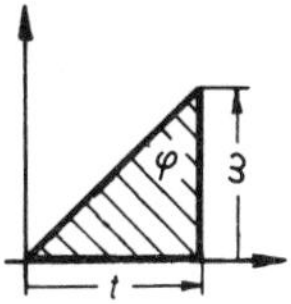	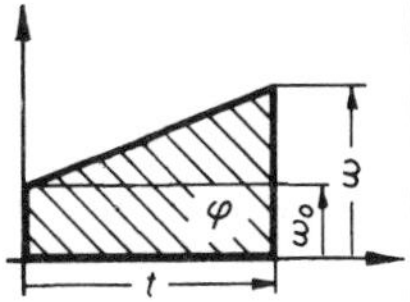	
L 24	$\varphi =$	ωt	$\frac{\omega t}{2} = \frac{\alpha t^2}{2} = \frac{\omega^2}{2\alpha}$	$\frac{t}{2}(\omega_0 + \omega) = \omega_0 t + \frac{1}{2}\alpha t^2$	– rad
L 25	$\omega =$	$\frac{\varphi}{t}$	$\sqrt{2\alpha\varphi} = \frac{2\varphi}{t} = \alpha t$	$\omega_0 + \alpha t = \sqrt{\omega_0^2 + 2\alpha\varphi}$	1/s m/m s rad/s
L 26	$\omega_0 =$	const.	0	$\omega - \alpha t = \sqrt{\omega^2 - 2\alpha\varphi}$	
L 27	$\alpha =$	0	$\frac{\omega}{t} = \frac{2\varphi}{t^2} = \frac{\omega^2}{2\varphi}$	$\frac{\omega - \omega_0}{t} = \frac{\omega^2 - \omega_0^2}{2\varphi}$	$1/s^2$ m/m s^2 rad/s^2
L 28	$t =$	$\frac{\varphi}{\omega}$	$\sqrt{\frac{2\varphi}{\alpha}} = \frac{\omega}{\alpha} = \frac{2\varphi}{\omega}$	$\frac{\omega - \omega_0}{\alpha} = \frac{2\varphi}{\omega_0 + \omega}$	s min h

Note

The shaded areas represent the angle of rotation φ covered during a time period t. (Angle of rotation $\varphi = 2\pi \times$ number of rotations resp. $360^\circ \times$ number of rotations)

KINEMATICS
Oscillations
L 7

Linear simple harmonic motion

A body supported by a spring performs a linear harmonic oscillation. For this kind of motion, quantities s, v, and a as functions of time are equal to the projections $\underline{s}$, $\underline{v}$, and $\underline{a_n}$ of a uniform rotation of a point.

uniform rotation	path (project.)	simple harmonic motion

position — position-time curve

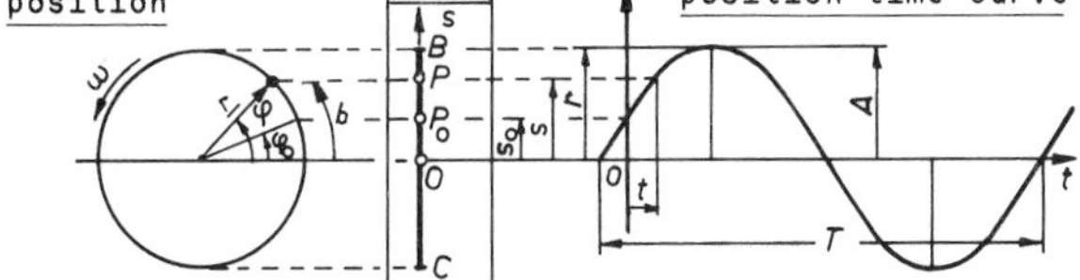

29 $\varphi = \omega t + \varphi_0$; $b = r(\omega t + \varphi_0)$ $\qquad s = A \sin(\omega t + \varphi_0)$

velocity — velocity-time curve

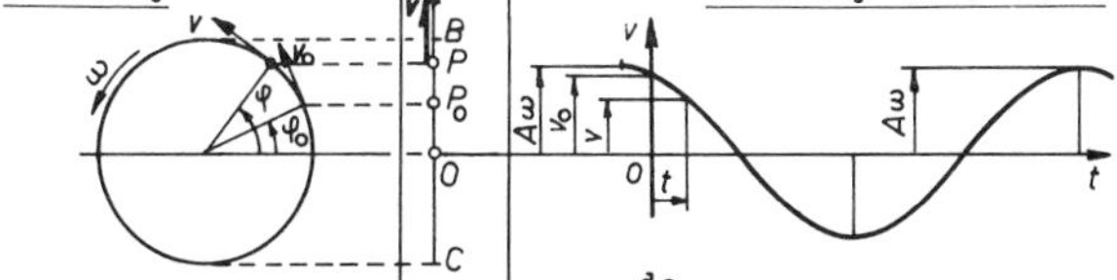

30 $v = r\,\omega$ $\qquad v = \dfrac{ds}{dt} = A\omega \cos(\omega t + \varphi_0)$

acceleration — acceleration-time curve

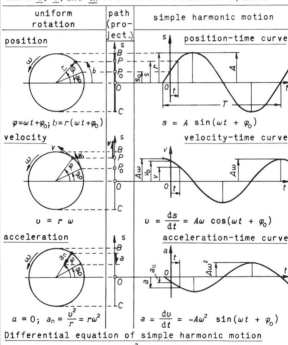

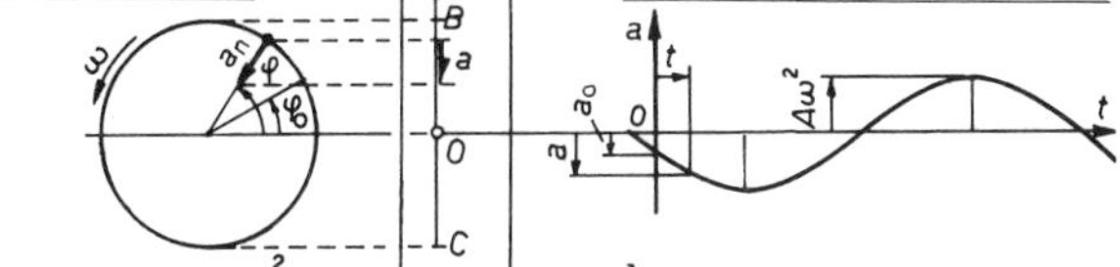

31 $a = 0$; $a_n = \dfrac{v^2}{r} = r\omega^2$ $\qquad a = \dfrac{dv}{dt} = -A\omega^2 \sin(\omega t + \varphi_0)$

Differential equation of simple harmonic motion

32 $$a = \frac{d^2 s}{dt^2} = -\omega^2 s$$

φ_0	angular	at time $t = 0$	s: displacement
φ	position	at time t	A: amplitude (max. displacem.)

a_n: centripetal accelerat. | r: radius of circle

$\underline{r}$: radius vector (origin: centre of circle; head: pos. of body)

B, C: extreme positions of oscillating point

KINEMATICS

Free fall and projection

L 8

Free fall and vertical projection

	unknown param.	free fall $v_0 = 0$	vertic. project. upwards ($v_0 > 0$), downwnds. ($v_0 < 0$)	EU
		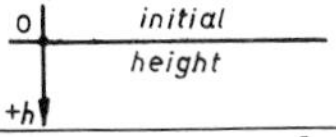	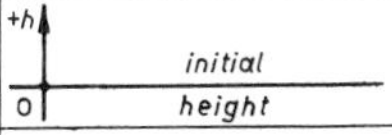	
33	$h =$	$\frac{g}{2}t^2 = \frac{v}{2}t = \frac{v^2}{2g}$	$v_0 t - \frac{g}{2}t^2 = \frac{v_0 + v}{2}\,t$	m, cm
34	$v =$	$g\,t = \frac{2h}{t} = \sqrt{2gh}$	$v_0 - g\,t = \sqrt{v_0^2 - 2gh}$	m/s, km/h
35	$t =$	$\frac{v}{g} = \frac{2h}{v} = \sqrt{\frac{2h}{g}}$	$\frac{v_0 - v}{g} = \frac{2h}{v_0 + v}$	s, min

Horizontal and angled projection

	unknown param.	horizont. proj. $v_0 > 0$	angled proj. upwards ($\alpha > 0$), downwds. ($\alpha < 0$) $v_0 > 0$	EU
		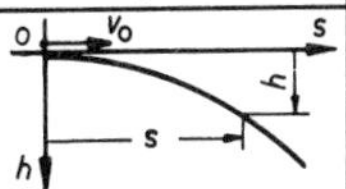	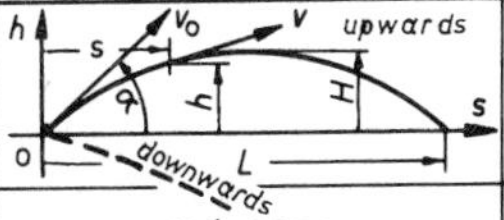	
36	$s =$	$v_0 t = v_0\sqrt{\frac{2h}{g}}$	$v_0 t \cos\alpha$	m, cm
37	$h =$	$\frac{g}{2}t^2$	$v_0 t \sin\alpha - \frac{g}{2}t^2$	m, cm
38	$v =$	$\sqrt{v_0^2 + g^2 t^2}$	$\sqrt{v_0^2 - 2gh}$	m/s, km/s

Range L and max. height H for an angled upwards project.

39	general	$L = \frac{v_0^2}{g}\sin 2\alpha$	$H = \frac{v_0^2}{2g}\sin^2\alpha$	m, cm
40	general	$t_L = \frac{2v_0}{g}\sin\alpha$	$t_H = \frac{v_0}{g}\sin\alpha$	s, min
	maxima	at $\alpha = 45^0$	at $\alpha = 90^0$	
41	maxima	$L_{max} = \frac{v_0^2}{g}$	$H_{max} = \frac{v_0^2}{2g}$	m, cm
42	maxima	$t_{L_{max}} = \frac{v_0\sqrt{2}}{g}$	$t_{H_{max}} = \frac{v_0}{g}$	s, min

α : angle of projection with respect to horizontal plane

t_H : time for height H | t_L : time for distance L

KINEMATICS

Motion on an inclined plane

L 9

Sliding motion on an inclined plane

	unknown param.	excluding friction $\mu = 0$	including friction $\mu > 0$
43	$a =$	$g \sin\alpha$	$g(\sin\alpha - \mu\cos\alpha)$
44			otherwise $g\,\dfrac{\sin(\alpha-\varrho)}{\cos\varrho}$
45	$v =$	$a\,t = \dfrac{2s}{t} = \sqrt{2as}$	
46	$s =$	$\dfrac{t^2}{2} = \dfrac{vt}{2} = \dfrac{v^2}{2a}$	
	α	$0 \ldots \alpha^*$	$\varrho_0 \ldots \alpha^*$

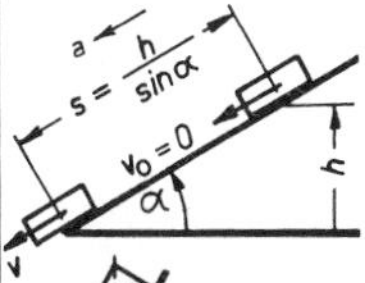

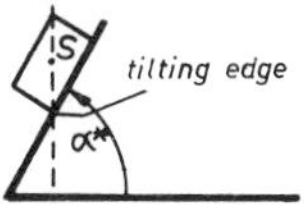

Rolling motion on an inclined plane

	unknown param.	excluding friction $f = 0$	including friction $f > 0$
47	$a =$	$\dfrac{g\,r^2}{r^2+k^2}\sin\alpha$	$g r^2\,\dfrac{\sin\alpha - \frac{f}{r}\cos\alpha}{r^2 + k^2}$
48	$v =$	see above l 45	
49	$s =$	see above l 46	
50	α	$0 \ldots \alpha_{max}$	α_{min}: $\tan\alpha_{min} = \dfrac{f}{r}$
51		$\tan\alpha = \mu_0\dfrac{r^2+k^2}{k^2}$	α_{max}: $\tan\alpha_{max} = \mu_0\,\dfrac{r^2 + k^2 + f\,r}{k^2}$

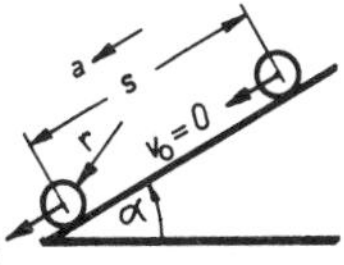

	ball	solid cylinder	pipe with low wall thickness
52	$k^2 = \dfrac{2}{5}r^2$	$k^2 = \dfrac{r^2}{2}$	$k^2 = \dfrac{r_1^2 + r_2^2}{2} \approx r^2$

α^*: tilting angle, where centre of gravity S vertically above tilting edge

μ : coefficient of sliding friction (see Z 20)

μ_0 : coefficient of static friction (see Z 20)

53 ϱ : angle of sliding friction ($\mu = \tan\varrho$)

54 ϱ_0 : angle of static friction ($\mu_0 = \tan\varrho_0$)

f : lever arm of rolling resistance (see k 70 and Z 7)

k : radius of gyration

Simple Conn-Rod mechanism

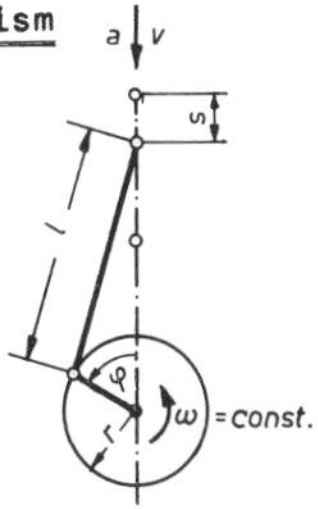

55 $s = r(1 - \cos\varphi) + \frac{\lambda}{2} r \sin^2\varphi$

56 $v = \omega r \sin\varphi(1 + \lambda\cos\varphi)$

57 $a = \omega^2 r(\cos\varphi + \lambda\cos 2\varphi)$

58 $\lambda = \frac{r}{l} = \frac{1}{4} \dots \frac{1}{6}$

59 $\varphi = \omega t = 2\pi n t$

(λ is called the crank ratio)

Scotch-Yoke mechanism

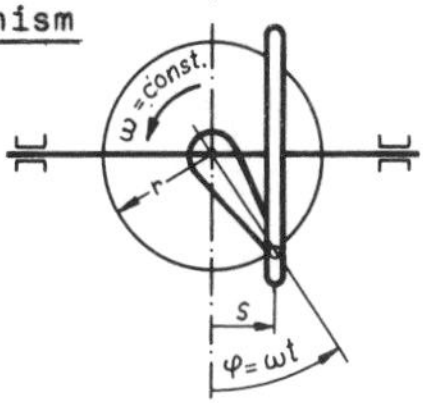

60 $s = r\sin(\omega t)$

61 $v = \omega r \cos(\omega t)$

62 $a = -\omega^2 r \sin(\omega t)$

63 $\omega = 2\pi n$

(motion is simple harmonic)

Cardan joint

For uniform drive the off-drive will be

non-uniform	uniform due to auxiliary shaft H	

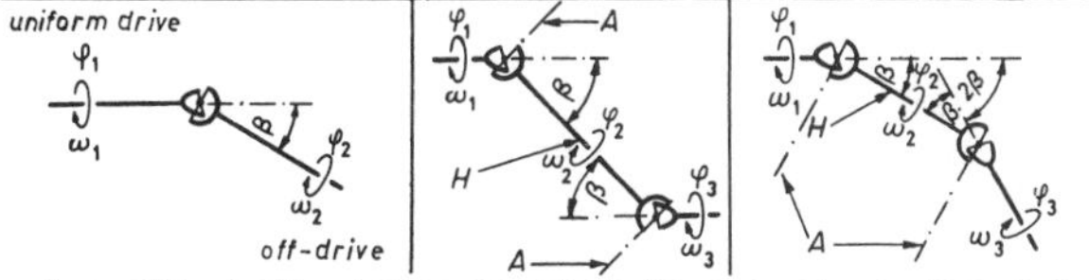

For all shafts being located in one plane the following relations apply:

64 $\tan\varphi_2 = \tan\varphi_1 \times \cos\beta$	$\tan\varphi_3 = \tan\varphi_1$	$\tan\varphi_3 = \tan\varphi_1$
65 $\omega_2 = \omega_1 \frac{\cos\beta}{1 - \sin^2\beta \times \sin^2\varphi_1}$	$\omega_3 = \omega_1$	$\omega_3 = \omega_1$

66 $\alpha_2 = \omega_1^2 \frac{\sin^2\beta \times \cos\beta \times \sin 2\varphi_1}{(1 - \sin^2\beta \times \sin^2\varphi_1)^2}$

	Both axes A of the auxiliary shaft joints must be parall. to each other

The more the angle of inclination β, increases, the more the max. acceleration α and the accelerating moment M_α become; therefore, in practice $\beta \leq 45^\circ$..

General

Dynamics deals with the forces acting on bodies in motion and with the terms "work, energy, and power".

The most important quantities of dynamics and their units

Mass m (is a base quantity, see preface)

Units: kg; Mg = t; g

1 kg is the mass of the international standard. A mass is measured by means of a steelyard.

Force (gravitational force) F

The force F is the product of mass m and acceleration a.

1 $$F = m a$$

The gravitational force W is the force acting on a mass m due to the earth's acceleration g:

2 $$W = m g$$

Being a gravitational force the weight W is measured by means of a spring balance.

Units: N; [kgf; lbf]

1 N is the force that, when acting on a body of a mass m = 1 kg for 1 s, accelerates this body to a final velocity of 1 m s^{-1} (i.e. accelerates this body at 1 m s^{-2}). 9·81 N [= 1 kgf] is the gravitational force acting on a mass of 1 kg due to the earth's attraction.

Work W

The mech. work is the product of force F and distance s, where the constant force F acts on a body in linear motion in a direction parall. to the distance s covered. ($W = F s$)

Units: N m = Joule = J = W s; [kgf m; ft lbf]

Where a force of 1 N acts over a distance of 1 m. it produces the work (energy) of 1 N m (J).

Power P

The power P is the derivative of work with respect to time. Where work (energy) increases or decreases linearily with time, power is the quotient of work and time. ($P = W/t$)

Units: W(Watt); [kgf m s^{-1}; H.P.]

Where for a period of 1 s an energy of 1 J is converted at a constant rate, the corresponding power is 1 W.

$$1\ \text{W} = 1\ \text{J/s}$$

DYNAMICS

Mass, Mass moment of inertia

M 2

Definition of the mass moment of inertia I

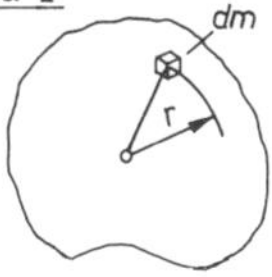

The mass mom. of inertia of a body about an axis has been defined as the sum of the products of mass-elements and the squares of their dist. from the axis.

3 $I = \sum r^2 \Delta m = \int r^2 \, dm$ kg m²

Steiner's theorem (Parallel axis theorem) (see also J 9)

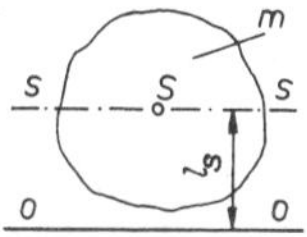

Where the mass moment of inertia of a body of mass m about an axis through its centre of gravity S-S is I_{SS}, the mass moment of inertia about a parallel axis O-O at a distance l_S will be:

4 $I_{OO} = I_{SS} + m \, l_S^2$ kg m²

Radius of gyration k

The radius of gyration of a body of mass m and mass moment of inertia I is the radius of an imaginary cylinder of infinitely small wall thickness having the same mass and the same mass moment of inertia as the body in question.

5 hence $k = \sqrt{\frac{I}{m}}$ m, cm, mm

Flywheel effect

6 Flywheel effect $W k^2 = m g k^2 = g I$ kg cm³ s⁻², N m²
(k^2 formulae see M 3)

Equivalent mass (for rolling bodies)

8 $m_{eq} = \frac{I}{k^2}$ kg

Basic formulae

	linear motion formulae	units	rotational motion formulae	units
9	$F_a = m a$	N, [kgf]	$M_a = I \alpha$	N m, [kgf m]
10	$W = F s$ (F=const.)	N m, [kgf m]	$W = M \varphi$ (M=const.)	N m, [kgf m]
11	$W_K = \frac{1}{2} m v^2$	J, [kgf m]	$W_K = \frac{1}{2} I \omega^2$	J, [kgf m]
12	$W_P = m g h$	J, [kgf m]	$\omega = 2 \pi n$	s^{-1}, min^{-1}
13	$W_F = \frac{1}{2} F \Delta l$	J, [kgf m]	$W_F = \frac{1}{2} M \Delta \beta$	W s, [kgf m]
14	$P = \frac{dW}{dt} = F v$	W, kW	$P = \frac{dW}{dt} = M \omega$	W, kW

For explanation of symbols see M 4

	Mass moment of inertia about axis $a-a$ (turning axis)	Mass moment of inertia about axis $b-b$ passing through centre of gravity S	type of body
m 15	$I = m r^2$	$I = \frac{1}{2} m r^2$	circular hoop 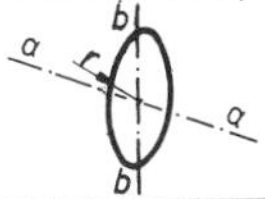
m 16	$k^2 = r^2$	$k^2 = \frac{1}{2} r^2$	
m 17	$I = \frac{1}{2} m r^2$	$I = \frac{m}{12}(3 r^2 + h^2)$	cylinder 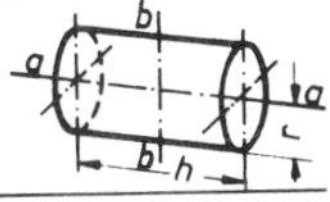
m 18	$k^2 = \frac{1}{2} r^2$	$k^2 = \frac{1}{12}(3r^2 + h^2)$	
m 19	$I = \frac{1}{2} m(R^2 + r^2)$	$I = \frac{m}{12}(3R^2 + 3r^2 + h^2)$	hollow cylinder 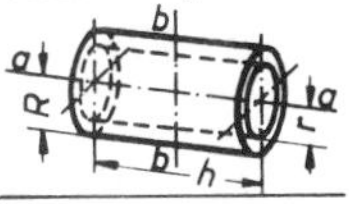
m 20	$k^2 = \frac{1}{2}(R^2 + r^2)$	$k^2 = \frac{1}{12}(3R^2 + 3r^2 + h^2)$	
m 21	$I = \frac{3}{10} m r^2$	$I = \frac{3}{80} m(4r^2 + h^2)$	cone 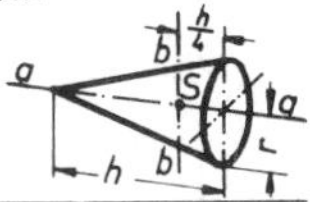
m 22	$k^2 = \frac{3}{10} r^2$	$k^2 = \frac{3}{80}(4r^2 + h^2)$	
m 23	$I = \frac{2}{5} m r^2$	$I = \frac{2}{5} m r^2$	sphere 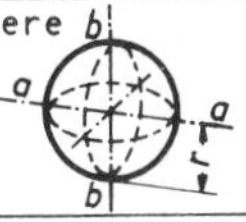
m 24	$k^2 = \frac{2}{5} r^2$	$k^2 = \frac{2}{5} r^2$	
m 25	$I = m(R^2 + \frac{3}{4} r^2)$	$I = m \frac{4R^2 + 5r^2}{8}$	torus 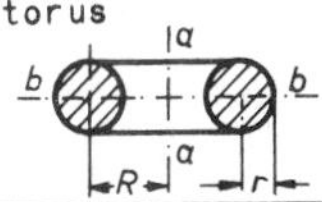
m 26	$k^2 = R^2 + \frac{3}{4} r^2$	$k^2 = \frac{1}{8}(4R^2 + 5r^2)$	
m 27	$I = \frac{1}{3} m l^2$	$I = \frac{m}{12}(d^2 + c^2)$	bar 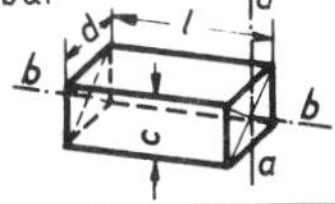
m 28	$k^2 = \frac{1}{3} l^2$	$k^2 = \frac{1}{12}(d^2 + c^2)$	

Total kinetic energy of a body

29 $$W_K = \frac{1}{2} m v_s^2 + \frac{1}{2} I_s \omega^2 \qquad \text{J, [kgf m]}$$

Kinetic energy of a rolling body - no sliding

30 $$W_K = \frac{1}{2} (m + m_{eq}) v_s^2 \qquad \text{J, [kgf m]}$$

31 $$v_s = \omega r \qquad \text{m/s, km/h}$$

Rotational torque

32 $$M = \frac{P}{\omega} = \frac{P}{2 \pi n} \qquad \text{N m, [kgf m]}$$

Transmission ratios

Transmission ratio

33 $$i = \frac{d_2}{d_1} = \frac{z_2}{z_1} = \frac{n_1}{n_2} = \frac{\omega_1}{\omega_2}$$

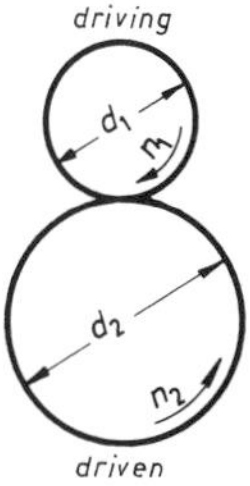

Torque ratio

34 $$\frac{\text{moment of force}}{\text{moment of load}} = \frac{M_F}{M_L} = \frac{1}{i \eta}$$

Efficiency

35 $$\eta = \frac{\text{work produced}}{\text{work applied}} = \frac{\text{output}}{\text{input}}$$

Overall efficiency for a series of transmissions

36 $$\eta = \eta_1 \times \eta_2 \times \eta_3 \times \dots$$

- m_{eq} : see m 8
- v_s : velocity of linear motion of centre of gravity
- F_a : accelerating force — N, [kgf]
- M_a : accelerating moment — N m, [kgf m]
- W_K : kinetic energy — J, [kgf m]
- W_P : potential energy — J, [kgf m]
- W_F : energy of helical spring under tension — J, [kgf m]
- Δl : extension of helical spring
- $\Delta \hat{\beta}$: angular defection of spiral spring (in radians)

Centrifugal force

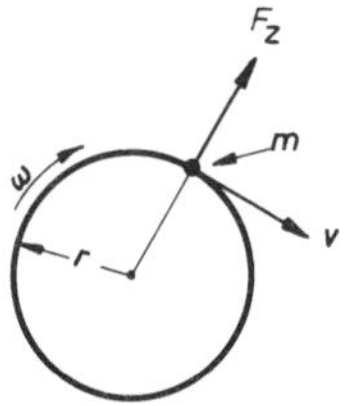

37 $F_z = m\omega^2 r = \frac{m v^2}{r}$ N, [kgf]

38 $= 4\pi^2 m n^2 r$ N, [kgf]

39 $v = 2\pi r n$ m/s, km/h

40 $\omega = 2\pi n$ 1/s, 1/min

Stresses in rotation bodies (appr. formulae)

Disc

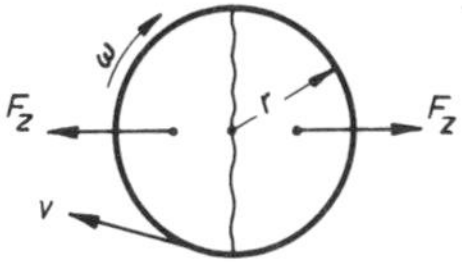

41 $\sigma_z = \frac{\omega^2 r^2 \varrho}{3} = \frac{v^2 \varrho}{3}$

N/m^2, $[kgf/cm^2]$

Ring

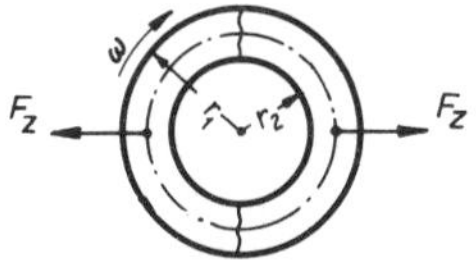

42 $\sigma_z = \frac{\omega^2 \varrho}{3} (r_1^2 + r_1 r_2 + r_2^2)$

N/m^2, $[kgf/cm^2]$

l_s :	distance from centre of gravity	m, cm, mm
e :	maximum pendulum swing	m, cm, mm
f :	instantaneous pendulum swing	m, cm, mm
F_z :	centrifugal force	N, [kgf, gf]
I_O :	mass moment of inertia about O	$kg\,m^2$, $[kgf\,m\,s^2]$
I_S :	mass moment of inertia about S	$kg\,m^2$, $[kgf\,m\,s^2]$
M_1 :	moment required to deflect spiral spring by 1 rad = 57·3°	N m, [kgf cm]
σ_z :	tensile stress	N/m^2, kgf/cm^2, $[kgf/mm^2]$
T :	period of oscillation (B to B' and back)	s, min
v_E :	velocity at E	m/s, cm/s, km/h
v_F :	velocity at F	m/s, cm/s, km/h
W_{KE} :	kinetic energy at E	N m, [kgf m]

Mechanical oscillation

General

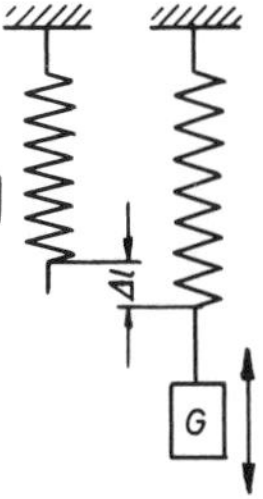

43 period $T = 2\pi\sqrt{\dfrac{m}{k}}$ s, min

44 stiffness $k = \dfrac{G}{\Delta l}$ N/m, [kgf/cm]

45 frequency $f = \dfrac{1}{T}$ (see L 1) s^{-1}, min^{-1}

46 angular velocity $\omega = 2\pi f = \sqrt{\dfrac{k}{m}}$ s^{-1}, min^{-1}

Critical speed n_c of shaft

47 $$n_c = \frac{1}{2\pi}\sqrt{\frac{c_q}{m}}$$

48 $$= 300\sqrt{\frac{10\ c_q\ \text{mm}}{9 \cdot 81\ \text{N}} \times \frac{\text{kg}}{m}} \quad \text{min}^{-1}$$

stiffness k_q of

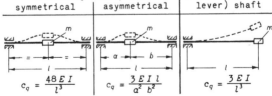

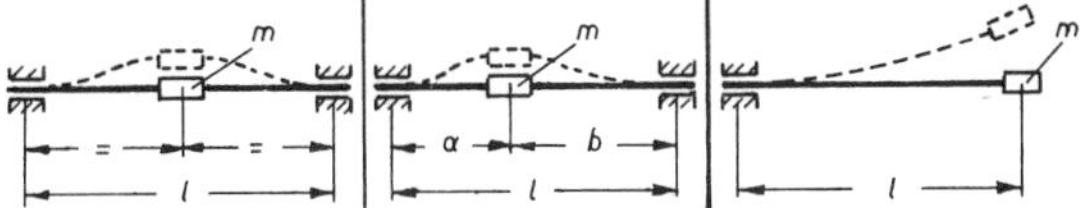

2-bearing shaft, load symmetrical	2-bearing shaft, load asymmetrical	overhung (cantilever) shaft
49 $c_q = \dfrac{48\,E\,I}{l^3}$	$c_q = \dfrac{3\,E\,I\,l}{a^2\,b^2}$	$c_q = \dfrac{3\,E\,I}{l^3}$

Δl: deflection or elongation
I : second moment of area of shaft cross section
m : mass. When calculating the critical speed the mass m (e.g. of a belt disc) is assumed to be concentrated at a single point. The mass of the shaft should be allowed for by a slight increase
c_q : stiffness for transverse oscillations

Pendulum

(Explanations see L 4)

Conical pendulum

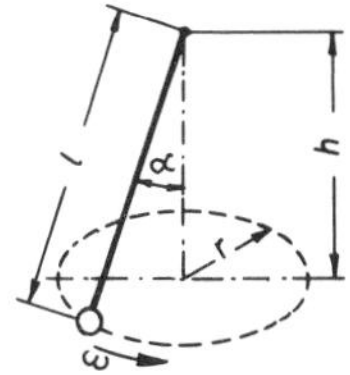

50 $T = 2\pi\sqrt{\dfrac{h}{g}} = 2\pi\sqrt{\dfrac{l\cos\alpha}{g}}$ s, min

51 $\tan\alpha = \dfrac{r\,\omega^2}{g} = \dfrac{r}{h}$

52 $\omega = \sqrt{\dfrac{g}{h}}$ | $h = \dfrac{g}{\omega^2}$ m, cm

Simple pendulum

The arm of a pendulum has no mass, the total mass is represented as a point.

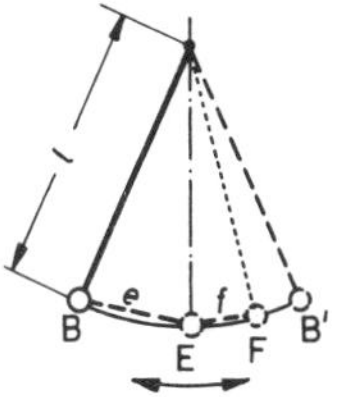

53 $T = 2\pi\sqrt{\dfrac{l}{g}}$ s, min

54 $v_E = e\sqrt{\dfrac{g}{l}}$ | $v_F = \sqrt{\dfrac{g}{l}(e^2 - f^2)}$ m/s, km/h

55 $W_{KE} = m\,g\,\dfrac{e^2}{2\,l}$ J, N m, [kgf cm]

Compound pendulum

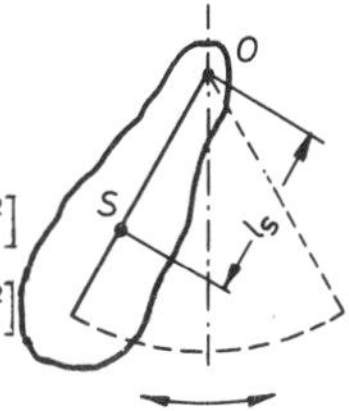

56 $T = 2\pi\sqrt{\dfrac{I_0}{m\,g\,l_S}} = 2\pi\sqrt{\dfrac{k_0^2}{g}}$ s, min

57 $I_0 = I_S + m\,l_S^2$ N m s², [kgf cm s²]

58 $I_S = m\,g\,l_S\left(\dfrac{T^2}{4\pi^2} - \dfrac{l_S}{g}\right)$ N m s², [kgf cm s²]

If a body with centre of gravity S is suspended from O, distant l_S from S, and the period of swing determined the mass moment of inertia about O can be calculated using m 58.

Torsional pendulum

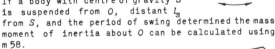

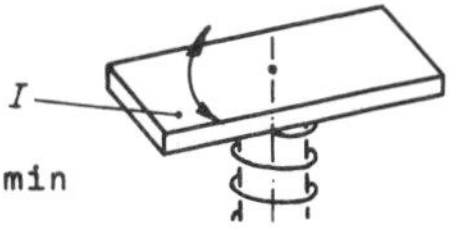

59 $T = 2\pi\sqrt{\dfrac{I}{M_1}}$ s, min

For explanation of symbols see M 5

Impact

When two bodies of mass m_1 and m_2 and velocities v_{11}
60 and v_{21} collide, the total momentum $p = m\,v$ will remain constant over the whole impact period (velocities become v_{12} and v_{22}):

61 $$p = m_1\,v_{11} + m_2\,v_{21} = m_1\,v_{12} + m_2\,v_{22}$$

Impact-direction

direct and concentric impact	velocities parallel to normal to surfaces at point of impact	normal to surfaces at point of impact through centre of gravity of both bodies
oblique and concentric impact	any random velocities	normal to surfaces at point of impact through centre of gravity of both bodies
oblique and excentric impact	any random velocities	any random normal to surfaces at point of impact

Types of impact

		elastic impact+)	plastic impact
	relative velocity	equal, before and after impact	equals zero after impact
62	velocity after direct and concentric impact	$v_{12} = \dfrac{v_{11}(m_1 - m_2) + 2m_2\,v_{21}}{m_1 + m_2}$	$v_{02} = \dfrac{m_1\,v_{11} + m_2\,v_{21}}{m_1 + m_2}$
63			
64		$v_{22} = \dfrac{v_{21}(m_2 - m_1) + 2m_1\,v_{11}}{m_1 + m_2}$	
65	coeff. of restit.	$\varepsilon = 1$	$\varepsilon = 0$

Coefficient of restitution ε

This indicates by what factor the relative velocities will vary before (v_{r1}) and after (v_{r2}) impact:

66 $$\varepsilon = \frac{v_{r2}}{v_{r1}}, \qquad \text{here} \qquad 0 \leqq \varepsilon \leqq 1$$

+) For an oblique, concentric, elastic impact the velocity vector v is split into a normal and a tangential component. The normal component v_n produces a direct impact (see above), the tangential component v_t has no effect on the impact.

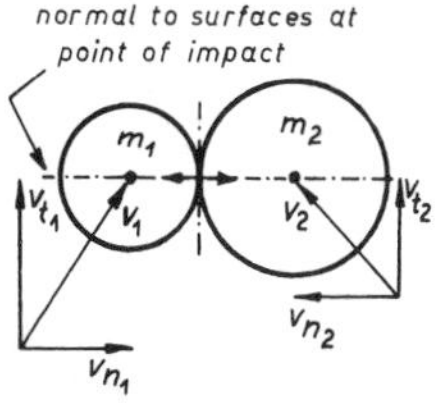

General

Hydraulics deals with the behaviour of liquids. With good approximation, liquids may be considered incompressible, which means that the influence of pressure changes on their densities is negligibly small.

Quantities

Pressure p see O 1

Density ρ see O 1

Dynamic viscosity η

(EU: Pascal-second Pa s $= \frac{kg}{m\ s} = \frac{N\ s}{m^2}$ [= 10 P])

The dynamic viscosity is a material constant, which is a function of pressure and temperature:

n 1 $$\eta = f(p, t)$$

The dependence on pressure can often be neglected. Hence

n 2 $$\eta = f(t) \quad \text{(for figures see Z 14)}$$

Kinematic viscosity ν (EU: m^2/s [= 10^4 St = 10^6 cSt])

The kinematic viscosity is the quotient of dynamic viscosity η and density ρ:

n 3 $$\nu = \frac{\eta}{\rho}$$

Hydrostatics

Pressure distribution in a fluid

n 4 $$p_1 = p_0 + g\,\rho\,h_1$$

n 5 $$p_2 = p_1 + g\,\rho\,(h_2 - h_1)$$

$$= p_1 + g\,\rho\,\Delta h$$

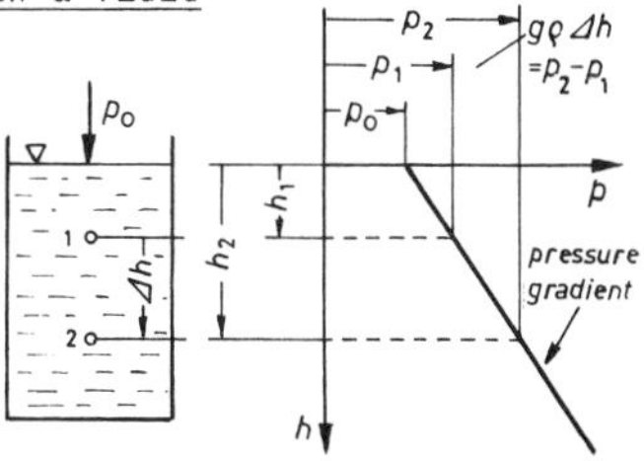

continued on N 2

Hydrostatic forces on plane surfaces

Hydrostatic force is the component acting on the surface which is caused by the weight of the fluid alone, i.e. without taking into account the atmospheric pressure p_o.

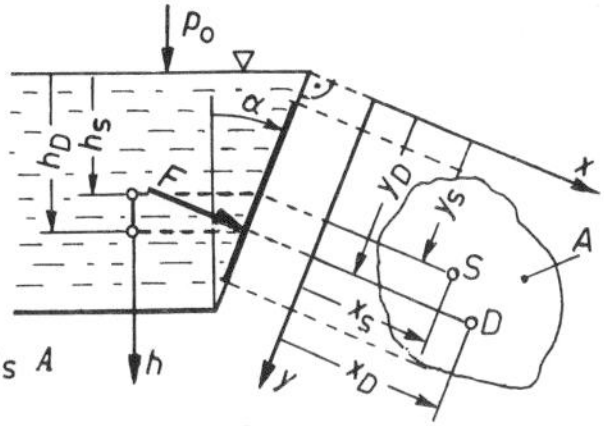

n 6 $$F = g\,\varrho\,y_S\,A\cos\alpha = g\,\varrho\,h_S\,A$$

n 7 $$y_D = \frac{I_x}{y_S\,A} = y_S + \frac{I_S}{y_S\,A} \quad ; \qquad x_D = \frac{I_{xy}}{y_S\,A} \qquad \text{m, mm}$$

Hydrostatic forces on curved surfaces

The hydrostatic force acting on the curved surface 1,2 is resolved into the horizontal component F_H and the vertical component F_V.

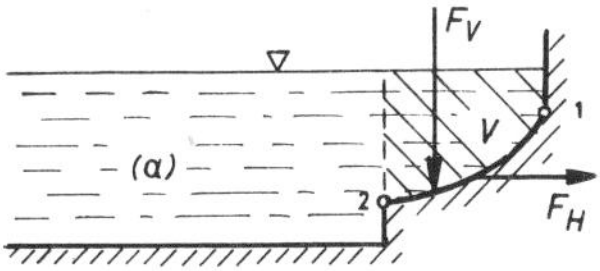

F_V is equal to the weight of the fluid in (a) or the equivalent weight of fluid (b), above the surface 1,2. The line of action runs through the centre of gravity of the volume V.

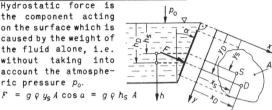

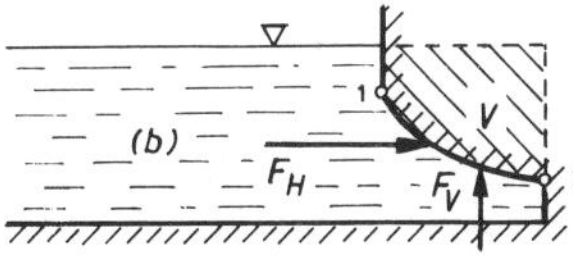

n 8 $$|F_V| = g\,\varrho\,V \qquad \text{N, kN}$$

F_H is equal to the hydrostatic pressure force acting on the projection of the considered surface 1,2 on the plane perpendicular to F_H. Calculation is accomplished by n 6 and n 7.

- S : centre of gravity of area A
- D : centre of pressure = point of action of force F
- I_x : second moment of area A in relation to axis x
- I_S : second moment of area A in relation to an axis running parallel to axis x through the centre of gravity (see J 10 and P 10)
- I_{xy} : centrifugal moment of area A in relation to axes x and y (see J 10).

Buoyancy

The buoyancy F_A is equal to the weight of the displaced fluids of densities ϱ and ϱ'.

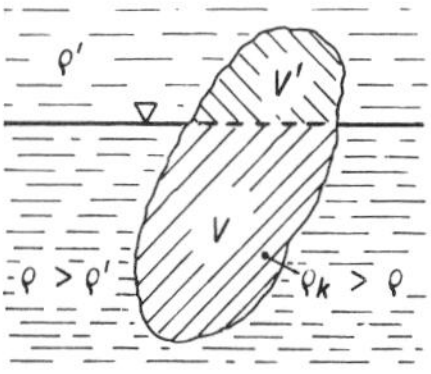

n 9 $$F_A = g\,\varrho\,V + g\,\varrho'\,V' \qquad \text{N, kN}$$

If the fluid of density ϱ' is a gas, the following formula is valid:

n 10 $$F_A \approx g\,\varrho\,V \qquad \text{N, kN}$$

With ϱ_k being the density of the body,

n 11 $\varrho > \varrho_k$	the body will float	in the heavier fluid
n 12 $\varrho = \varrho_k$	" " " remain suspended	in the heavier fluid
n 13 $\varrho < \varrho_k$	" " " sink	in the heavier fluid

Determination of density ϱ of solid and liquid bodies

Solid body of greater density than the fluid used	Solid body of smaller density than the fluid used	For fluids first determine F_1 and m of a deliberate body in a fluid of known density ϱ_b. This yields:
n 14 $\varrho = \varrho_F \dfrac{1}{1 - \dfrac{F}{m\,g}}$	n 15 $\varrho = \varrho_F \dfrac{1}{1 + \dfrac{F_H - F}{m\,g}}$	n 16 $\varrho = \varrho_b \dfrac{1 - \dfrac{F}{m\,g}}{1 - \dfrac{F_1}{m\,g}}$

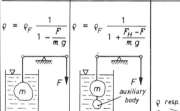

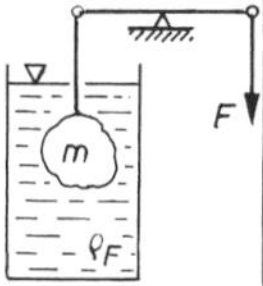

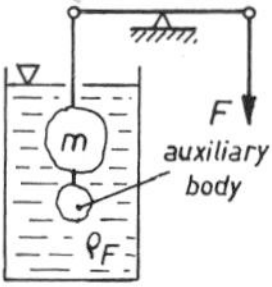

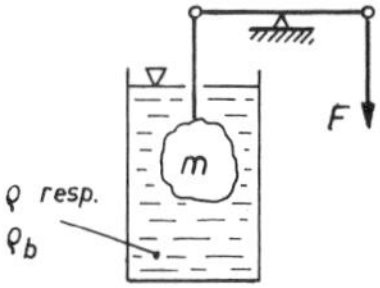

m : mass of the body remaining suspended in the fluid
F : equilibrium force necessary
F_H : equilibrium force necessary in the preliminary trial for the auxiliary body alone
ϱ_F : density of the fluid used

Hydrodynamics
(of a steady flow)

Continuity equation

Rule of conservation of mass:

n 17 $A_1 \upsilon_1 \varrho_1 = A \upsilon \varrho = A_2 \upsilon_2 \varrho_2$

n 18 $= \dot{m} = \dot{V} \varrho \qquad \frac{\mathrm{g}}{\mathrm{s}}, \ \frac{\mathrm{kg}}{\mathrm{s}}$

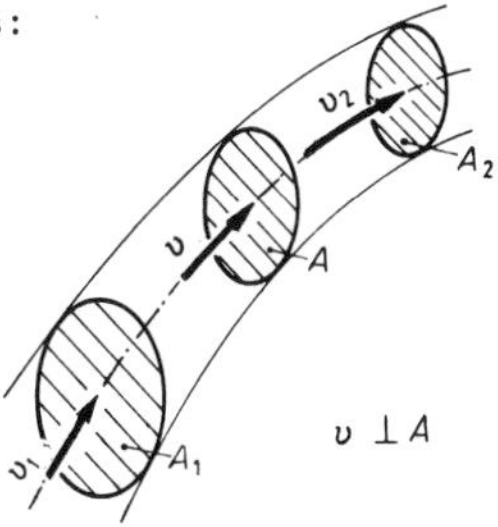

Rule of conservation of volume:

n 19 $\dot{V} = A \upsilon \qquad \frac{\mathrm{m}^3}{\mathrm{s}}, \ \frac{\mathrm{cm}^3}{\mathrm{s}}$

$\upsilon \perp A$

Bernoulli's equation (Rule of conservation of energy)

No friction (ideal fluid):

n 20 $\frac{p_1}{\varrho} + g z_1 + \frac{\upsilon_1^2}{2} = \frac{p}{\varrho} + g z + \frac{\upsilon^2}{2} = \frac{p_2}{\varrho} + g z_2 + \frac{\upsilon_2^2}{2} \qquad \frac{\mathrm{J}}{\mathrm{kg}}$

$\frac{p}{\varrho}$: pressure energy per unit mass

$g z$: potential energy per unit mass

$\frac{\upsilon^2}{2}$: kinetic energy per unit mass

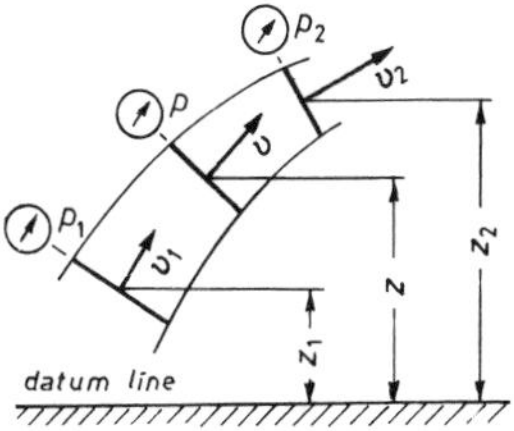

Including losses (real fluid):

n 21 $\frac{p_1}{\varrho} + g z_1 + \frac{\upsilon_1^2}{2} = \frac{p_2}{\varrho} + g z_2 + \frac{\upsilon_2^2}{2} + w_{R\,1,2} \qquad \frac{\mathrm{J}}{\mathrm{kg}}$

υ : velocity

$w_{R\,1,2}$: resistance losses along path from 1 to 2

Power P of an hydraulic machine

n 22 $$P = \dot{m}\, w_{t\,1,2} \qquad \text{kW, W}$$

technical work per unit mass:

n 23 $$w_{t\,1,2} = \frac{1}{\rho}(p_2-p_1) + g(z_2-z_1) + \frac{1}{2}(v_2^2-v_1^2) - w_{R\,1,2} \qquad \frac{\text{J}}{\text{kg}}$$

n 24 for hydraulic machines: $w_{t\,1,2} < 0$

n 25 for pumps: $w_{t\,1,2} > 0$

Momentum equation

For a fluid flowing through a stationary reference volume the following vector equation is valid:

n 26 $$\Sigma\vec{F} = \dot{m}(\vec{v_2} - \vec{v_1}) \qquad \text{N, kN}$$

$\Sigma\vec{F}$ is the vector sum of the forces acting on the fluid contained in the reference volume. These can be:
- volume forces (e.g. weight)
- pressure forces
- friction forces.

$\vec{v_2}$ is the exit velocity of the fluid leaving the reference volume

$\vec{v_1}$ is the entrance velocity of the fluid entering the reference volume.

Angular-momentum equation

In a steady state rotational flow a torque M is exerted on the fluid flowing through the reference volume, given by:

n 27 $$M = \dot{m}(v_{2,u} \times r_2 - v_{1,u} \times r_1) \qquad \text{N m}$$

$v_{2,u}$ and $v_{1,u}$ are the circumferential components of exit velocity out of and entrance velocity into the reference volume.

r_2 and r_1 are the radii associated with v_2 and v_1.

Friction losses in pipe flow

n 28 Friction loss per unit mass $w_{R\,1,2} = \Sigma(\zeta \cdot a \frac{v^2}{2})$, hence

n 29 Pressure loss $\Delta p_V = \varrho\, w_{R\,1,2}$

Determination of coefficient of resistance ζ and coefficient of shape a:

	circular pipes	non circular pipes
n 30	$Re = \frac{v\,d\,\varrho}{\eta}$	$Re = \frac{v\,d_h\,\varrho}{\eta}$

n 31 Where $Re < 2000$, the flow is laminar.
n 32 Where $Re > 3000$, the flow is turbulent.
Where $Re = 2000...3000$, the flow can be either laminar or turbulent.

	Flow (circular)		Flow (non circular)	
	laminar	turbulent *)	laminar	turbulent *)
n 33	$\zeta = \frac{64}{Re}$	$\zeta = f(Re, \frac{k}{d})$	$\zeta = \varphi \frac{64}{Re}$	$\zeta = f(Re, \frac{k}{d_h})$

n 34 $a = \frac{l}{d}$ for straight pipes | $a = \frac{l}{d_h}$ for straight pipes

n 35 $a = 1$ for fittings, unions and valves.

Determination of coefficient φ

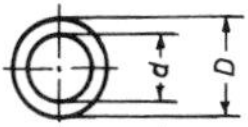

n 36 For annular cross sections:

D/d	1	3	5	7	10	30	50	70	100	∞
φ	1·50	1·47	1·44	1·42	1·40	1·32	1·29	1·27	1·25	1·00

n 37 For rectangular cross sections:

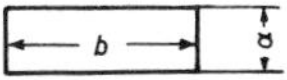

a/b	0	0·1	0·2	0·3	0·4	0·5	0·6	0·7	0·8	1·0
φ	1·50	1·34	1·20	1·10	1·02	0·97	0·94	0·92	0·90	0·89

n 38 d : internal diameter of pipe | l : length of pipe
$d_h = 4\,A/U$: hydraulic diameter
A : cross section perpendicular to fluid flow
U : wetted circumference
k/d and k/d_h : relative roughness
k : mean roughness (see Z 9)

*) ζ is taken from diagram Z 8

Flow of liquids from containers

Base apertures

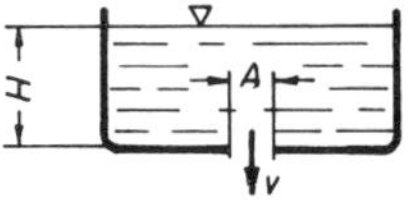

39 $$v = C_v \sqrt{2gH}$$

40 $$Q = C_d A \sqrt{2gH}$$

Small lateral apertures

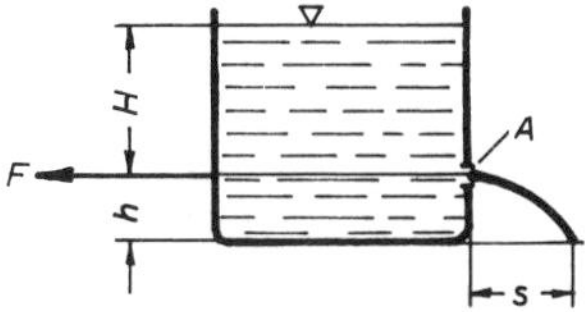

41 $$v = C_v \sqrt{2gH}$$

42 $$s = 2\sqrt{Hh}$$

43 $$Q = C_d A \sqrt{2gH}$$

44 $$F = \rho Q v$$

Large lateral apertures

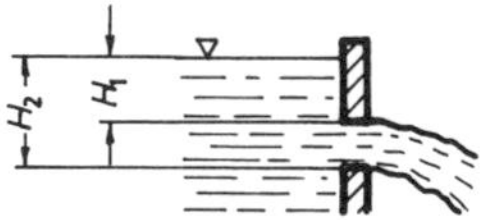

45 $$Q = \frac{2}{3} C_d b \sqrt{2g} \left(H_2^{3/2} - H_1^{3/2}\right)$$

Excess pressure on surface of liquid

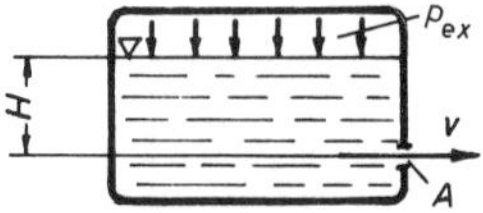

46 $$v = C_v \sqrt{2\left(gH + \frac{p_{ex}}{\rho}\right)}$$

47 $$Q = C_d A \sqrt{2\left(gH + \frac{p_{ex}}{\rho}\right)}$$

Excess pressure applied to an outlet point

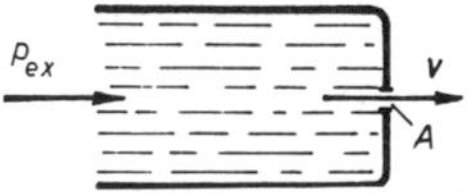

48 $$v = C_v \sqrt{2 \frac{p_{ex}}{\rho}}$$

49 $$Q = C_d A \sqrt{2 \frac{p_{ex}}{\rho}}$$

v : outlet velocity — m/s, km/h

p_{ex}: pressure in excess of atmosph. pressure — N/m^2, kgf/cm^2

C_d : discharge coefficient ($C_d = C_c \times C_v$)

C_c : contraction coeff. ($C_c = 0{\cdot}62$ for sharp edge apert.)
($C_c = 0{\cdot}97$ for well rounded apert.)

C_v : velocity coefficient (for water $C_v = 0{\cdot}97$)

b : width of aperture — m, cm

F : reaction force — N, kgf

Q : volume of outlet flow — m^3/s, m^3/h

HEAT

Thermal variables of state

O 1

Thermal variables of state are pressure p, temperature t, and density ϱ or volume per unit mass (specific volume), respectively.

Pressure p (EU: $N/m^2 = Pa$, bar)

Pressure is the quotient of force F and area A:

o 1 $$p = \frac{F}{A}$$

The absolute pressure can be interpreted as the total result of the impacts of the molecules on the wall. The pressure measured with a pressure gauge is the pressure differential Δp in relation to the ambient pressure p_u. A state of pressure means $\Delta p > 0$, vacuum means $\Delta p < 0$. Thus, the absolute pressure p can be expressed by:

o 2 $$p = p_u + \Delta p$$

Temperature T, t (Base quantity) see explanations at front of book.

The unit of temperature T is the Kelvin K, defined by equation

o 3 $$1\ K = \frac{T_{TR}}{273 \cdot 16}$$

where T_{TR} is the temperature of pure water at the triple point. In addition to the Kelvin scale the centigrade scale is also used. The temperature of this scale has been internationally defined by:

o 4 $$t = \left(\frac{T}{K} - 273 \cdot 15\right) {}^{\circ}C \; ; \quad T = \left(\frac{t}{{}^{\circ}C} + 273 \cdot 15\right) K$$

Density ϱ (EU: kg/m^3)

Density is the quotient of mass m and volume V:

o 5 $$\varrho = \frac{m}{V}$$

Volume per unit mass (specific volume) υ (EU: m^3/kg)

Specific volume is the quotient of the volume V and the mass m:

o 6 $$\upsilon = \frac{V}{m} = \frac{1}{\varrho}$$

Molecular volume V_m (EU: m^3/mol)

Molecular volume is the quotient of volume V and number of moles contained in the volume:

o 7 $$V_m = \frac{V}{n}$$

Amount of substance n (Base quantity) see explanations at front of book.

Heating of solid and liquid bodies

Heat (thermal energy) Q (EU: J)

Heat is energy exchanged between systems of different temperatures, where these systems interact with each other through diathermal walls.

Heat per unit mass q (EU: J/kg)

The heat per unit mass is the quotient of heat Q and mass m:

8 $$q = \frac{Q}{m}$$

Specific heat c EU: J/(kg K)

The specific heat c denotes the amount of heat Q to be supplied to or extracted from a substance of mass m to change its temperature by a difference t:

9 $$c = \frac{Q}{m\,\Delta t} = \frac{q}{\Delta t}$$

The specific heat is a function of temperature. For figures see Z 1...Z 5.

Latent heats per unit mass l (EU: J/kg) Values see Z10

The supply or extraction of latent heat causes a body to change its state without changing its temperature. The following latent heats exist:

10	l_f	Latent heat of	fusion	is the heat necessary to convert a	solid body of the fusing temperature into a fluid	of the same temperature
11	l_d	Latent heat of	vapourisation	is the heat necessary to convert a	fluid of the boiling temperature (dependent on pressure) into dry saturated vapour	of the same temperature
12	l_s	Latent heat of	sublimation	is the heat necessary to convert a	solid body of a temperature below its triple temperature at the sublimation temperature (dependent on pressure) directly into dry saturated vapour	of the same temperature

HEAT

Heating of solid and liquid bodies

Expansion of solid bodies

A solid body changes its dimensions due to temperature changes. With α being the coefficient of linear expansion (for figures see Z 11) the following formulae are valid for:

o 13 Length: $l_2 = l_1 \left[1 + \alpha(t_2 - t_1)\right]$

o 14 $\Delta l = l_2 - l_1 \approx l_1 \alpha(t_2 - t_1)$

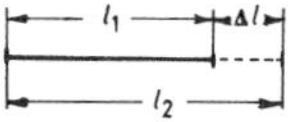

o 15 Area: $A_2 \approx A_1 \left[1 + 2\alpha(t_2 - t_1)\right]$

o 16 $\Delta A = A_2 - A_1 \approx A_1 2\alpha(t_2 - t_1)$

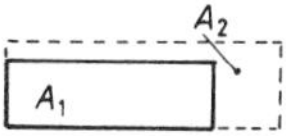

o 17 Volume: $V_2 \approx V_1 \left[1 + 3\alpha(t_2 - t_1)\right]$

o 18 $\Delta V = V_2 - V_1 \approx V_1 3\alpha(t_2 - t_1)$

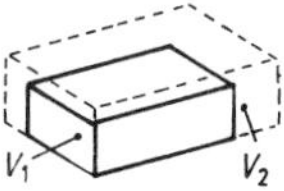

Expansion of liquid bodies

With β being the coefficient of volume expansion (for figures see Z 11) the following formulae apply:

o 19 $V_2 = V_1 \left[1 + \beta(t_2 - t_1)\right]$

o 20 $\Delta V = V_2 - V_1 = V_1 \beta(t_2 - t_1)$

Bending due to heat A

Bimetallic strips are subject to bending due to heat. Bending occurs towards the side of the metal with the lower coefficient of expansion. With α_b being the "specific thermal bending" the bending due to heat can be calculating by (α_b approx. 14×10^{-6}/K, for exact values see manufacturers catalogues):

o 21 $A = \dfrac{\alpha_b L^2 \Delta t}{s}$

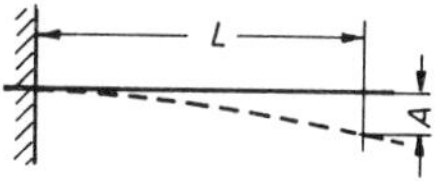

l_1 : length at $t = t_1$	A_1 : area at $t = t_1$	
l_2 : length at $t = t_2$	A_2 : area at $t = t_2$	
V_1 : volume at $t = t_1$	t_1 : temperature prior to	heating
V_2 : volume at $t = t_2$	t_2 : temperature after	heating
s : thickness	Δt : temperature difference	

General equation of state of ideal gases

The state of ideal gas is determined by two thermal variables of state. Thus, the third variable can be calculated using the general equation of state. With R being the characteristic gas constant (different values for different gases, see Z 12) the equation reads as follows:

o 22 $$p\,v = R\,T \quad \text{or} \quad p\,V = m\,R\,T \quad \text{or} \quad p = \varrho\,R\,T$$

If the gas constant is related to the mole volume, the equation reads

o 23 $$p\,V_m = R_m T$$

where $R_m = 8314 \cdot 3$ J/(kmol K) is the universal gas constant (valid for all ideal gases). R and R_m are related by

o 24 $$R_m = M\,R$$

where M is the molecular mass (see Z 12).

Thermal state of real (non ideal) gases and vapours

The thermal state of real gases and vapours is calculated using special equations or diagrams.

Changes of state

Changes of state are caused by interactions of the system with the surroundings. These interactions are calculated using the 1st and the 2nd law:

	1st law for closed systems	1st law for open systems	2nd law for all systems
o 25 o 26 o 27	$q_{1,2} + w_{1,2} = u_2 - u_1$	$q_{1,2} + w_{t\,1,2} = h_2 - h_1 + \Delta e$	$q_{1,2} = \int_1^2 T\,ds$

In these formulae, energy input is positive (i.e. $q_{1,2}$, $w_{1,2}$, $w_{t\,1,2}$) and energy output negative.

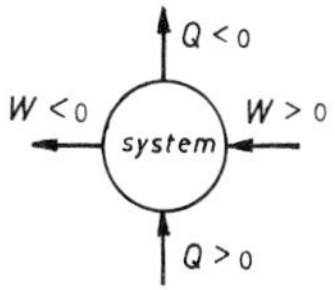

h : enthalpy per unit mass | s: entropy per unit mass
u : internal energy per unit mass
$w_{1,2}$: external work done per unit mass } see O 7
$w_{t\,1,2}$: continuous external work done per unit mass } see O 7
Δe : changes in kinetic or potential energies

Changes of state of ideal gases

The table on page O 6 shows the relations for different changes of state, which have been developed from formulae o 25 to o 27. The following explanations apply: Each change of state may be represented by an equation

o 28
$$p\upsilon^n = \text{const.}$$

The various exponents n are given in column 1.

c_{pm} and c_{vm} are the mean specific heats for constant pressure and constant volume, respectively, in the temperature range between t_1 and t_2. There, the following relations apply (values for c_{pm} see Z 13):

o 29
$$c_{pm} = c_{pm}\Big|_{t_1}^{t_2} = \frac{c_{pm}\Big|_0^{t_2} t_2 - c_{pm}\Big|_0^{t_1} t_1}{t_2 - t_1}$$

o 30
$$c_{vm} = c_{vm}\Big|_{t_1}^{t_2} = c_{pm}\Big|_{t_1}^{t_2} - R$$

o 31
$$\gamma = \gamma_m = \gamma_m\Big|_{t_1}^{t_2} = c_{pm}\Big|_{t_1}^{t_2} \Big/ c_{vm}\Big|_{t_1}^{t_2}$$

The change of entropy occurring during the change of state is given by:

o 32
$$s_2 - s_1 = c_{pm}\ln\left(\frac{T_2}{T_1}\right) - R\,\ln\left(\frac{p_2}{p_1}\right) = c_{vm}\ln\left(\frac{T_2}{T_1}\right) + R\,\ln\left(\frac{\upsilon_2}{\upsilon_1}\right)$$

Changes of state of real gases and vapours

The table below shows the relations for different changes of state, which have been developed from formulae o 25 to o 27. The thermal variables of state, p, υ, T as well as the properties, u, h, s are generally taken from appropriate diagrams.

	Change of state, constant quantity	external work $w_{1,2} = \int_1^2 p\,d\upsilon$	continuous external work $w_{t\,1,2} = -\int_1^2 \upsilon\,dp$	heat per unit mass $q_{1,2}$
o 33	isochoric, $\upsilon = \text{const.}$	0	$\upsilon(p_2 - p_1)$	$u_2 - u_1 = (h_2 - h_1) - \upsilon(p_2 - p_1)$
o 34	isobaric, $p = \text{const.}$	$p(\upsilon_1 - \upsilon_2)$	0	$h_2 - h_1$
o 35	isothermal, $T = \text{const.}$	$(u_2 - u_1) - T(s_2 - s_1) = (h_2 - h_1) - T(s_2 - s_1) - (p_2\upsilon_2 - p_1\upsilon_1)$	$(h_2 - h_1) - T(s_2 - s_1)$	$T(s_2 - s_1)$
o 36	isentropic, $s = \text{const.}$	$u_2 - u_1 = (h_2 - h_1) - (p_2\upsilon_2 - p_1\upsilon_1)$	$h_2 - h_1$	0

HEAT

Ideal gases in open and closed systems

process details, exponent	relation between state 1 and state 2	closed system, reversible $w_{1,2} = -\int_1^2 p\, dv$	open system, reversible $w_{t\,1,2} = \int_1^2 v\, dp$	heat transfer per unit mass $q_{1,2}$	p–v–diagram	T–s–diagram
isochoric v = const. $n = \infty$ (o 37)	$\frac{p_2}{p_1} = \frac{T_2}{T_1}$	0	$v(p_2 - p_1)$ $= R(T_2 - T_1)$	$c_{vm}(T_2 - T_1)$	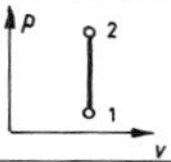	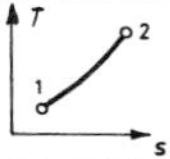
isobaric p = const. $n = 0$ (o 38)	$\frac{v_2}{v_1} = \frac{T_2}{T_1}$	$p(v_1 - v_2)$ $= R(T_1 - T_2)$	0	$c_{pm}(T_2 - T_1)$	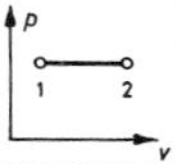	less curved than isochoric 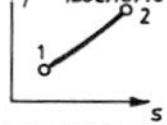
isothermal T = const. $n = 1$ (o 39)	$\frac{p_2}{p_1} = \frac{v_1}{v_2}$	$R\,T \ln\left(\frac{v_1}{v_2}\right)$ $= R\,T \ln\left(\frac{p_2}{p_1}\right)$	$w_{1,2}$	$-w_{1,2}$	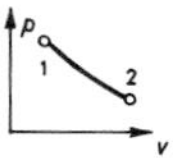	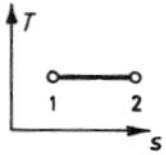
isentropic s = const. $n = \gamma$ (o 40)	$\frac{p_2}{p_1} = \left(\frac{v_1}{v_2}\right)^{\gamma}$ $\frac{p_2}{p_1} = \left(\frac{T_2}{T_1}\right)^{\frac{\gamma}{\gamma-1}}$ $\frac{v_2}{v_1} = \left(\frac{T_1}{T_2}\right)^{\frac{1}{\gamma-1}}$	$u_2 - u_1 = c_{vm}(T_2 - T_1)$ $= c_{vm}\,T_1\left[\left(\frac{p_2}{p_1}\right)^{\frac{\gamma-1}{\gamma}} - 1\right]$ $= \frac{1}{\gamma-1} R\,T_1\left[\left(\frac{p_2}{p_1}\right)^{\frac{\gamma-1}{\gamma}} - 1\right]$	$h_2 - h_1 = c_{pm}(T_2 - T_1)$ $= c_{pm}\,T_1\left[\left(\frac{p_2}{p_1}\right)^{\frac{\gamma-1}{\gamma}} - 1\right]$ $= \frac{\gamma}{\gamma-1} R\,T_1\left[\left(\frac{p_2}{p_1}\right)^{\frac{\gamma-1}{\gamma}} - 1\right]$	0	steeper than isothermal 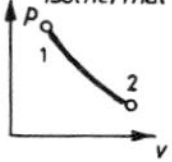	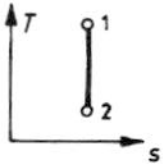
polytropic process n = const. (o 41)	$\frac{p_2}{p_1} = \left(\frac{v_1}{v_2}\right)^{n}$ $\frac{p_2}{p_1} = \left(\frac{T_2}{T_1}\right)^{\frac{n}{n-1}}$ $\frac{v_2}{v_1} = \left(\frac{T_1}{T_2}\right)^{\frac{1}{n-1}}$	$\frac{1}{n-1} R(T_2 - T_1)$ $= \frac{1}{n-1} R\,T_1\left[\left(\frac{p_2}{p_1}\right)^{\frac{n-1}{n}} - 1\right]$	$\frac{n}{n-1} R(T_2 - T_1)$ $= \frac{n}{n-1} R\,T_1\left[\left(\frac{p_2}{p_1}\right)^{\frac{n-1}{n}} - 1\right]$	$c_{vm}\frac{n-\gamma}{n-1}(T_2 - T_1)$	drawn to fit existing process	

p–υ diagram

For reversible processes the area between the curve of the variation of state and the υ-axis represents the external work per unit mass, the area between the curve and the p-axis represents the continuous external work per unit mass.

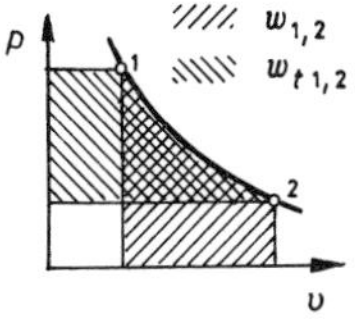

T–s diagram

For reversible processes the area between the curve and the s-axis represents the heat transfer per unit mass.

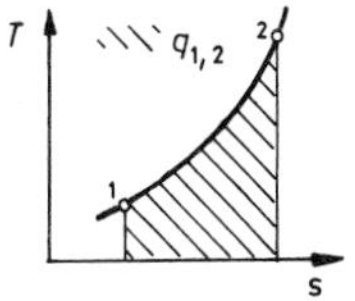

Total transfer of heat

The heat added to or removed from a closed system during a single variation of state is:

o 42 $$Q_{1,2} = m\,q_{1,2} \qquad \text{J}$$

The heat flow continuously added to or removed from an open system is:

o 43 $$\Phi_{1,2} = \dot{Q}_{1,2} = \dot{m}\,q_{1,2} \qquad \text{W}$$

where $\dot{m}$ is the mass flow (EU: kg/s).

Total transfer of work

The external work added to or done by a closed system during a single variation of state is:

o 44 $$W_{1,2} = m\,w_{1,2} \qquad \text{J}$$

The external power continuously added to or done by an open system is given by:

o 45 $$P_{1,2} = \dot{m}\,w_{t\,1,2} \qquad \text{W}$$

Mass m of a mixture of components $m_1,\ m_2,\ \ldots$

o 46 $$m = m_1 + m_2 + \ldots + m_n = \sum_{i=1}^{i=n} m_i$$

Mass fractions ξ_i of a mixture

o 47 $$\xi_i = \frac{m_i}{m} \quad \text{and} \quad \sum_{i=1}^{i=n} \xi_i = 1$$

Number of moles n of a mixture of components $n_1,\ n_2,\ \ldots$

o 48 $$n = n_1 + n_2 + \ldots + n_n = \sum_{i=1}^{i=n} n_i$$

Mole fractions ψ_i of a mixture

o 49 $$\psi_i = \frac{n_i}{n} \quad \text{and} \quad \sum_{i=1}^{i=n} \psi_i = 1$$

Equivalent molecular mass M of a mixture

For the molecular mass the following formulae apply:

o 50 $$M_i = \frac{m_i}{n} \quad \text{and} \quad M = \frac{m}{n}$$

where the equivalent molecular mass M of the mixture can be calculated as follows:

o 51 $$M = \sum_{i=1}^{i=n} (M_i \times \psi_i) \quad \text{and} \quad \frac{1}{M} = \sum_{i=1}^{i=n} \left(\frac{\xi_i}{M_i} \right)$$

Conversion between mass- and mole-fractions

o 52 $$\xi_i = \frac{M_i}{M} \psi_i$$

Pressure p of the mixture and partial pressures p_i of the components

o 53 $$p = \sum_{i=1}^{i=n} p_i \quad \text{where} \quad p_i = \psi_i \times p$$

continued on page O 9

Continuation of page O 8

Volume fractions r_i of a mixture

o 54 $$r = \frac{V_i}{V} = \psi_i \quad \text{and} \quad \sum_{i=1}^{i=n} r_i = 1$$

Here, by partial volume V_i we mean the volume the component would occupy at the temperature T and the total pressure p of the mixture. For ideal gases the following formulae apply:

o 55 $$V_i = \frac{m_i R_i T}{p} = \frac{n_i R_m T}{p} \quad \text{and} \quad \sum_{i=1}^{i=n} V_i = V$$

Internal energy u and enthalpy h of a mixture

o 56 $$u = \sum_{i=1}^{i=n} (\xi_i \times u_i) \quad ; \qquad h = \sum_{i=1}^{i=n} (\xi_i \times h_i)$$

Using these formulae, the temperature of the mixture can be determined, for real gases and vapours by using diagrams, and for ideal gases as follows:

o 57	internal energy	$t = \frac{c_{v_{m1}} \times t_1 m_1 + c_{v_{m2}} \times t_2 m_2 + \ldots + c_{v_{mn}} \times t_n m_n}{c_{v_m} \times m}$
o 58	enthalpy	$t = \frac{c_{p_{m1}} \times t_1 m_1 + c_{p_m} \times t_2 m_2 + \ldots + c_{p_{mn}} \times t_n m_n}{c_{p_m} \times m}$

where the specific heats of the mixture are determined as follows:

o 59 $$c_{v_m} = c_{p_m} - R$$

o 60 $$c_{p_m} = \sum_{i=1}^{i=n} (\xi_i \times c_{p_{m_i}})$$

Due to the temperature difference between two points heat flows from the point of higher temperature towards the point of lower temperature. The following kinds of heat transmission must be distinguished:

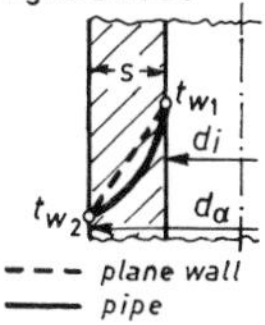

Conduction

o 61 in a plane wall: $\Phi = \dot{Q} = \lambda A \dfrac{t_{w_1} - t_{w_2}}{s}$

o 62 in the wall of a pipe: $\Phi = \dot{Q} = \lambda A_m \dfrac{t_{w_1} - t_{w_2}}{s}$

The mean logarithmic area is

o 63 $A_m = \pi d_m L$; where $d_m = \dfrac{d_a - d_i}{\ln\left(\dfrac{d_a}{d_i}\right)}$

L : length of the pipe

Convection

By heat convection we mean the heat transfer in a fluid. Due to their flow the molecules as carriers of the mass are also the carriers of the heat. Where the flow originates by itself, the convection is called natural convection. The convection taking place in a flow is called forced convection.

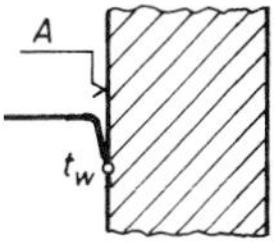

o 64 $\Phi = \dot{Q} = \alpha A (t - t_w)$

Radiation

This kind of heat transmission does not require mass as a carrier (e.g. the radiation of the sun through space). For calculations formula o 64 is used.

Heat transfer

By heat transfer we mean the combined result of the different processes contributing to the heat transmission:

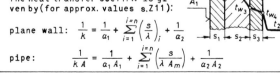

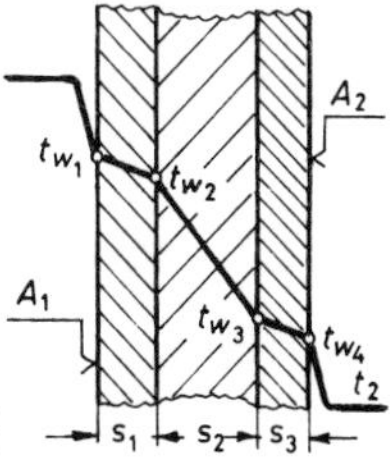

o 65 $\Phi = \dot{Q} = k A (t_1 - t_2)$

The heat transfer coeff. k is given by (for approx. values s. Z 11):

o 66 plane wall: $\dfrac{1}{k} = \dfrac{1}{\alpha_1} + \sum_{i=1}^{i=n} \left(\dfrac{s}{\lambda}\right)_i + \dfrac{1}{\alpha_2}$

o 67 pipe: $\dfrac{1}{k A} = \dfrac{1}{\alpha_1 A_1} + \sum_{i=1}^{i=n} \left(\dfrac{s}{\lambda A_m}\right) + \dfrac{1}{\alpha_2 A_2}$

λ : thermal conductivity (for values see Z 1...Z 5)
α : heat transfer coeff. (for calculation see O 12)

Heat exchanger

A heat exchanger transmits heat from one fluid to another. The heat flow may be calculated by:

o 68 $$\Phi = \dot{Q} = k\,A\,\Delta t_m .$$

Here, Δt_m is the logarithmic mean temperature difference. The following formula applies for both parallel-flow and counterflow heat exchangers:

o 69 $$\Delta t_m = (\Delta t_{great} - \Delta t_{small}) \Big/ \ln \frac{\Delta t_{great}}{\Delta t_{small}}$$

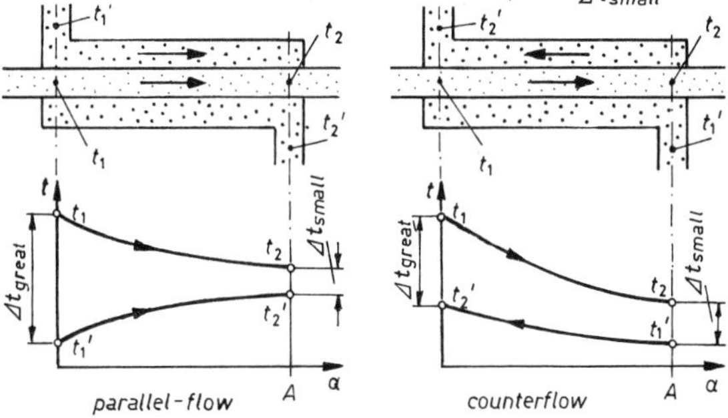

In counterflow operation Δt_{great} and Δt_{small}, can occur on the opposite ends of the exchanger, to that shown in the figure.

Symbols used on page O 12:

A_1 : surface of the smaller body
A_2 : surface of the bigger body
d : inside diameter of pipe
D : outer diameter of pipe
Gr : Grashof'number
H : height of plate
L : length of plate
w : velocity
C_1 and C_2 : radiation constants of the surfaces exchanging radiation (for values see Z 12)
$C_s = 5{\cdot}67 \times 10^{-8}$ W/(m² K⁴) : radiation constant of the black body
o 70 Pr : Prandtl-number; $Pr = (\eta\, c_p)/\lambda$
$\Delta t = |t_W - t_\infty|$: absolute temperature difference between wall and fluid in the thermally not affected region
t_∞: ambient temperature
ν : kinematic viscosity ($\nu = \eta/\varrho$)
η : dynamic viscosity
η_{Fl}: dynamic viscosity at mean temperature of fluid
η_W : dynamic viscosity at wall temperature
λ : thermal conductiv. of fluid (for values s. Z 5, Z 6)
β : volume expansion coefficient (see Z 11 and o 77)
β^* : temperature factor

Calculation of heat transfer coefficient α [1]

For free convection (according to Grigull)

o 71	on a vertical plate	$\alpha = \dfrac{Nu\,\lambda}{H}$	$Nu = 0{\cdot}55\sqrt[4]{Gr\,Pr}$, for $1700 < Gr\,Pr < 10^8$
o 72			$Nu = 0{\cdot}13\sqrt[3]{Gr\,Pr}$, for $Gr\,Pr > 10^8$
o 73			$Gr = \dfrac{g\,\beta\,\Delta t\,H^3}{\nu^2} = \dfrac{g\,\beta\,\Delta t\,\varrho^2\,H^3}{\eta^2}$
o 74	on a horizontal plate	$\alpha = \dfrac{Nu\,\lambda}{D}$	$Nu = 0{\cdot}41\sqrt[4]{Gr\,Pr}$, for $Gr\,Pr > 10^5$
o 75			$Gr = \dfrac{g\,\beta\,\Delta t\,D^3}{\nu^2} = \dfrac{g\,\beta\,\Delta t\,\varrho^2\,D^3}{\eta^2}$

o 76 Fluid properties must be related to reference temperature $\quad t_B = \dfrac{t_W + t_\infty}{2}$

o 77 The expansion coefficient of gases is: $\beta_{gas} = 1/T_\infty$

For forced convection inside pipes (accord. to Hausen)

o 78
$$\alpha = Nu\,\lambda/d$$

o 79	flow	laminar $Re<2000$	$Nu = \left[3{\cdot}65 + \dfrac{0{\cdot}0668\left(Re\,Pr\,\dfrac{d}{L}\right)}{1 + 0{\cdot}045\left(Re\,Pr\,\dfrac{d}{L}\right)^{2/3}}\right]\left(\dfrac{\eta_{Fl}}{\eta_W}\right)^{0,14}$
o 80			if $10^4 > Re\,Pr\dfrac{d}{L} > 10^{-1}$, where $Re = \dfrac{v\,d\,\varrho}{\eta}$
o 81		turbulent $Re>3000$	$Nu = 0{\cdot}116(Re^{2/3} - 125)Pr^{1/3}\left[1 + \left(\dfrac{d}{L}\right)^{2/3}\right]\left(\dfrac{\eta_{Fl}}{\eta_W}\right)^{0,14}$
			if $2320 < Re < 10^6$; $0{\cdot}6 < Pr < 500$; $1 > L/d < \infty$

With the exception of η_W all material values are related to the mean temperature of the fluid.

For gases factor $(\eta_{Fl}/\eta_W)^{0,14}$ must be omitted.

For radiation (heat transfer coefficient: α_{Str})

o 82
$$\alpha_{Str} = \beta^* C_{1,2}$$

o 83	between	parallel	surfaces	$\beta^* = \dfrac{T_1^4 - T_2^4}{T_1 - T_2}$	$C_{1,2} = \dfrac{1}{\dfrac{1}{C_1} + \dfrac{1}{C_2} - \dfrac{1}{C_s}}$
o 84					
o 85		enveloping			$C_{1,2} \approx \dfrac{1}{\dfrac{1}{C_1} + \dfrac{A_1}{A_2}\left(\dfrac{1}{C_2} - \dfrac{1}{C_s}\right)}$

[1] α in $J/(m^2\,s\,K)$ or $W/(m^2\,K)$

For explanation of symbols see O 11

STRENGTH

General terms

P 1

Stress

Stress is the ratio of applied force F and cross section A.

Tensile and compressive stresses occur at right angles to the cross section.

p 1 $$\sigma \text{ or } f = \frac{F}{A} \qquad \text{N/mm}^2$$

In calculations	tensile	stresses are usually	positive
	compressive		negative

Shear stresses act along to the cross section.

p 2 $$\tau \text{ or } q = \frac{F}{A} \qquad \text{N/mm}^2$$

Stress-strain diagrams (tensile test)

Materials with

yield point (e.g. mild steel)	plastic yield (e.g. aluminium alloy)

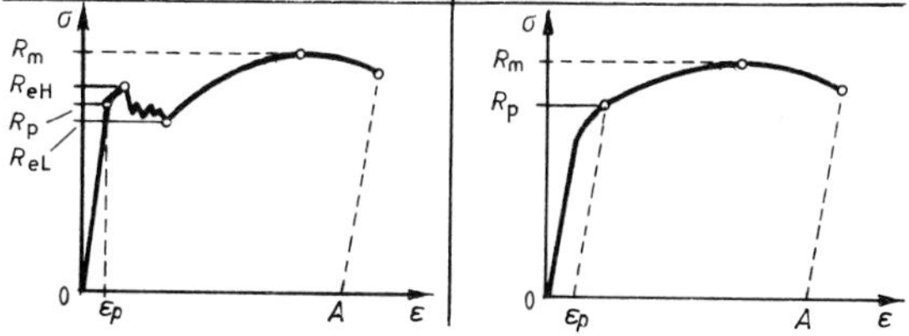

Notation: The standard symbols are from BS 18 and DIN 50145.

p 3 $R_m = \frac{F}{S_0}$; $\left[\sigma_B = \frac{F}{A_0}\right]$: tensile stress, where

F : tensile force

S_0 ; $[A_0]$: original cross section (of unloaded [specimen)

p 4 $\varepsilon = \frac{\Delta L}{L_0} \times 100\%$; $\left[\varepsilon = \frac{\Delta l}{l_0} \times 100\%\right]$: strain, where

L_0 ; $[l_0]$: original length (of unloaded specimen)

ΔL ; $[\Delta l]$: change in length of loaded specimen

continued on P 2

STRENGTH
General terms

continued from P 1 (stress-strain diagram)

R_p ; $[\sigma_p]$: Proof stress or yield strength (offset)
The limit of proportionality in sometimes known as the elastic limit.
$\varepsilon_p \simeq 0{\cdot}01\% \Longrightarrow R_{p\,0{\cdot}01}$; $[\sigma_p]$

Yield point (ferrous metals)
R_{eH} ; $[\sigma_{So}]$: upper yield stress or upper yield point
R_{eL} ; $[\sigma_{Su}]$: lower yield stress or lower yield point.

Proof stress (non-ferrous metals)
$\varepsilon_p \simeq 0{\cdot}2\% \Longrightarrow R_{p\,0{\cdot}2}$; $[\sigma_{p\,0{\cdot}2}]$

p 5 $R_m = \dfrac{F_m}{S_o}$; $\left[\sigma_B = \dfrac{F_{max}}{A_o}\right]$: tensile strength

p 6 $A = \dfrac{\Delta L}{L_o} \times 100\%$; $\left[\delta = \dfrac{\Delta l_{max}}{l_o} \times 100\%\right]$: percentage elongation after fracture.
For specimens with circular cross sections, percentage elongations may be quoted, based on gauge lengths, e.g. A_5 ; $[\delta_5]$ is based on a gauge length of $5 \times \sqrt{\dfrac{4\,S_o}{\pi}}$ mm.

Permissible stress (allowable stress)
Must be below the elastic limit R_p , thus

the permissible stress is: $\sigma_t = \dfrac{R_m}{\nu}$

R_m: yield strength of material
ν : safety factor, always greater than 1.

Ultimate safety factor (against fracture)	Proof safety factor (against yield or 0·2 proof)
$\nu = 2 \ldots 3 \ldots 4$	$\nu = 1{\cdot}2 \ldots 1{\cdot}5 \ldots 2$

Loads

type	nature of stress	load diagram
I	dead	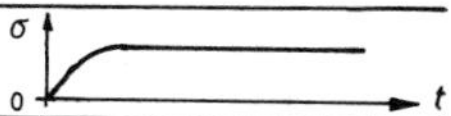
II	undulating	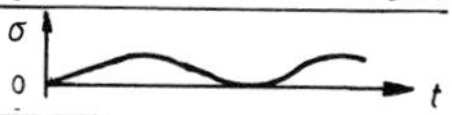
III	alternating	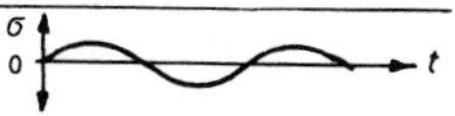

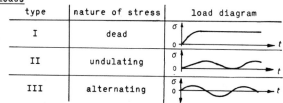

Modulus of elasticity E: The relationship between σ and ε (Hooke's law) is applicable to the elastic range, i.e. below the elastic limit (see Z 18/19 for values of E). E is known as "Young's modulus".

p 7 $\sigma = E \times \varepsilon = E \times \Delta l / l_0$; $E = \sigma / \varepsilon = \sigma \times l_0 / \Delta l$

Tensile and compressive stresses σ_t and σ_c

p 8 $\sigma_t = \frac{F_t}{A} \leqslant p_t$; $\sigma_c = \frac{F_c}{A} \leqslant p_c$

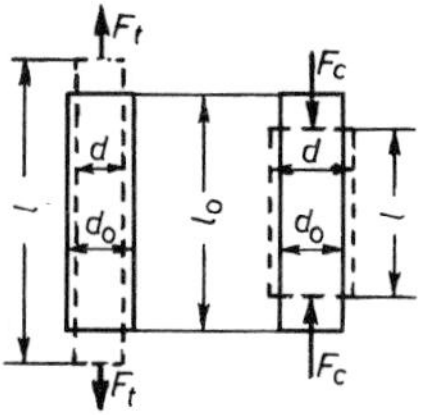

Strain ε under tension

p 9 $\varepsilon = \frac{\Delta l}{l_0} = \frac{l - l_0}{l_0} = \frac{\sigma_t}{E} = \frac{F_t}{E \times A}$

Compressive strain ε_c under compression

p10 $\varepsilon_c = \frac{\Delta l}{l_0} = \frac{l_0 - l}{l_0} = \frac{\sigma_c}{E} = \frac{F_c}{E \times A}$ | $E \times A$ = tensile or compressive stiffness.

Transverse contraction under tension (Poisson's ratio)

For circular cross section

p11 $\mu = \frac{\varepsilon_{cross}}{\varepsilon_{along}}$; where $\varepsilon_{along} = \frac{l - l_0}{l_0}$ and $\varepsilon_{cross} = \frac{d_0 - d}{d_0}$

For most metals Poisson's ratio can be assumed to be $\mu = 0{\cdot}3$.

Thermal stresses: Tensile or compressive stress is caused by restricting thermal expansion (see also o13/14):

p12 $\sigma_{th} = E \times \varepsilon_{th} = E \times \alpha \times \Delta t$ $(\varepsilon_{th} = \alpha \times \Delta t)$

Δt is the temperature difference between the unstressed original state and the state considered.

$\Delta t > 0$ tensile stress, positive

$\Delta t < 0$ compressive stress, negative.

For prestressed members subjected to thermal stress the total strain comprises:

p13 $\varepsilon_{tot} = \varepsilon_{el} + \varepsilon_{th} = F/(E \times A) + \alpha \Delta t$; $\varepsilon_{el} = F/(E \times A)$

Tensile and compressive stresses in thin-wall cylinders (boiler formula):

Hoop stress $\sigma = p\,d/(2s)$

p14 Tensile stress $\sigma = p_i\,d_i/(2s)$

p15 Compressive stress $\sigma = -p_a\,d_a/(2s)$

(p14, p15) valid for $\frac{d_a}{d_i} \leqslant 1{\cdot}2$

p_i and p_a: internal and external pressures

d_i and d_a: inside and outside diameters

$s = 0{\cdot}5(d_a - d_i)$: wall thickness

Tensile stresses in rotating bodies: see M 5.

Tensile stress in a shrunk-on ring (approx. formulae)

Shrunk-on ring on a rotating shaft:

The shrinkage force F_H of the ring must be at least twice the centripetal force F_C.

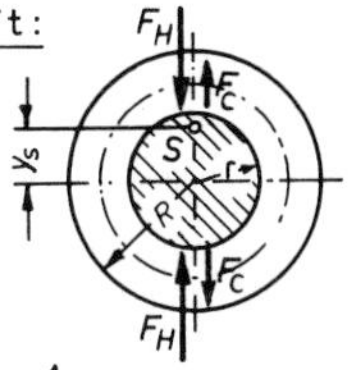

p 16 $$F_H \geqslant 2\,F_C$$

p 17 $$F_C = m\;y_s\;\omega^2 = 4\,\pi^2 m\;y_s\;n^2$$

p 18 $$y_s = \frac{4}{3\,\pi} \times \frac{R^3 - r^3}{R^2 - r^2}$$

p 19 Cross section $$A = \frac{F_H}{2\,p_t}$$

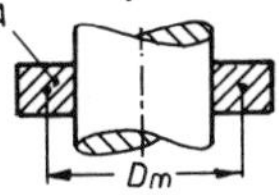

p 20 Shrinkage allowance $$A = \frac{1}{E}\,D_m \times p_t$$

(λ = outside dia. of shaft − inside dia. of ring)

Shrunk-on ring for clamping

Split, rotating clamped parts.

F_C comprises:

Centripetal force F_{CR} for ring
Centripetal force F_{CM} for clamped parts,

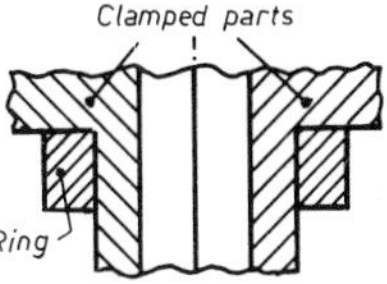

p 21 or $$F_H \geqslant 2(F_{CR} + F_{CM})\;;$$ then as p 19 and p 20

Energy of deformation U (Strain energy)

The energy stored in a deformed component is:

p 22 $$U = w\,V\;;\quad \text{where}$$

p 23 $$w = \frac{1}{2}\,\sigma\,\varepsilon = \frac{1}{2}\,E\,\varepsilon^2 = \frac{\sigma^2}{2\,E}$$; V : volume of component

Limit cross section for similar types of stress

Where a tension (or compression) force is applied at a point within the dotted core area, only tension (or compr.) forces will occur over the whole cross section. If applied at any other point, bending stress, i.e. simultaneous tension and compr. stress will occur.

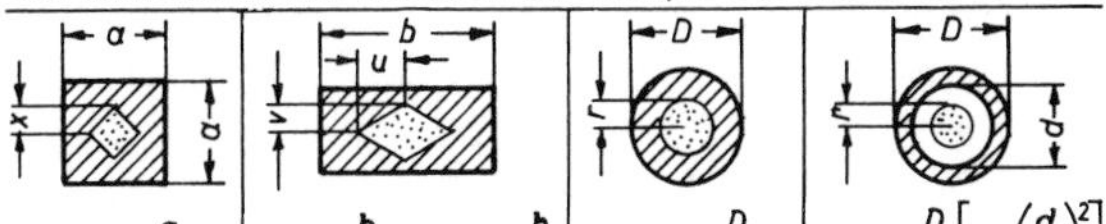

p 24 $x = \dfrac{a}{6}$ | $u = \dfrac{b}{6}$; $v = \dfrac{h}{6}$ | $r = \dfrac{D}{8}$ | $r = \dfrac{D}{8}\left[1+\left(\dfrac{d}{D}\right)^2\right]$

S : centre of mass of half ring (see K 7)
D_m : mean diameter ($D_m = R + r$)

Explanation

All external loads on a beam (including support reactions and its own weight) produce internal forces and moments which stress the material. By taking a section through the beam at a point x it is possible to show the internal loads: Vertical shear forces V and bending moments M.

End loads P and torsion T are considered separately.

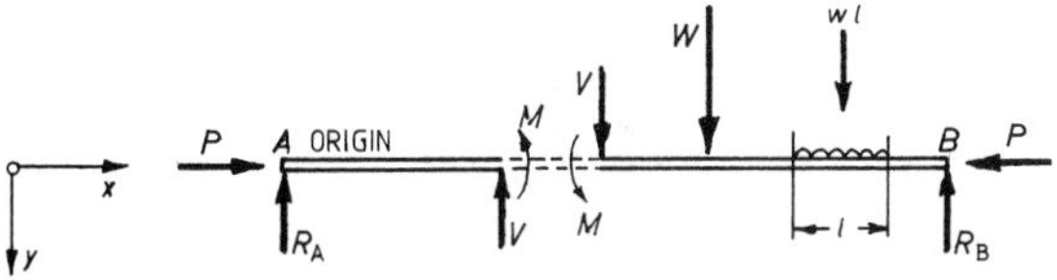

Referring to the x-y plane (z axis is at right angles):

Forces in direction of	x-axis	produce	end loads	P
	y-axis		shear forces	V
Moments about the	z-axis		bending moments	M
	x-axis		torsion	T

Always consider the left-hand side of the section.

In each part of the beam there must be equilibrium between all external and internal forces and moments:

Considered separately:

5/26 $$V + \sum_{i=1}^{n} V_i = 0 \qquad P + \sum_{i=1}^{n} P_i = 0$$

7/28 $$M + \sum_{i=1}^{n} M_i = 0 \qquad T + \sum_{i=1}^{n} T_i = 0$$

Method of calculation

1. Calculate the reactions.
2. Section the beam at the following places:
   - 2.1 Points of action of point loads W and beginning and end of disributed loads w.
   - 2.2 Points where the beam axis changes direction or the cross section changes.
   - 2.3 Any other convenient places.

continued on P 6

continued from P 5

3. Find the forces and moments on the left hand side of the section as in p 25...p 28.
4. Plot shear force and bending moment diagrams.

Relations between w, V and M at any point x

9/30 $$\frac{dM}{dx} = V \qquad \Big| \qquad \frac{dV}{dx} = -w$$

Rules:
M is a maximum when $V = 0$
In sections with no loads V = constant.

Example: Simply supported beam with end load. (Fixed at A). The reactions are:

$R_A = 2{\cdot}5$ kN; $P_A = 3$ kN; $R_B = 1{\cdot}5$ kN

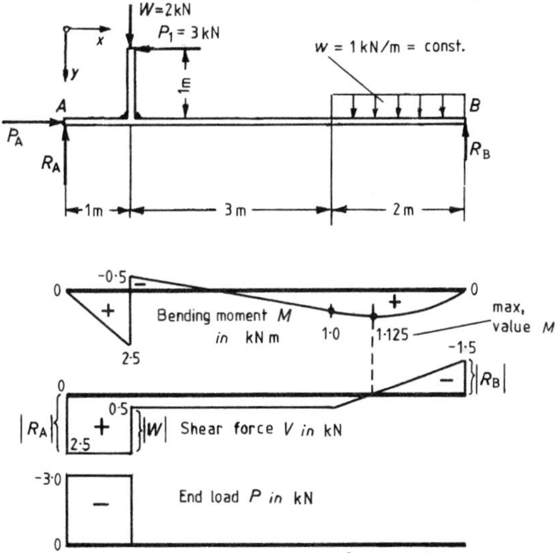

Calculations see P 7

Continued from P 6

Range	from equation p...	
$0 < x < 1\,m$	27	$M - R_A \times x = 0\,; \quad M = R_A \times x = 2{\cdot}5\,kN \times x$ *)
	26	$V - R_A = 0\,; \quad V = R_A = 2{\cdot}5\,kN = \text{const.}$
	29	or $V = \frac{d}{dx}(2{\cdot}5\,kN \times x) = 2{\cdot}5\,kN = \text{const.}$
	25	$P + P_A = 0\,; \quad P = -P_A = -3\,kN$
$1\,m < x < 4\,m$	27	$M - R_A \times x + W(x - 1m) + P_1 \times 1m = 0\,; \quad M = 0{\cdot}5\,kN\,x - 1\,kN\,m$ *)
	26	$V - R_A + W = 0\,; \quad V = R_A - W = 2{\cdot}5\,kN - 2\,kN = 0{\cdot}5\,kN = \text{const.}$
	29	or $V = \frac{d}{dx}(0{\cdot}5\,kN \times x - 1\,kN\,m) = 0{\cdot}5\,kN = \text{const.}$
	25	$P + P_A - P_1 = 0\,; \quad P = P_1 - P_A = 3\,kN - 3\,kN = 0$
$4\,m < x < 6\,m$	27	$M - R_A \times x + W(x - 1\,m) + P_1 \times 1\,m + w\frac{(x - 4\,m)^2}{2} = 0$ $M = -9\,kNm + 4{\cdot}5\,kN \times x - 0{\cdot}5\frac{kN}{m} \times x^2$ **)
	26	$V - R_A + W + w(x - 4m) = 0\,; \quad V = R_A - W - w(x-4m) = 2{\cdot}5kN - 2kN - 1\frac{kN}{m}(x-4m) = 4{\cdot}5kN - 1\frac{kN}{m}x$ *)
	29	or $V = \frac{d}{dx}(-9\,kN\,m + 4{\cdot}5\,kN\,x - 0{\cdot}5\frac{kN}{m}x^2) = 4{\cdot}5\,kN - 1\frac{kN}{m}\,x$
	25	$P + P_A - P_1 = 0\,; \quad P = P_1 - P_A = 3\,kN - 3\,kN = 0$

*) Straight line

**) Parabola

continued on P 8

Continued from P 7

Example:

Curved cantilever beam (r = const.)

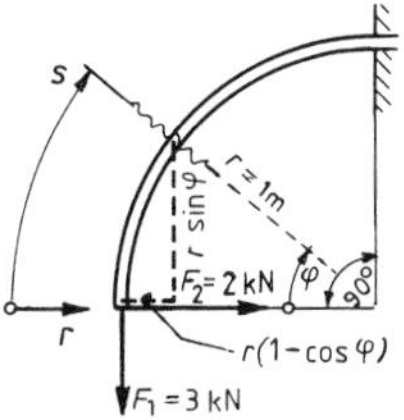

The limits are: $0 \leqslant \varphi \leqslant 90^\circ$

or: $0 \leqslant s \leqslant r\,\dfrac{\pi}{2}$

Bending moment:

p 31 $M + F_1\,r(1-\cos\varphi) + F_2\,r\sin\varphi = 0$

p 32 $M = -F_1\,r + F_1\,r\cos\varphi - F_2\,r\sin\varphi$

At the section φ, F_1 and F_2 are resolved into tangential and radial components.

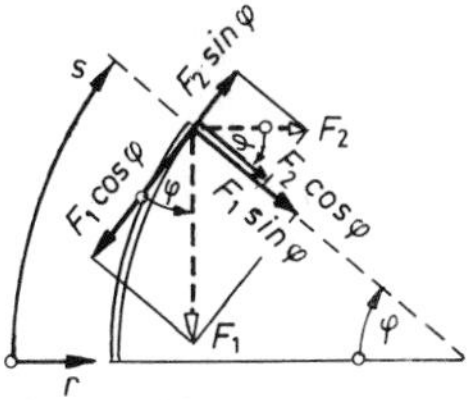

Shear force (radial):

p 33 $F_q + F_1\sin\varphi + F_2\cos\varphi = 0$

p 34 $F_q = -F_1\sin\varphi - F_2\cos\varphi$; or from p 30:

p 35 $F_q = \dfrac{dM}{ds} = \dfrac{1}{r} \times \dfrac{dM}{d\varphi}$ (because $s = r\varphi$; $ds = r\,d\varphi$)

p 36 $= \dfrac{1}{r} \times \dfrac{d(-F_1\,r + F_1\,r\cos\varphi - F_2\,r\sin\varphi)}{d\varphi} = -F_1\sin\varphi - F_2\cos\varphi$

Normal force (tangential):

p 37 $F_n - F_1 \times \cos\varphi + F_2 \times \sin\varphi = 0$

p 38 $F_n = F_1 \times \cos\varphi - F_2 \times \sin\varphi$

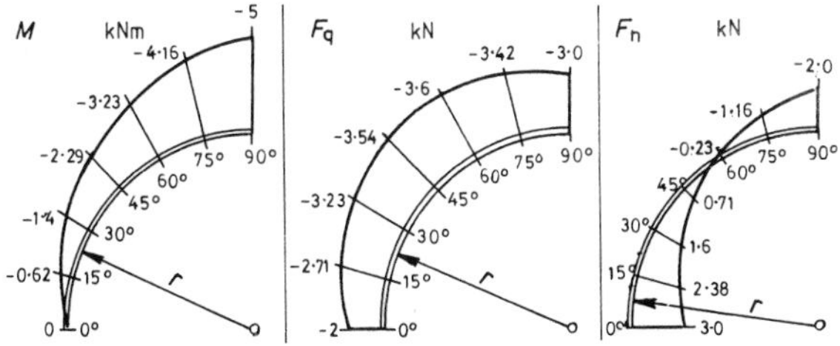

For a graphical method of determining bending moment see K 4.

Maximum bending stress

p 39 $\sigma_{bt\,max} = \frac{M\,y_{max}}{I_{xx}}$

p 40 $= \frac{M}{Z_{min}} \leqslant p_b$

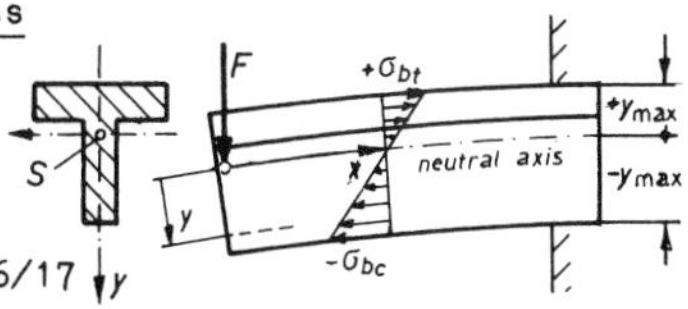

Values for p_b see Z16/17

$+y_{max}$ (tens)	distance from surface fibre to the x-axis passing through the centroid S or neutral axis
$-y_{max}$ (comp)	

I_{xx} : Second moment of area about axis S_z or about plane of neutral axis.

Bending stress at distance y from the neutral axis

p 41 $\sigma_b = \frac{M}{I}\,y$

Section modulus Z_{min}

p 42 $Z_{min} = \frac{I}{y_{max}}$

Second moments of area

Axial second moment of area see J 10 and table P 10
Polar second moment of area see J 10
Product moment see J 10.

Principal second moments of area and principal axes

The principal second moments of area $I_1 = I_{max}$ and $I_2 = I_{min}$ are applicable to asymmetric sections, when the principal axes are rotated through the angle φ_0.

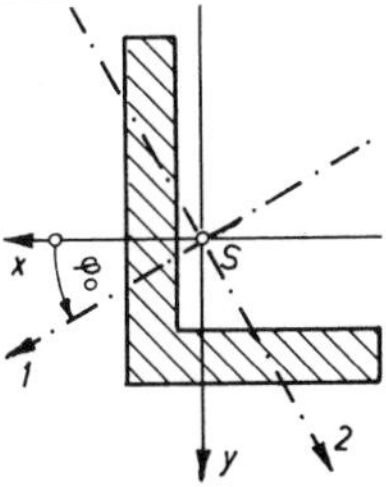

p 43 $I_{\substack{1\\2}} = I_{\substack{max\\min}} = \frac{1}{2}(I_y + I_x) \pm \frac{1}{2}\sqrt{(I_y - I_x)^2 + 4 I_{xy}^2}$

p 44 $\tan 2\varphi_0 = \frac{2 I_{xy}}{I_y - I_x}$

For calculations of I_{xy} see J 10/11.

The principal axes are always perpendicular to each other.

The axis of symmetry of a symmetrical section is one principal axis e.g. $I_1 = I_x$.

Values of I and Z for some common sections

(see p 41 and p 42)

For position of centroid S (or neutral axis) see K 7

	I_x and I_y	Z_x and Z_y	Cross section A
p 45 p 46	$I_x = \frac{b\,d^3}{12}$ $I_y = \frac{d\,b^3}{12}$	$Z_x = \frac{b\,h^2}{6}$ $Z_y = \frac{h\,b^2}{6}$	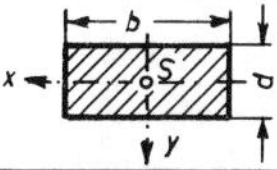
p 47	$I_x = I_y = \frac{\pi\,D^4}{64}$	$Z_x = Z_y = \frac{\pi\,D^3}{32} \triangleq \frac{D^3}{10}$	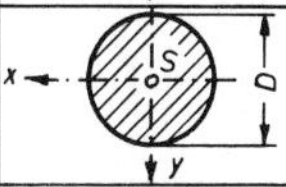
p 48	$I_x = I_y = \frac{\pi}{64}(D^4 - d^4)$	$Z_x = Z_y$ $= \frac{\pi}{32} \times \frac{D^4 - d^4}{D} \triangleq \frac{D^4 - d^4}{10D}$	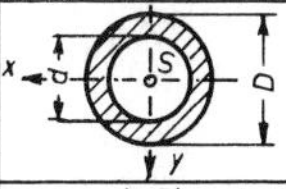
p 49 p 50 p 51 p 52	$I_x = I_y = 0{\cdot}06014\,s^4$ $= 0{\cdot}5412\,R^4$	$Z_x = 0{\cdot}1203\,s^3$ $= 0{\cdot}6250\,R^3$ $Z_y = 0{\cdot}1042\,s^3$ $= 0{\cdot}5413\,R^3$	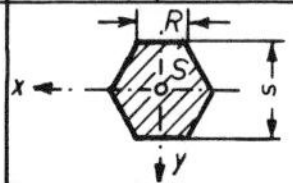
p 53 p 54	$I_x = \frac{\pi\ a\ b^3}{4}$ $I_y = \frac{\pi\ a^3\ b}{4}$	$Z_x = \frac{\pi\,a\,b^2}{4}$ $Z_y = \frac{\pi\,a^2\,b}{4}$	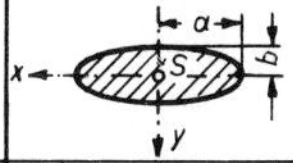
p 55 p 56	$I_x = \frac{b\ h^3}{36}$ $I_y = \frac{b^3\ h}{48}$	$Z_x = \frac{b\,h^2}{24}$ $Z_y = \frac{b^2\,h}{24}$	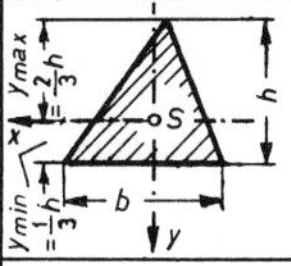
p 57 p 58 p 59	$I_x = \frac{h^3}{36} \times \frac{(a+b)^2 + 2ab}{a+b}$ $y_{max} = \frac{h}{3} \times \frac{2a+b}{a+b}$ $y_{min} = \frac{h}{3} \times \frac{a+2b}{a+b}$	$Z_x = \frac{h^2}{12} \times \frac{(a+b)^2 + 2ab}{2a+b}$	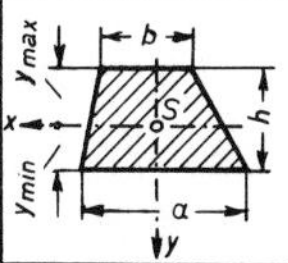

Steiner's theorem

(Parallel axis theorem for second moments of area).

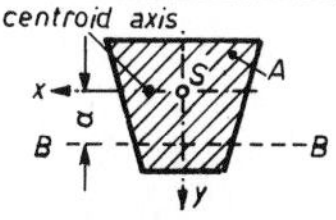

p 60 $I_{B\text{-}B} = I_x + A\ a^2$

Beams of uniform cross section

Equation of the elastic curve

The following apply to each section of the beam (see P 5, Method of calculation, Item 2):

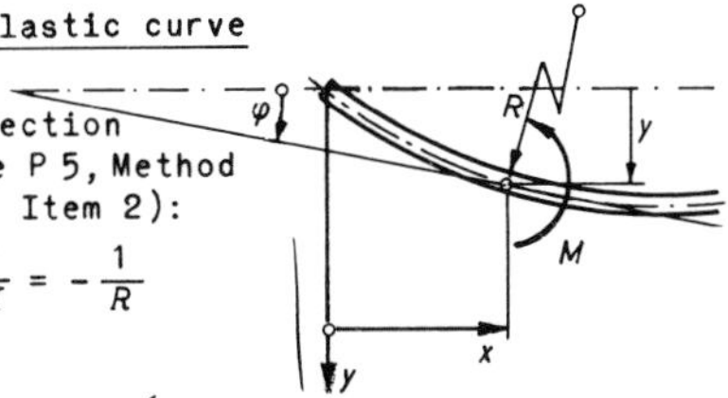

p 61 $$\frac{d^2y}{dx^2} = y'' = -\frac{M}{EI} = -\frac{1}{R}$$

p 62 $$E \times I \times y'' = -M$$

p 63 $$E \times I \frac{dy}{dx} = E \times I \times y' = -\int M\,dx + C_1$$

p 64 $$E \times I \times y = -\iint M\,dx \times dx + C_1 \times x + C_2$$

R : radius of curvature of the elastic curve at point x.

p 65 $y' = \tan\varphi$: inclination of the tangent to the elastic curve at point x.

y = deflection of beam at point x.

C_1 and C_2 are constants of integration and are determined from known factors.

e.g. $y = 0$ at the support.

$y_i = y_{i+1}$ at junction between sections i and $(i+1)$.

$y' = 0$ at the support of a cantilever beam and at the centre of a beam with symmetrical loading.

$y'_i = y'_{i+1}$ at the junction between sections i and $(i+1)$.

Strain energy due to bending U:

For a beam of length l:

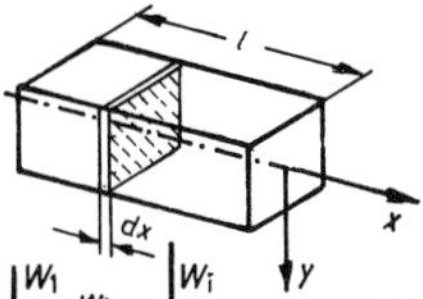

p 66 $$U = \frac{1}{2}\int_0^l \frac{M^2}{EI}\,dx$$

A beam with discontinuous loads with may be divided into n length:

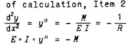

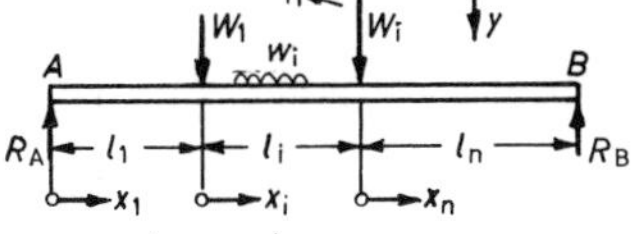

p 67 $$U_{tot} = \frac{1}{2E}\left(\int_{x_1=0}^{x_1=l_1} \frac{M^2}{I_1}\,dx_1 + \ \ldots\ + \int_{x_n=0}^{x_n=l_n} \frac{M^2}{I_n}\,dx_n\right)$$

p 68 p 69 p 70 p 71

Loading	Support reactions	M_{max} at point (...)	Deflection, y / Slope, $y' = \tan\varphi$	Deflection at C, y_c / Max. y_m
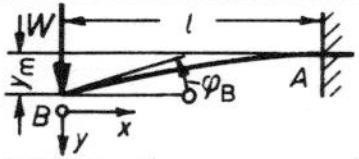	$R_A = W$ $M_A = W\,l$	$W\,l$ (A)	$y = \frac{W\,l^3}{6\,E\,I}\left(2 - 3\frac{x}{l} + \frac{x^3}{l^3}\right)$ $\tan\varphi_B = -\frac{W\,l^2}{2\,E\,I}$	$y_m = \frac{W\,l^3}{3\,E\,I}$
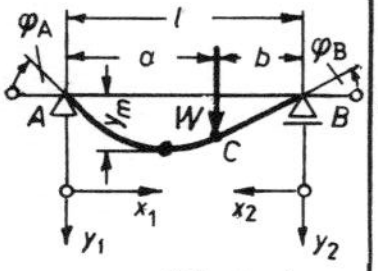 Note $a > b$	$R_A = W\frac{b}{l}$ $R_B = W\frac{a}{l}$	$W\frac{a\,b}{l}$ (C)	$y_1 = \frac{W\,l^3}{6\,E\,I}\frac{a}{l}\frac{b^2}{l^2}\frac{x_1}{l}\left(1 + \frac{l}{b} - \frac{x_1^2}{a\,b}\right)$ $y_2 = \frac{W\,l^3}{6\,E\,I}\frac{b}{l}\frac{a^2}{l^2}\frac{x_2}{l}\left(1 + \frac{l}{a} - \frac{x_2^2}{a\,b}\right)$ $\tan\varphi_A = \frac{y_C}{2a}\left(1 + \frac{l}{b}\right)$ $\tan\varphi_B = \frac{y_C}{2b}\left(1 + \frac{l}{a}\right)$	$y_c = \frac{W\,l^3}{3\,E\,I}\frac{a^2}{l^2}\frac{b^2}{l^2}$ $y_m = y_c\frac{l+b}{3b}\sqrt{\frac{l+b}{3a}}$ y_m at point $x_1 = a\sqrt{\frac{l+b}{3a}}$
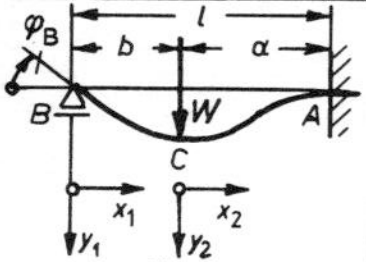 Note $a > b$	$R_A = W - R_B$ $R_B = W\frac{a^2}{l^2}\left(\frac{a}{l} + \frac{3b}{2l}\right)$ $M_A = W\frac{l+b}{2}\frac{a}{l}\frac{b}{l}$	M_A (A) when $b = 0{\cdot}414\,l$: $0{\cdot}171Wl$ (C) $-0{\cdot}171Wl$ (A)	$y_1 = \frac{R_B\,l^3}{6\,E\,I}\left(3\frac{x_1}{l} - \frac{x_1^3}{l^3}\right) - \frac{W\,a^2\,x_1}{2\,E\,I}$ $y_2 = \frac{W\,a^3}{6\,E\,I}\left(2 - 2\frac{x_2}{a} + \frac{x_2^3}{a^3}\right) - \frac{R_B\,l^3}{6\,E\,I}\left(2 - 3\frac{x_1}{l} + \frac{x_1^3}{l^3}\right)$ $\tan\varphi_B = W\,a^2 b/(4\,E\,I\,l)$	$y_c = \frac{R_B\,l^3}{6EI}\left(3\frac{b}{l} - \frac{b^3}{l^3}\right) - \frac{W\,a^2\,b}{2\,E\,I}$
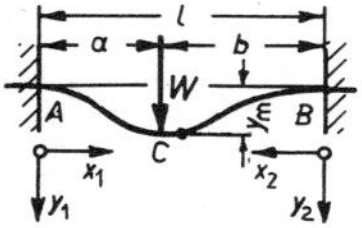 Note $a < b$	$R_A = W\frac{b^2}{l^2}\left(3 - 2\frac{b}{l}\right)$ $R_B = W\frac{a^2}{l^2}\left(3 - 2\frac{a}{l}\right)$ $M_A = W\,a\,b^2/l^2$ $M_B = W\,b\,a^2/l^2$	$-Wa\frac{b^2}{l^2}$ (A) $-Wb\frac{a^2}{l^2}$ (B) $2Wl\frac{a^2\,b^2}{l^2\,l^2}$ (C)	$y_1 = \frac{W}{6\,E\,I}\frac{b^2}{l^2}\left(3ax_1^2 - 3x_1^3 + 2\frac{b}{l}x_1^3\right)$ $y_2 = \frac{W}{6\,E\,I}\frac{a^2}{l^2}\left(3bx_2^2 - 3x_2^3 + 2\frac{a}{l}x_2^3\right)$ $\tan\varphi_A = \tan\varphi_B = 0$	$y_m = \frac{2Wl^3}{3EI}\frac{b^3}{l^3}\frac{a^2}{b^2}\left(\frac{l}{2b+l}\right)^2$ y_m at point $x_1 = \frac{2l\cdot b}{2b + l}$ $y_c = \frac{W}{3\,E\,I}\frac{a^3\,b^3}{l^3}$

STRENGTH

Deflection of beams in bending

p 75 p 74 p 73 p 72

Loading	Support reactions	M_{max} at point (...)	Deflection, y / Slope, $y' = \tan\varphi$	Deflection at C, y_C; Max. y_m
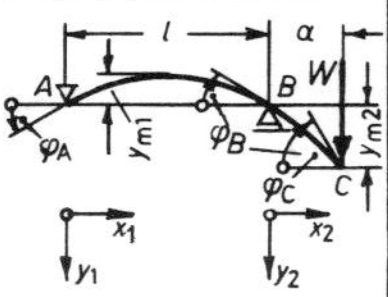	$R_A = W\frac{a}{l}$ $R_B = W\left(1+\frac{a}{l}\right)$	$-Wa$ (B)	$y_1 = \frac{W a l^2}{6EI}\left(\frac{x^3}{l^3} - \frac{x}{l}\right)$ $y_2 = \frac{W}{6EI}(2a\,l\,x_2 + 3a\,x_2^2 - x_2^3)$ $\tan\varphi_A = -\frac{W a l}{6EI}$; $\tan\varphi_B = \frac{W a l}{3EI}$ $\tan\varphi_C = \frac{W a}{6EI}(2l + 3a)$	$y_{m_1} = -\frac{W a l^2}{9\sqrt{3}\,EI}$ y_{m_1} at point $x_1 = 0{\cdot}577\,l$ $y_{m_2} = y_C = \frac{W a^2}{3EI}(l+a)$
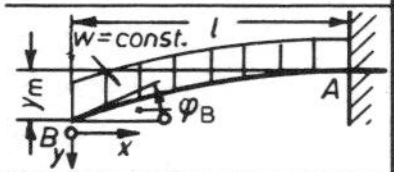	$R_A = w\,l$ $M_A = \frac{1}{2}w\,l^2$	$\frac{1}{2}w\,l$ (A)	$y = \frac{w l^4}{24EI}\left(\frac{x^4}{l^4} - 4\frac{x}{l} + 3\right)$ $\tan\varphi_B = -\frac{w l^3}{6EI}$	$y_m = \frac{w l^4}{8EI}$
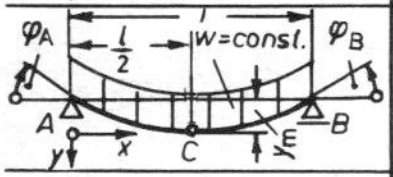	$R_A = R_B = \frac{w\,l}{2}$	$\frac{w l^2}{8}$ (C)	$y = \frac{w l^4}{24EI}\frac{x}{l}\left(1 - 2\frac{x^2}{l^2} + \frac{x^3}{l^3}\right)$ $\tan\varphi_A = \frac{w l^3}{24EI} = -\tan\varphi_B$	$y_m = \frac{5 w l^4}{384EI}$
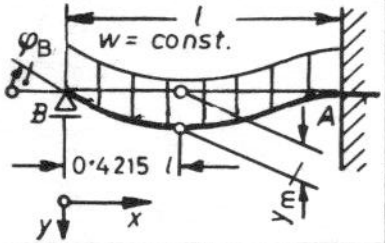	$R_B = \frac{3}{8}w\,l$ $R_A = \frac{5}{8}w\,l$ $M_A = \frac{1}{8}w\,l^2$	M_A (A)	$y = \frac{w l^4}{48EI}\left(\frac{x}{l} - 3\frac{x^3}{l^3} + 2\frac{x^4}{l^4}\right)$ $\tan\varphi_B = \frac{w l^3}{48EI}$	$y_m = \frac{w l^4}{185EI}$ y_m at point $x = 0{\cdot}4215\,l$
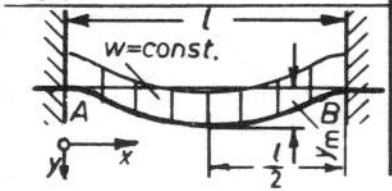	$R_A = w \times l/2$ $R_B = w \times l/2$ $M_A = w \times l^2/12$ $M_B = w \times l^2/12$	$-\frac{w l^2}{12}$ (A, B) $\frac{w l^2}{24}$ (C)	$y = \frac{w l^4}{24EI}\left(\frac{x^2}{l^2} - 2\frac{x^3}{l^3} + \frac{x^4}{l^4}\right)$ $\tan\varphi_A = \tan\varphi_B = 0$	$y_m = \frac{w l^4}{384EI}$

Formulae for y and y_m do not allow for shear deflection.

Mohr's analogy

Graphical method

1. Determine the bending moment curve by constructing a link polygon (see also K 4).

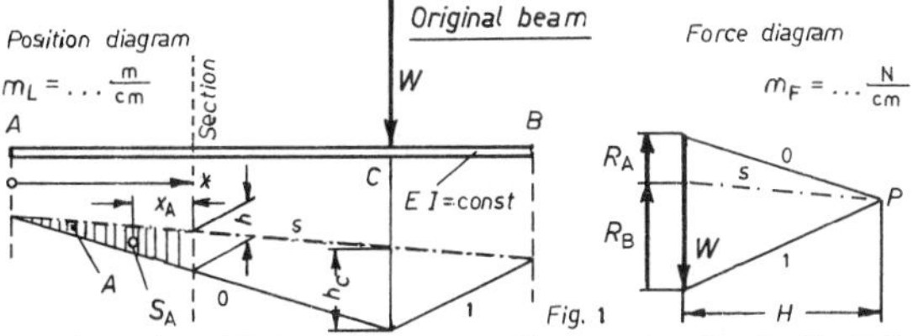

Fig. 1

2. Construct a link polygon as the equivalent distributed load w^* on the "equivalent beam". Another link polygon will give the tangents to the elastic curve.

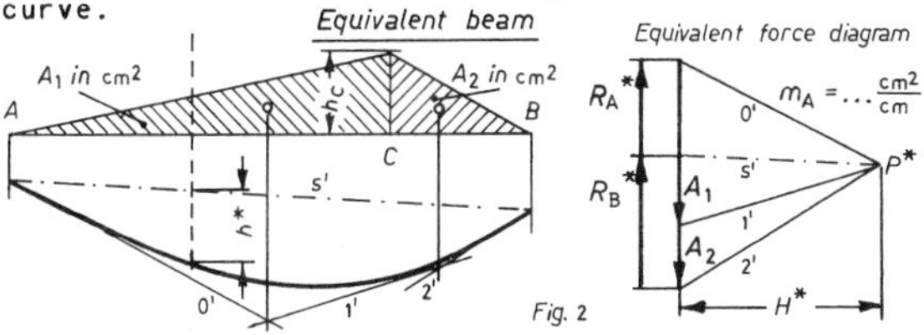

Fig. 2

3. Deflection of the original beam at point x:

p 76 $$y = h^* \frac{H H^*}{E I} m_F \, m_A \, m_L^3 .$$

Slope at support A and B:

p 77 $$\tan \varphi_A = R_A^* \frac{H}{E I} m_F \, m_A \, m_L^2 \text{ resp. } \tan \varphi_B = R_B^* \frac{H}{E I} m_F \, m_A \, m_L^2 .$$

Mathematical method

1. Calculate the equivalent support reaction R_A^* of the "equivalent beam" carrying the equivalent distributed load $w^x = A_1 + \dots + A_n$ (see fig. 2). p 78
2. Calculate the equivalent bending moment M^* and the equivalent shear force V^* at the point x:

p 79 $M^* = R_A^* z - A \, z_A$; $V^* = R_A^* - A$ (see fig. 1+2)

x_A : Distance between c of g of equivalent distributed load A and the section x.

p 80 3. Deflection $y = M^*/EI$; Slope $y = V^*/EI$

continued on P 15

continued from P 14 (Mohr's analogy)

Choice of equivalent beam

The supports of the equivalent beam must be such that its maximum equivalent bending moment M^*_{max} coincides with the point of maximum deflection in the original beam.

	Original beam	Equivalent beam
Simple beam		
Cantilever beam		

Beam of varying cross section

Fig. 1
Original beam e.g. shaft

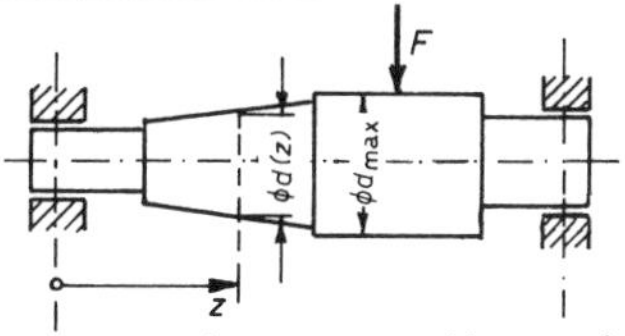

Plot the bending moment curve as the equivalent distributed load $w^*(z)$ on a uniform equivalent beam of cross section equal to the maximum second moment of area $I_{x\max}$ of the original beam. (See P 14, item 1.).
Plot $w^*(z)$ according to the ratio $\frac{I_{x\max}}{I_x(z)}$:

p 81

Fig. 2
Equivalent beam of the shaft in Fig. 1

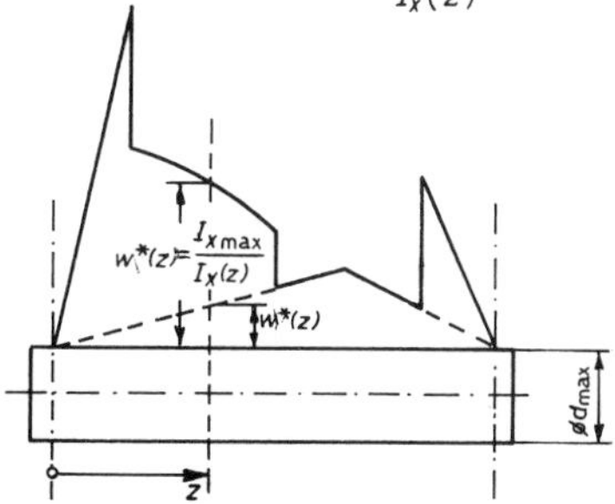

Then calculate according to P 14 (items 2 and 3 or p 78 ... p 80.

Beams of uniform strength

	maximum section dimension	typical dimension x = resp. y =	maximum deflect. f	type of beam
p 82	$h=\sqrt{\frac{6\,W\,l}{b\,p_{bt}}}$	$\sqrt{\frac{6\,W\,x}{b\,p_{bt}}}$	$\frac{8\,W}{b\,E}\left(\frac{l}{h}\right)^3$	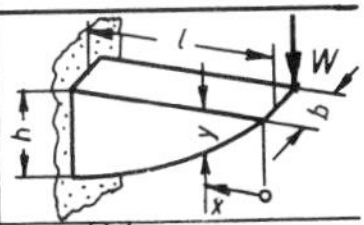 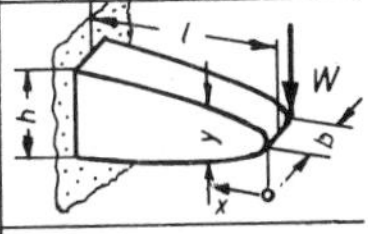
p 83	$b=\frac{6\,W\,l}{h^2\,p_{bt}}$	$\frac{6\,W\,x}{h^2\,p_{bt}}$	$\frac{6\,W}{b\,E}\left(\frac{l}{h}\right)^3$	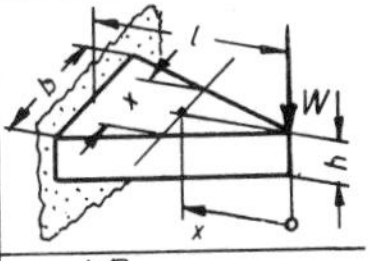
p 84	$h=\sqrt{\frac{3\,w\,l^2}{b\,p_{bt}}}$	$x\sqrt{\frac{3\,w\,l}{b\,l\,p_{bt}}}$	$\frac{3\,w\,l}{b\,E}\left(\frac{l}{h}\right)^3$	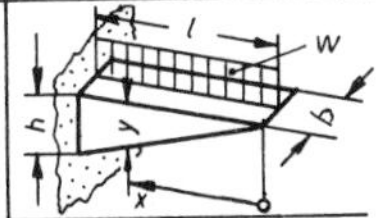
p 85	$b=\frac{3\,w\,l^2}{h^2\,p_{bt}}$	$\frac{3\,w\,l\,x^2}{h^2\,l\,p_{bt}}$		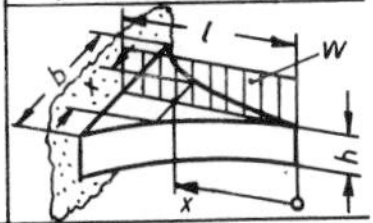
p 86	$h=\sqrt{\frac{3\,w\,l^2}{4\,b\,p_{bt}}}$	$\sqrt{\frac{3\,w\,l^2}{4\,b\,p_{bt}}\left(1-\frac{4x^2}{l^2}\right)}$	$\frac{w\,l^4}{64\,E\,I}$	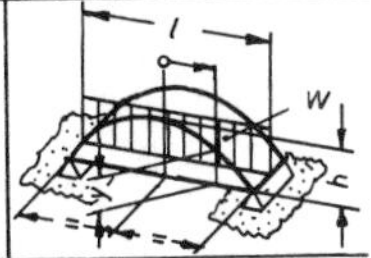
p 87	$d=\sqrt[3]{\frac{32\,W\,l}{\pi\,p_{bt}}}$	$\sqrt[3]{\frac{32\,W\,x}{\pi\,p_{bt}}}$	$\frac{192}{5}\times\frac{W\,l^3}{E\,\pi\,d^4}$	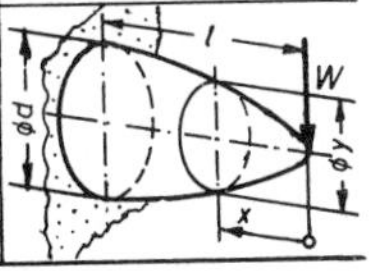

W : point load — kN

w : uniformly distributed load — kN/m

p_{bt} : permissible bending stress — N/mm^2 (see Z 17)

STRENGTH

Statically indeterminate beams

Convert a statically indeterminate beam (fig. 1) into a statically determinate one (fig. 2) by replacing one support by its support reaction (R_C in fig. 2).

Divide into two separate beams or subsystems. Determine the deflections at the point of the statically indeterminate support (see P 11 to P 15) from each subsystem, in terms of R_C.

Since no deflection can occur at support C:

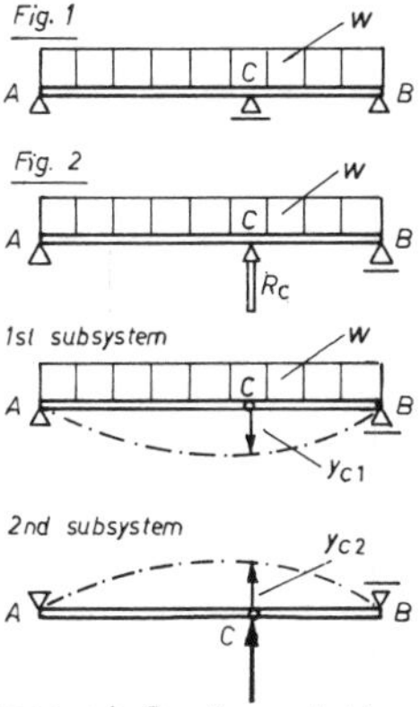

p 88

$$|y_{C1}| = |y_{C2}|$$

Hence, calculate the support force at C, R_C and then the remaining support reactions.

Method of solution for simple static. indeterm. beams

Statically indetermin. beam	Statically determ. beam	1st subsystem	2nd subsystem

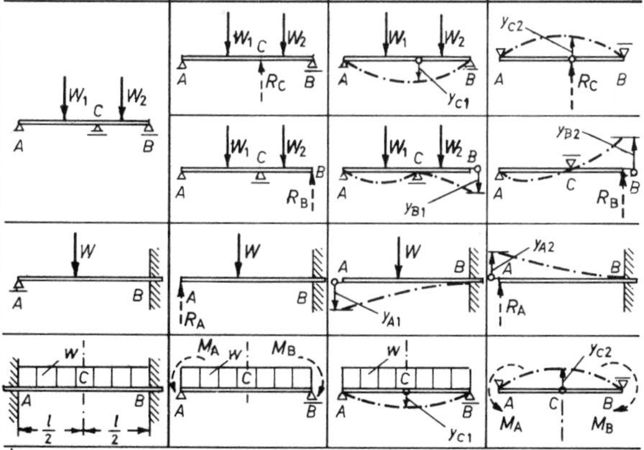

↑↱ : statically indeterm. support reactions and moments

STRENGTH

Shear

Hooke's law for shear stress

p 89 q or $\tau = G\,\gamma$

G : shear modulus
γ : shear strain

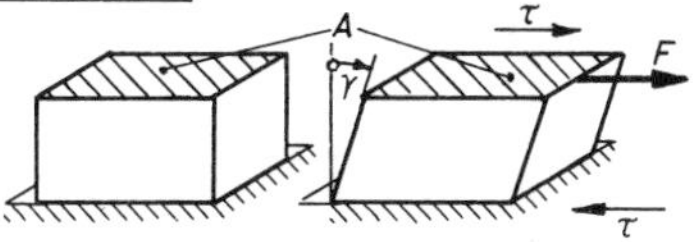

Relation between shear modulus and modulus of elasticity or Young's modulus

p 90 $G = \dfrac{E}{2(1+\mu)} \doteq 0{\cdot}385\,E$; corresp. to P 3 with $\mu = 0{\cdot}3$

Mean shear stress q_f or τ_f

p 91 q_f or $\tau_f = \dfrac{F}{A} \leqslant p_q$

Permissible shear strength p_q (for values see Z 16)

type of load (see P 2)		$p_q \doteq$
	dead	$p_t/1{\cdot}5$
	undulating	$p_t/2{\cdot}2$
	alternating	$p_t/3{\cdot}0$

Ultimate shear stress q_s

p 92 $q_s = \dfrac{Q_{max}}{A}$

for ductile metals: $q_s = 0{\cdot}8\,R_m$
for cast iron: $q_s = R_m$

Shearing force F

applied by guillotine shears	applied by cutting tools (punch...)
$Q \doteq 1{\cdot}2\,q_s\,l\,s$	$Q \doteq 1{\cdot}2\,q_s\,l_p\,s$

p 93 l : length of cut; l_p : perimeter of cut

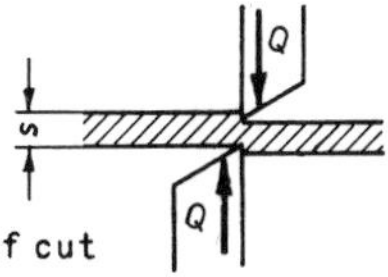

Theorem of related shear stresses

The shear stresses on two perpendicular faces of an element are equal in magnitude, perpendicular to their common edge and act either towards it or away from it.

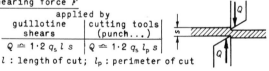

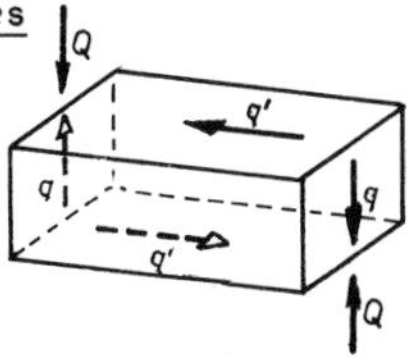

p 94 $q = q'$

q : transverse shear stress (transverse to beam axis) resulting from shear forces Q

q' : axial shear stress (parallel to beam axis) "complementary shear"

Axial shear stress due to shear forces

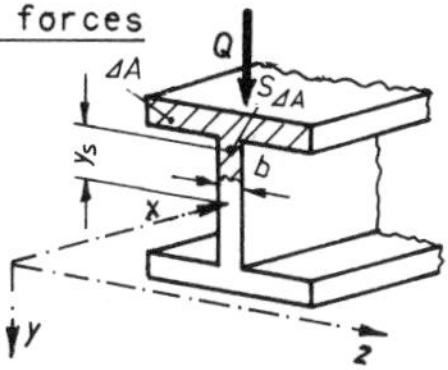

p 95 $$q = \frac{Q\,M}{b\,I} = q'$$

p 96 $$M = \Delta A\; y_s$$

$q = 0$ for $\sigma_b = \sigma_{b\,max}$.
q_{max} occurs when $\sigma_b = 0$.
i.e. on neutral axis.

Max. shear stress for different cross sections

$$q_{max} = q'_{max} = k\,\frac{Q}{A}$$

	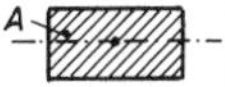	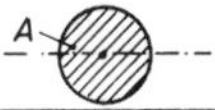	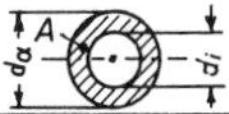
p 97 k	$\frac{3}{2}$	$\frac{4}{3}$	$\frac{4}{3} \times \frac{d_a^2 + d_a\,d_i + d_i^2}{d_a^2 + d_i^2}$ for thin-wall tubes: 2 ($d_a \approx d_i$)

Strain energy u due to shear

p 98 $$u = \frac{1}{2}\,q\,\gamma = \frac{q^2}{2\,G}$$

Shear deflection of a beam

p 99 $$y = \varkappa\,\frac{M}{G\,A} + C = \varkappa\,\frac{2{\cdot}6\,M}{E\,A} + C$$

Determine the constant C from known factors, e.g. $y = 0$ at the supports.

p100 The factor $\varkappa = A \int\limits_{(A)} \left[\frac{M}{b\,I}\right]^2 dA$ allows for the form of cross section. For example:

	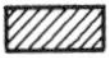		I 80	I 240	I 500
$\varkappa =$	1·2	1·1	2·4	2·1	2·0

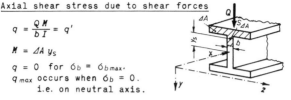

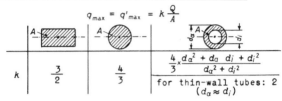

Q : shear force at point x of the beam
M : bending moment at the section A
I : second moment of area of the total cross section A about the z axis
b : width of section at point y
$S_{\Delta A}$: centre of area of section ΔA

STRENGTH
Torsion

General

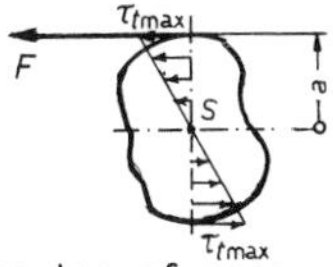

101 Shear stress due torsion $\Big\}\ \tau_t = \dfrac{T\,a}{J} \leqslant p_{qt}$

102 Torque $\quad T = \dfrac{P}{\omega} = \dfrac{P}{2\,\pi\,n} = F\,a$

P : power
a : distance from surface fibre to centre of mass
J : torsion constant; formulae see P 21 (attention: torsion constant is not the polar moment of inertia; only for circular cross section $J = I_p$, $a = D/2$)

Bars of circular cross section

Angle of twist φ (see e 5)

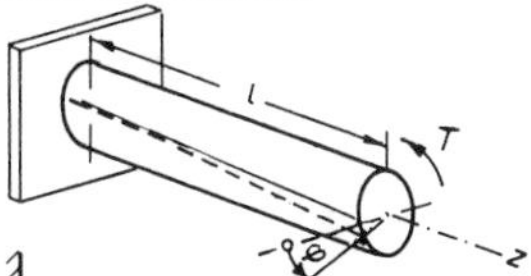

104 $\varphi = \dfrac{T\,l}{I_p\,G} = \dfrac{180^\circ}{\pi} \times \dfrac{T\,l}{I_p\,G}$

Stepped shafts:

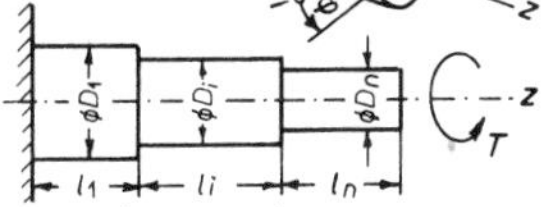

105 $\varphi = \dfrac{T}{G} \times \sum\limits_{i=1}^{n} \dfrac{l_i}{I_{pi}}$

106 $\quad = \dfrac{180^\circ}{\pi} \times \dfrac{T}{G} \times \sum\limits_{i=1}^{n} \dfrac{l_i}{I_{pi}}$

	polar moment of inertia I_p	max. shear stress τ_t	cross section
107	$\dfrac{\pi\,D^4}{32}$	$\simeq 5{\cdot}1 \dfrac{T}{D^3}$	
108	$\dfrac{\pi}{32}(D^4 - d^4)$	$\simeq 5{\cdot}1\ T\,\dfrac{D}{D^4 - d^4}$	

Bars of non-circular, solid or thin-wall hollow section

109 Angle of twist $\varphi = \dfrac{T\,l}{I_t\,G} = \dfrac{180^\circ}{\pi} \times \dfrac{T\,l}{I_t\,G}$

	torsion constant J	position and magnitude of τ_t	cross section
110	$c_1\,h\,b^3$	in 1: $\tau_{t1} = \tau_{t\max} = \dfrac{c_2\,T}{c_1\,h\,b^2}$ in 2: $\tau_{t2} = c_3\,\tau_{t\max}$ in 3: $\tau_{t3} = 0$	$h/b \geqslant 1$ 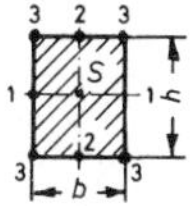

$h/b =$	1	1,5	2	3	4	6	8	10	∞
c_1	0·141	0·196	0·229	0·263	0·281	0·298	0·307	0·312	0·333
c_2	0·675	0·852	0·928	0·977	0·990	0·997	0·999	1·000	1·000
c_3	1·000	0·858	0·796	0·753	0·745	0·743	0·743	0·743	0·743

continued on P 21

	torsion constant J	position and magnitude of τ_t	cross section
p 111 p 112	$\frac{a^4}{46\cdot19} \simeq \frac{h^4}{26}$	at 1: $\tau_{t1} = \tau_{t\,max} = \frac{20\,T}{a^3} \approx \frac{13\,T}{h^3}$ at 2: $\tau_{t2} = 0$	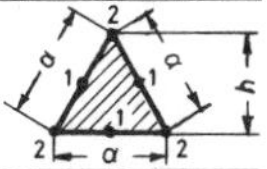
p 113 p 114 p 115	$0\cdot1154\ s^4$ $= 0\cdot0649\ d^4$	at 1: $\tau_t = \tau_{t\,max} = 5\cdot297\,\frac{T}{s^3}$ $= 8\cdot157\,\frac{T}{d^3}$	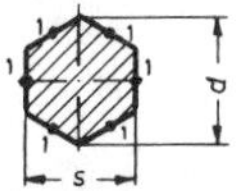
p 116 p 117	$\frac{\pi}{16}\times\frac{D^3\,d^3}{D^2+d^2}$	at 1: $\tau_{t1} = \tau_{t\,max} \simeq 5\cdot1\,\frac{T}{D\,d^2}$ at 2: $\tau_{t2} = \tau_{t\,max}\,\frac{d}{D}$	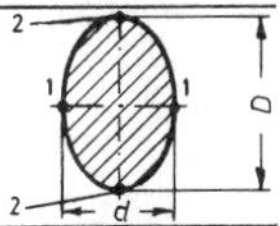
p 118 p 119 p 120	$\frac{\pi}{16}\times\frac{n^3(d^4-d_i^4)}{n^2+1}$ $D/d = D_i/d_i = n \geqslant 1$	at 1: $\tau_{t1} = \tau_{t\,max} \simeq 5\cdot1\,\frac{T\,d}{n(d^4-d_i^4)}$ at 2: $\tau_{t2} = \tau_{t\,max}\,\frac{d}{D}$	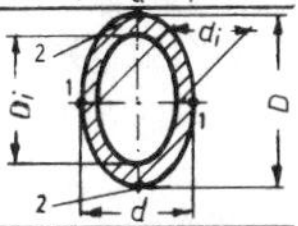
p 121 p 122 p 123	with varying wall thickness: $\frac{4\,A_m^2}{\sum_{i=1}^{n}\frac{s_i}{t_i}}$	at 1: $\tau_{t1} = \tau_{t\,max} = \frac{T}{2\,A_m\,t_{min}}$ at i: $\tau_{t2} = \frac{T}{2\,A_m\,t_i}$	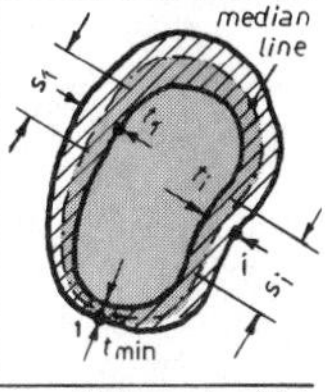
p 124	with thin, uniform wall: $\frac{4\,A_m^2\,t}{s_m}$	$\tau_t = \frac{T}{2\,A_m\,t}$	
p 125	$\frac{\eta}{3}\sum_{i=1}^{n} b_i^3\,h_i$	$\tau_{t\,max} = \frac{T\,b_{max}}{I_t}$ midway along the long side h of the rectangular section of max. thickness b_{max} (e.g. position 1 in the sketch).	profiles built up of rectang. cross sections 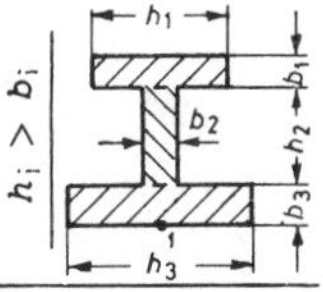

Föppl's factor:

	I $n=3$	L $n=2$	⊥ $n=2$
η	$\simeq 1\cdot3$	$\simeq 1\cdot0$	$1\cdot12$
		⊏ $n=3$	+ $n=2$
η		$1 < \eta < 1\cdot3$	$1\cdot17$

A_m : area enclosed by the median line
s : length of the median line
t (t_{min}): wall thickness, (min. wall thickn.) [t_i = const
s_i : part length of median line when wall thickness

Euler's formula

Applies for elastic instability of struts. Minimum load P_e at which buckling occurs:

127 $$P_e = \pi^2 \frac{E\, I_{min}}{l_k^2}$$

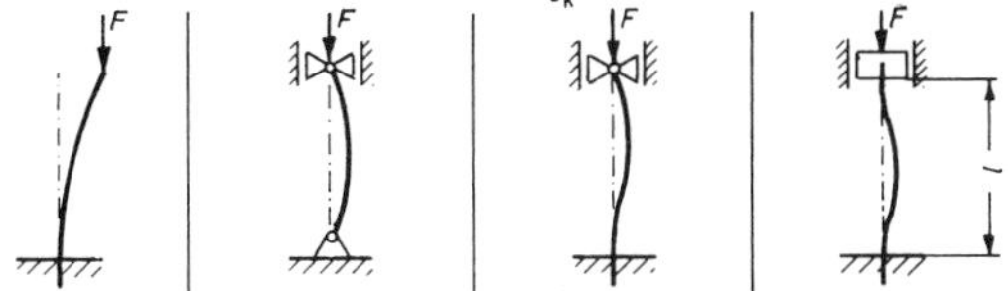

$l_k = 2\,l$	$l_k = l$	$l_k = 0{\cdot}707\,l$	$l_k = 0{\cdot}5\,l$

128 Permissible working load $F = P_e / \nu_k$

129 Slenderness ratio $\lambda = \frac{l}{k} = l\sqrt{\frac{A}{I_{min}}}$

Limit 1 based on $R_{p\,0{\cdot}01}$	Limit 2 based on $R_{p\,0{\cdot}2}$
130 $\lambda_{lim\,0{\cdot}01} = \pi\sqrt{\frac{E}{R_{p\,0{\cdot}01}}}$	$\lambda_{im\,0{\cdot}2} = \pi\sqrt{\frac{E}{R_{p\,0{\cdot}2}}}$

Tetmajer formula

Valid in the range $R_{p\,0{\cdot}01} \leqslant \frac{P_e}{A} \leqslant R_{p\,0{\cdot}2}$

Material of strut fails due to bending and compression

131 $$p_t = \frac{F}{A} = a - b\,\lambda + c\,\lambda^2 = p_c\,\nu_k$$

Material	a	b N/mm²	c	valid for λ =
mild steel ASTM-A 283 Gr. C	289	0·818	0	60...100
mild steel ASTM-A 440	589	3·818	0	60...100
cast iron ASTM-A 48 A 25 B	776	12·000	0·053	5... 80
timber	30	0·20	0	2...100
beach or oak	38	0·25	0	0...100

Calculation method

First determine the minimum second moment of area using the Euler formula:

132 $$I_{min} = \frac{P_e\, l_k^2}{\pi^2 E}\;;\quad P_e = \nu_k \times F.$$

Then select a suitable cross section, e. g. circular tube, solid rectangle, etc., and find I and A.

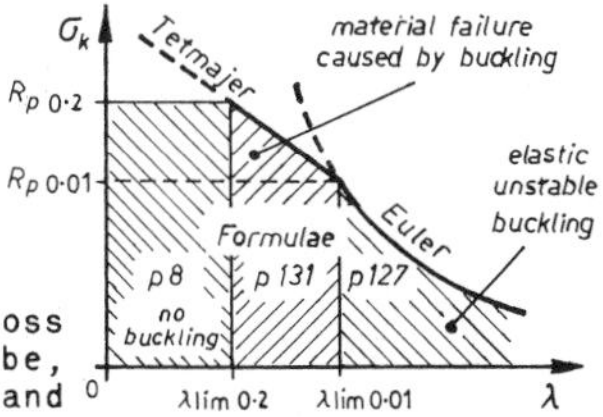

continued on P 23

Continued from P 22

$\nu_k = 3 \ldots 5$ in the Tetmajer range

$\nu_k = 4 \ldots 6$	in the Euler range	for large	structures
$\nu_k = 6 \ldots 8$		for small	

Present	slenderness ratio	λ	calculate from	p 129
Limiting		$\lambda_{\lim 0{\cdot}01}$ and $\lambda_{\lim 0{\cdot}2}$		p 130

Determine buckling and compressive stress as follows:

If $\lambda \geq \lambda_{\lim 0{\cdot}01}$	use p 127
$\lambda_{\lim 0{\cdot}01} > \lambda \geq \lambda_{\lim 0{\cdot}2}$	use p 131
$\lambda < \lambda_{\lim 0{\cdot}2}$	use p 8.

If $P_e < F \nu_k$, redesign with larger cross section.

Method of buckling coefficient ω (DIN 4114)

Specified for building and bridge construction, steelwork and cranes.

p 133 Buckling coefficient $\left.\right\}$ $\omega = \dfrac{p_c}{p_k} = \dfrac{\text{permissible compress. stress}}{\text{buckling stress}}$

p 134 $\sigma_\omega = \omega \dfrac{F}{A} \leq p_c$ where $\omega = f(\lambda)$

λ	Buckling coefficient ω for mild steel ASTM-A 283 Grade C	mild steel ASTM-A 440	alum. alloy AA 2017	cast iron ASTM-A48 A25B
20	1·04	1·06	1·03	1·05
40	1·14	1·19	1·39	1·22
60	1·30	1·41	1·99	1·67
80	1·55	1·79	3·36	3·50
100	1·90	2·53	5·25	5·45
120	2·43	3·65	7·57	not valid in this range
140	3·31	4·96	10·30	
160	4·32	6·48	13·45	
180	5·47	8·21	17·03	
200	6·75	10·31	21·02	

Calculation method: Estimate ω and choose cross section calculate A, I_{min} and λ from p 134. Then from table read off ω. Repeat the calculation with the appropriate new value of ω, until the initial and final values are identical.

Combination of direct stresses

Bending in two planes with end loads

The stresses σ arising from bending and end loads must be added together.

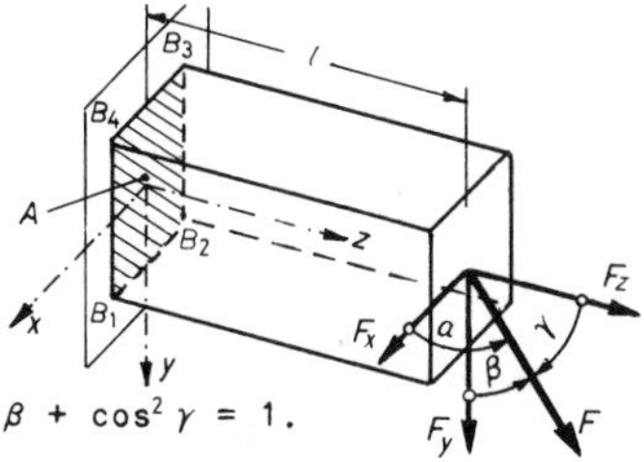

p 135 $F_x = F \cos\alpha$

p 136 $F_y = F \cos\beta$

p 137 $F_z = F \cos\gamma$

p 138 where $\cos^2\alpha + \cos^2\beta + \cos^2\gamma = 1$.

For any point $P(x,y)$ on the cross section $B_1\,B_2\,B_3\,B_4$ the resultant normal stress is in the z-direction:

p 139 $$\sigma = \frac{F_z}{A} - \frac{F_y\,l}{I_x}\,y - \frac{F_x\,l}{I_y}\,x$$

Note the sign of x and y. If F_z is a compressive force, α, β and γ will be in different quadrants. For the sign of cosine functions see E 3.

Long beams in compression should be examined for buckling.

Neutral axis of $\sigma = 0$ is the straight line:

p 140 $$y = -\frac{F_x}{F_y}\times\frac{I_x}{I_y}\,x + \frac{F_z}{F_y}\times\frac{I_x}{A\,l}$$

which intercepts the axes at:

p 141 $$x_0 = \frac{F_z}{F_x}\times\frac{I_y}{A\,l}\,; \qquad y_0 = \frac{F_z}{F_y}\times\frac{I_x}{A\,l}$$

With asymmetrical cross section F, resolve in the directions of the principal axes (see P 9).

Bending in one axis with end load

Either F_x or F_y in formulae p 139 ... p 141 is zero.

Bending with	tension	displaces the neutral axis towards the	compr.	zone
	compression		tension	

Stress in curved beams ($R < 5h$)

The direct force F_n and bending moment M_x (see P 8) act at the most highly stressed cross section A.

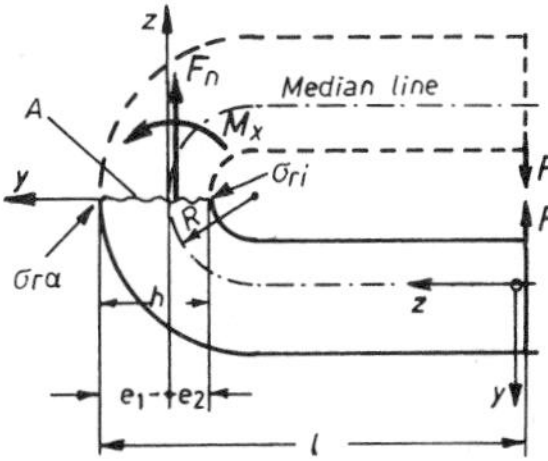

For the stress distribution over the cross section:

142 $$\sigma_t = \frac{F_n}{A} + \frac{M_x}{A\,R} + \frac{M_x\,R}{C} \times \frac{y}{R + y}$$

The stresses at the inner and outer surfaces are:

143 $$\sigma_{ra} = \frac{F_n}{A} + \frac{M_x}{A\,R} + \frac{M_x\,R}{C} \times \frac{|e_1|}{R + |e_1|} \leqslant p_t$$

144 $$\sigma_{ri} = \frac{F_n}{A} + \frac{M_x}{A\,R} - \frac{M_x\,R}{C} \times \frac{|e_2|}{R - |e_2|} \leqslant p_t$$

Formulae for coefficient C:

145 $$C = b\,R^3 \left(\ln \frac{1 + \frac{d}{2R}}{1 - \frac{d}{2R}} - \frac{d}{R} \right)$$

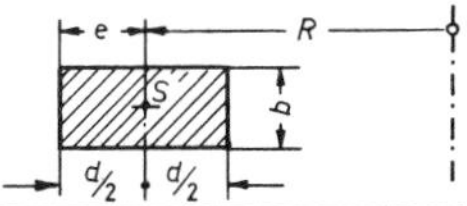

146 $$C = e^2 \pi R^2 \frac{1 - \sqrt{1 - \left(\frac{e}{R}\right)^2}}{1 + \sqrt{1 - \left(\frac{e}{R}\right)^2}}$$

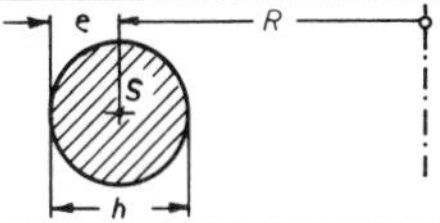

147 $$C = R^4 \left[\frac{a-b}{h} \left(1 + \frac{a e_1 + b e_2}{R(a-b)} \right) \times \ln \frac{1 + \frac{e_1}{R}}{1 - \frac{e_2}{R}} - \frac{a-b}{R} - \frac{(a+b)\,h}{2R^2} \right]$$

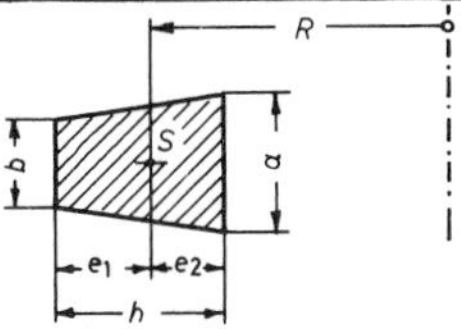

Position of the center of mass, see K 7

148 $$C = R^4 \left[\frac{b}{3h} \left(3 + 2\frac{h}{R} \right) \ln \frac{3 + \frac{2h}{R}}{3 - \frac{h}{3}} - \frac{b}{R} - \frac{b\,h}{2R^2} \right]$$

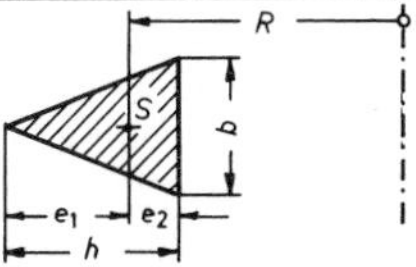

Combination of shear stresses

The stresses arising from shear and torsion at any cross section must be added vectorially.

The maximum shear stress τ_{res} occurs at point 1 and acts in the cross sectional plane. A complementary shear acts perpendicular to it.

	Maximum torsional stress τ_{res} in point 1	cross section
p 149	$\frac{5{\cdot}1\,T}{d^3} + \frac{1{\cdot}7\,F}{d^2} \leqslant p_{qt}$	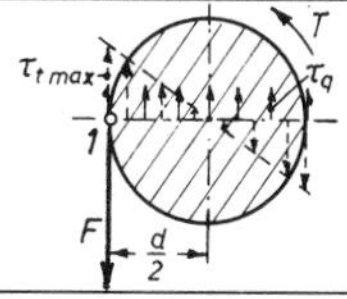
	where $T = F\frac{d}{2}$:	
p 150	$4{\cdot}244 \times \frac{F}{d^2} \leqslant p_{qt}$	
p 151	$\frac{5{\cdot}1\,T\,D}{D^4 - d^4} + 1{\cdot}7F \times \frac{D^2 + Dd + d^2}{D^4 - d^4} \leqslant p_{qt}$	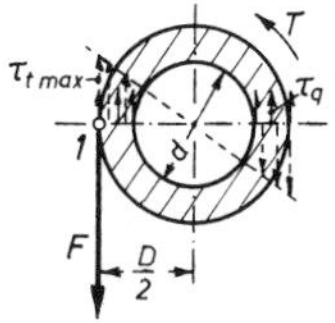
	where $T = F\frac{D}{2}$:	
p 152	$F\,\frac{4{\cdot}244\,D^2 + 1{\cdot}7d(D+d)}{D^4 - d^4} \leqslant p_{qt}$	
	For thin wall tubes:	
p 153	$\frac{5{\cdot}1\,T\,D}{D^4 - d^4} + \frac{2{\cdot}55\,F}{D^2 - d^2} \leqslant p_{qt}$	
p 154	$2{\cdot}55\,F \times \frac{2D^2 + d^2}{D^4 - d^4} \leqslant p_{qt}$	
p 155	$\frac{c_2}{c_1} \times \frac{T}{b^2\,h} + \frac{1{\cdot}5\,F}{b\,h} \leqslant p_{qt}$	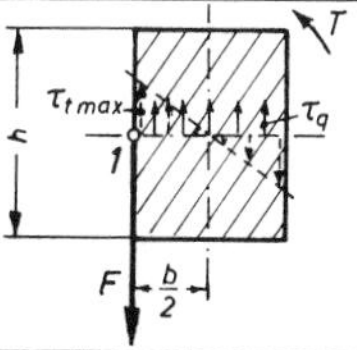
	where $T = F\frac{b}{2}$:	
p 156	$\frac{F}{2\,b\,h}\left(\frac{c_2}{c_1} + 3\right) \leqslant p_{qt}$	

p_{qt} : permissible shear stress (see Z 16)
τ : shear stress
τ_q : calculated maximum torsional shear stress
F : force producing torsion
T : torque produced by F
for c_1 and c_2 see P 20

Combination of direct and shear stresses

Material strength values can only be determined for single-axis stress conditions. Therefore, multi-axis stresses are converted to single-axis equivalent stresses σ_V (see P 29). The following then applies, according to the type of load:

$$\sigma_V \leq p_t \quad \text{or} \quad p_c \quad \text{or} \quad p_{bt}$$

Stresses in two dimensions

An element is subject to

direct stress	shear stress
σ_z in z direction	$\tau_{zy} = \tau$ } in y-z plane
σ_y in y direction	$\tau_{yz} = \tau$ } in y-z plane

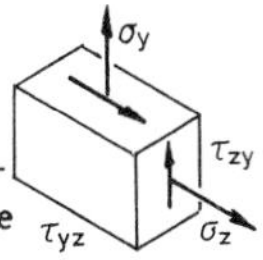

By rotating the element through the angle φ_σ the mixed stresses can be converted to tensile and compressure stresses only, which are called

Principal stresses

p 157 $$\sigma_1, \sigma_2 = 0{,}5(\sigma_z + \sigma_y) \pm 0{,}5\sqrt{(\sigma_z - \sigma_y)^2 + 4\tau^2}$$

Direction of the highest principal stress σ_1 at angle of rotation φ_σ from the original position is:

p 158 $$\tan 2\varphi_\sigma = \frac{2\tau}{\sigma_z - \sigma_y} \text{ *)}$$ (where the shear stress is zero).

Rotating the element through the angle φ_τ gives the

Maximum shear stresses

p 159 $$\tau_{max}, \tau_{min} = \pm 0 \cdot 5\sqrt{(\sigma_z - \sigma_y)^2 + 4\tau^2} = \pm 0 \cdot 5(\sigma_1 - \sigma_2)$$

The direct stresses act simultaneously:

p 160 $$\sigma_M = 0 \cdot 5(\sigma_z + \sigma_y) = 0 \cdot 5(\sigma_1 + \sigma_2)$$

Direction of the maximum shear stress τ_{max} is:

p 161 $$\tan 2\varphi_\tau = -\frac{\sigma_z - \sigma_y}{2\tau} \text{ *)}$$

The Principal stresses and Maximum shear stresses lie at 45° to each other.

*) The solution gives 2 angles. It means that both the Principal stresses and the Maximum shear stresses occur in 2 directions at right angles.

Stress in three dimensions

The stress pattern can be replaced by the

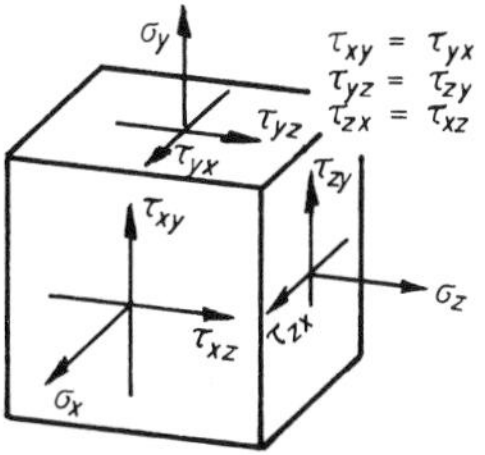

Principal stresses σ_1, σ_2, σ_3

They are the 3 solutions of the equation:

162 $\sigma^3 - R\,\sigma^2 + S\,\sigma - T = 0$

163 where $R = \sigma_x + \sigma_y + \sigma_z$

164 $S = \sigma_x\,\sigma_y + \sigma_y\,\sigma_z + \sigma_z\,\sigma_x - \tau_{xy}^2 - \tau_{yz}^2 - \tau_{zx}^2$

165 $T = \sigma_x\sigma_y\sigma_z + 2\tau_{xy}\,\tau_{yz}\tau_{zx} - \sigma_x\,\tau_{yz}^2 - \sigma_y\,\tau_{zx}^2 - \sigma_z\,\tau_{xy}^2$

Solve the cubic equation p 162 for σ_1, σ_2 and σ_3 as follows:

Put equation p 162 = y (or substitute y for 0 on the right hand side), then plot $y = f(\sigma)$. The points of intersection with the zero axis give the solution. Substitute these values in p 162 and obtain more accurate values by trial and interpolation.

The case where $\sigma_1 > \sigma_2 > \sigma_3$ gives the Maximum shear stress $\tau_{max} = 0{\cdot}5(\sigma_1 - \sigma_3)$.

Bending and torsion in shafts of circular cross section

According to the theory of maximum strain energy:

166 Equivalent stress $\sigma_E = \sqrt{f_{bt}^{\,2} + 3(\alpha_0\, f_{qt})^2} \leqslant p_{bt}$

167 Equivalent moment $M_E = \sqrt{M_b^{\,2} + 0{\cdot}75(\alpha_0\, T)^2}$

To find the diameter of the shaft, determine the necessary section modulus Z from:

168 $Z = \dfrac{M_E}{p_{bt}}$

---

f_{bt} : tensile stress due to bending
f_{qt} : shear stress due to torsion
M : bending moment
T : torque
α_0 : according to P 29

		Based on theory of maximum strain energy *) direct stress	shear stress	equivalent stress
three dimensional stress		tension: $\sigma_1 > 0$: $\sigma_{ED} = \sigma_1 = \sigma_{max}$ compr. : $\sigma_3 < 0$: $\sigma_{ED} = \sigma_3 = \sigma_{min}$	$\sigma_{ES} = 2\,\tau_{max} = \sigma_1 - \sigma_3$	$\sigma_E = \sqrt{0{,}5\left[(\sigma_1-\sigma_2)^2 + (\sigma_2-\sigma_3)^2 + (\sigma_3-\sigma_1)^2\right]}$
two dimensional stress		tension: $\sigma_1 > 0$: $\sigma_{ED} = \sigma_1 = \sigma_{max}$ $= 0{\cdot}5\left[(\sigma_z+\sigma_y) + \sqrt{(\sigma_z-\sigma_y)^2 + 4(a_0\,\tau)^2}\right]$ compr. : $\sigma_2 < 0$: $\sigma_{ED} = \sigma_2 = \sigma_{min}$ $= 0{\cdot}5\left[(\sigma_z+\sigma_y) - \sqrt{(\sigma_z-\sigma_y)^2 + 4(a_0\,\tau)^2}\right]$	$\sigma_{ES} = 2\,\tau_{max} = \sigma_1 - \sigma_2$ $= \sqrt{(\sigma_z-\sigma_y)^2 + 4(a_0\,\tau)^2}$	$\sigma_E = \sqrt{\sigma_1^2 + \sigma_2^2 - \sigma_1\,\sigma_2}$ $= \sqrt{\sigma_z^2 + \sigma_y^2 - \sigma_z\,\sigma_y + 3(a_0\,\tau)^2}$
loads I, II, III for σ and τ	equal	$a_0 = 1$	$a_0 = 1$	$a_0 = 1$
	un-equal	$a_0 = \dfrac{p_t\ \text{I, II, III}}{p_q\ \text{I, II, III}} = \dfrac{\sigma_{lim}\ \text{I, II, III}}{\tau_{lim}\ \text{I, II, III}}$	$a_0 = \dfrac{p_t\ \text{I, II, III}}{2\,p_q\ \text{I, II, III}} = \dfrac{\sigma_{lim}\ \text{I, II, III}}{2\,\tau_{lim}\ \text{I, II, III}}$	$a_0 = \dfrac{p_t\ \text{I, II, III}}{1{\cdot}73\,p_q\ \text{I, II, III}} = \dfrac{\sigma_{lim}\ \text{I, II, III}}{1{\cdot}73\,\tau_{lim}\ \text{I, II, III}}$
used for	type of stress and material	tension, bending, torsion of brittle materials: cast iron glass stone	compr. of brittle and ductile mater. Tens., bending, tors. of steel having pronounced yield point	all stressing of ductile materials: rolled, forged and cast steel; aluminium, bronze
	expected failure	cut off fracture	sliding fracture, flowing, deformat.	perman.-, sliding-, cut off fract.; flowing, deformat.

*) Give the best agreement with test results.

σ_{lim}, τ_{lim} are the typical values for the materials, e. g. R_m; q_s | σ_1, σ_2, σ_3 see P 27 + P 28

Lead screws see K 11

Fixing bolts

Bolted joints (approx. calculation)

Prestressed

axial working load F_A	shear load F_S
q 1 $A_3 \simeq A_S = \dfrac{F_{max}}{p_t}$	calculation for friction effect:
q 2 $F_{max} = (1{\cdot}3 \ldots 1{\cdot}6)\, F_A$ (using load-extension diagram, see P 1)	$A_3 \simeq A_S = \dfrac{F_{C\,req}}{(0{\cdot}25 \ldots 0{\cdot}5)\sigma_{0{\cdot}2}}$
q 3 $p_t = (0{\cdot}25 \ldots 0{\cdot}5)\sigma_{0{\cdot}2}$ (allowing for torsion and safety factor)	$F_{K\,erf} = \dfrac{\nu F_S}{\mu\, m\, n}$ (for values of μ see Z 7)

High-stress bolted joints see VDI 2230 (Verein Deutscher Ingenieure)

Bracket attachment (precise calculation not possible)

Practical assumption: that the centre of pressure is the point of rotation, e. g. $a \simeq h/4$.

q 4

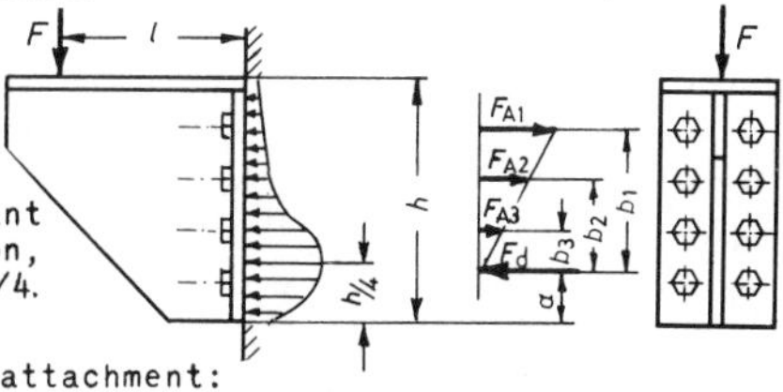

For a rigid attachment:

q 5 $$F\, l = F_{A1}\, b_1 + F_{A2}\, b_2 + \ldots F_{An}\, b_n$$

q 6 $$F_{A1} : F_{A2} : \ldots F_{An} = b_1 : b_2 \ldots b_n$$

Allow for the extra shear load $F_S = F$. There must be compressive stress over the whole attachment plane, when under load.

A_3 : core cross section

A_S : stress cross section $\left(A_S = \dfrac{\pi}{4} \left(\dfrac{d_2 + d_3}{2} \right)^2 \right)$

$F_{C\,req}$: required clamping force

m : no. of bolts

n : no. of joint faces

e.g.

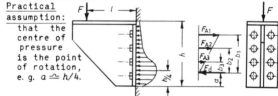

ν : safety factor against slipping $[\nu = 1{\cdot}5 \ldots (2)]$

$\sigma_{0{\cdot}2}$: proof stress

p_t : permissible stress

d_2 : outside diam. of bolt

d_3 : core diameter of bolt

Axles and shafts (approx. calculation)[1]

Axles

	Axle	required section modulus for bending	solid axle of circular cross sect. ($Z \simeq d^3/10$)	permissible bending stress [3]
q 7	fixed [2]			$p_{bt} = \frac{\sigma_{btU}}{(3\ldots5)}$
q 8		$Z = \frac{M}{p_{bt}}$	$d = \sqrt[3]{\frac{10\,M}{p_{bt}}}$	
q 9	rotating			$p_{bt} = \frac{\sigma_{btA}}{(3\ldots5)}$

Shafts

	Stress	diameter for solid shaft	permissible torsional stress [3]
q10	pure torsion		$p_{qt} = \frac{\tau_{tU}}{(3\ldots5)}$
q11		$d = \sqrt[3]{\frac{5\,T}{p_{qt}}}$	
q12	torsion and bending		$p_{qt} = \frac{\tau_{tU}}{(10\ldots15)}$

Bearing stress

q13 on shaft extension $\Big\}\ f_{bm} = \frac{F}{d\,b} \leqslant p_b$

(p_b see Z 18)

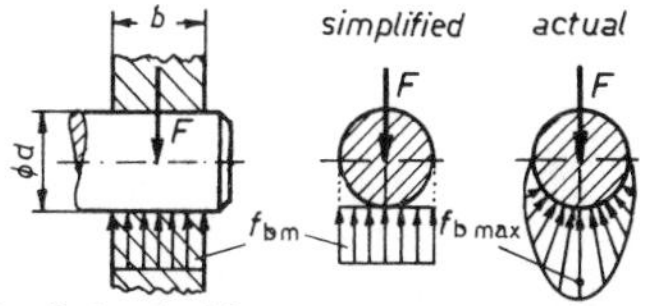

Shear due to lateral load: Calculation unnecessary when

$l > d/4$	for all shafts	with	circular	cross
$l > 0{\cdot}325\,h$	for fixed axles		rectangular	sections

Deflection due to bending see P 12
due to torsion see P 20

Vibrations see M 6.

---

1) For precise calculation see DIN 15017
2) Formulae are restricted to load classes I + II (P 2)
3) p_{bt} and p_{qt} allow for stress concentration-, roughness-, size- (s. DIN 15017), safety-factor and comb. stresses

---

l : arm of force F
M, T : bending moment, torque
f_{bm} (p_b) : mean (permissible) bearing stress (see Z 18)
$f_{b\,max}$ see q 47 for hydrodynamic. lubr. plain bearings,
σ_{btU}, σ_{btA}, τ_{tU} : values see Z 16. [other cases see Z 18

Friction-locked joints

Proprietary devices (e.g. annular spring, Doko clamping device, Spieth sleeve, etc.): see manufacturer's literature.
For interference fits see DIN 7190 (graph. method).

Clamped joint

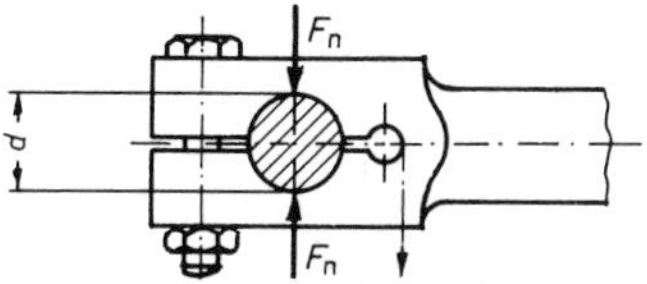

q 14 $$F_n = \frac{T\,\nu}{\mu\,d}$$

Taper joint

q 15 Taper from $\tan\alpha = \dfrac{D - d}{l}$

For tapered shaft extensions see DIN 1448, 1449, (254).
Approximate formula for axial force F_A on the nut:

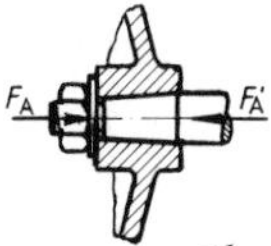

q 16 $$F_A = \frac{2\,T\,\nu}{\mu\,d_m}\tan\left(\frac{\alpha}{2} + \varrho\right)$$

q 17 $$d_m = \frac{D + d}{2}$$

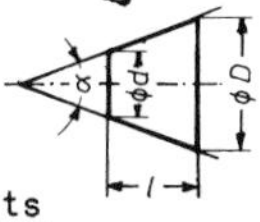

Specially machined joints

Proprietary splined fittings, hub to shaft e.g. polygon: see manufacturer's literature.

Plain key (approx. calculation)
Calculation is based on the bearing pressure on the side of the keyway in the weaker material. Allowing for the curvature of the shaft and the chamfer r_1, the bearing height of the key can be taken approx. as t_2.
The bearing length l is to transmit a torque T:

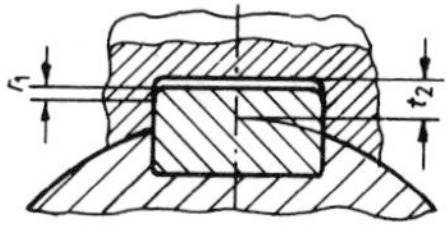

q 18 $$l = \frac{2\,T}{d\,t_2\,p_b}$$

Dimensions to DIN 6885, preferably Sheet 1. Allowance for fillets with form A.
For precise calculations refer to Mielitzer, Forschungsvereinigg. Antriebstechn. e.V. Frankfurt/M Forschungsheft 26, 1975.

Continued on Q 4

Continued from Q 3

Splined shaft

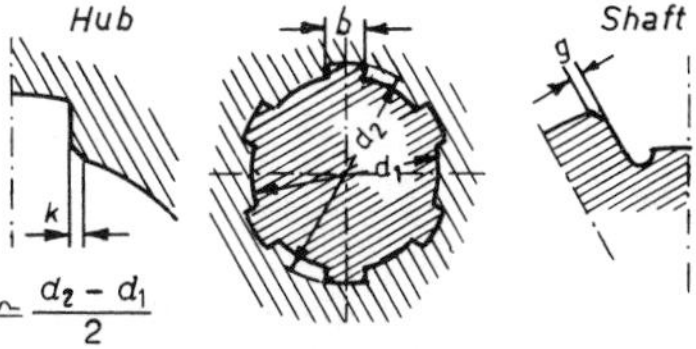

q 19 $$l = \frac{2\,T}{d_m\,h\,\varphi\,n\,p_b}$$

q 20 $$d_m = \frac{d_1 + d_2}{2}$$

q 21 $$h = \frac{d_2 - d_1}{2} - g - k \approx \frac{d_2 - d_1}{2}$$

The load is not shared equally between the splines so allowance for unequal bearing is made with the factor φ:

Type of location	φ
shaft located	0·75
hub located	0·9

For cross section dimensions refer to DIN 5462...5464.

Hub dimensions

Use diagram on page Q 5 to determine dimensions of hub.

Example: Find the length L and radial thickness s of a hub needed to transmit a torque T of 3000 N m, made in cast steel fitted with a plain key.

1. Determine the appropriate range "hub length L, CS/St, group e", follow the boundary lines to T = 3000 N m.
   Result: L = (110...140) mm.
2. Determine the appropriate range "hub thickness s, CS/St, group 1", follow the boundary lines to T = 3000 N m.
   Result: s = (43...56) mm.

F_n : normal force of transmitting surface
l : bearing length of joint
n : no. of splines
μ : coefficient of sliding friction (see Z 7)
ν : safety factor (see Q1)
ϱ : angle of friction (ϱ = arc tan μ)
p_b : permissible bearing pressure. For approximate calculation:

material	p_b in N/mm^2
CI (cast iron)	40... 50
CS (cast steel), St (steel)	90...100

Diagram to obtain hub geometry for Q 4

These empirical values are valid for steel shafts made of steel to BS 970, 060.A22, but not for special cases (such as high centrifugal force, etc). Increase *L* when there are other forces or moments being carried.

Type of joint	Ranges for hub length *L* CI	Ranges for hub length *L* CS, St	thickness *s* CI	thickness *s* CS, St
shrink-fit, interference fit, taper	a	d	g	k
keyed, clamped	b	e	h	l
splined shaft to DIN 5463 (medium)	c	f	i	m

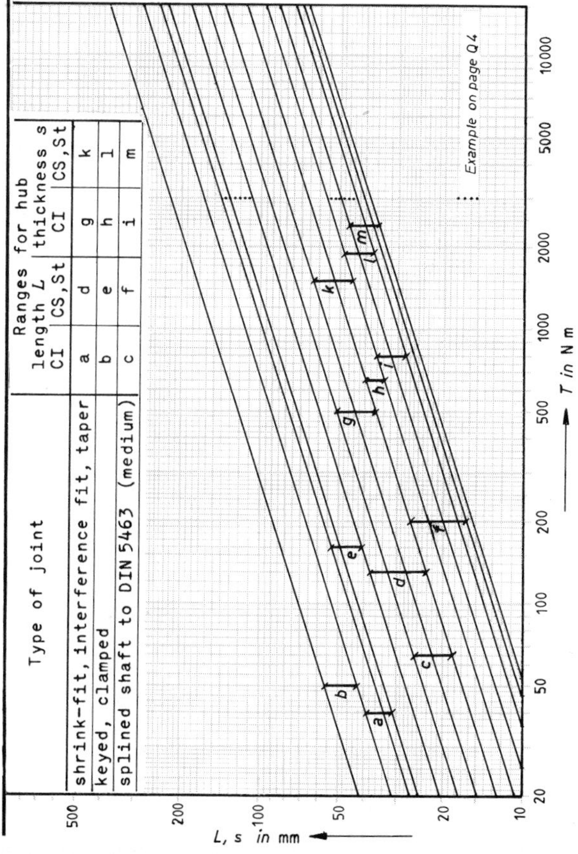

Spring rate R and spring work W (Strain energy)

		Characteristic general	const.
q 22	R	$\frac{dF}{ds}$	$\frac{F}{s}$
q 23	W_1	$\int_0^{s_1} F\, ds$	$F\,\frac{s_1}{2}$

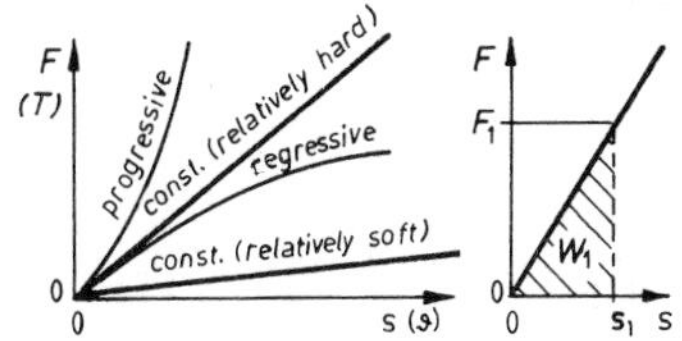

Springs in

		parallel	series
q 24	s_{tot} =	$s_1 = s_2 = s_3 \ldots = s_i$	$s_1 + s_2 + s_3 \ldots + s_i$
q 25	F_{tot} =	$F_1 + F_2 + F_3 \ldots + F_i$	$F_1 = F_2 = F_3 \ldots = F_i$
q 26	R	$R_{tot} = R_1 + R_2 + R_3, \ldots + R_i$	$\frac{1}{R_{tot}} = \frac{1}{R_1} + \frac{1}{R_2} + \ldots \frac{1}{R_i}$

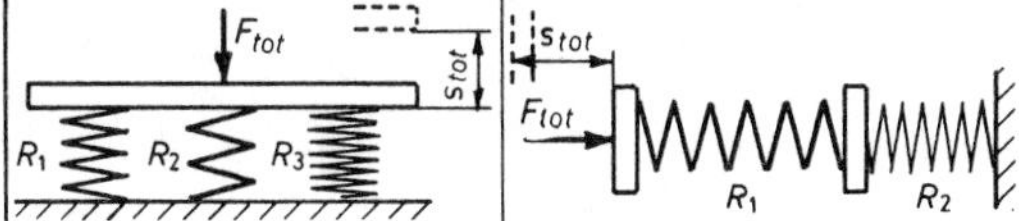

Springs in tension and compression

e.g. ring spring (Belleville spring)

Springs in bending

Rectangular, trapezoidal, triangular springs

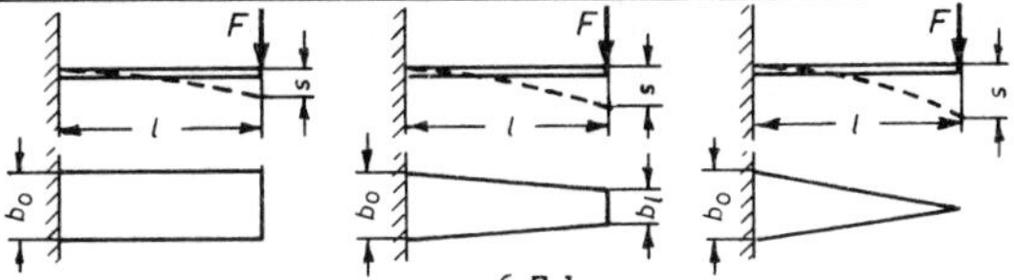

q 27 bending stress $f_{bt} = \frac{6\,F\,l}{b_0\,h^2}$

q 28 permissible load $F = \frac{b_0\,h^2\,p_{bt}}{6\,l}$

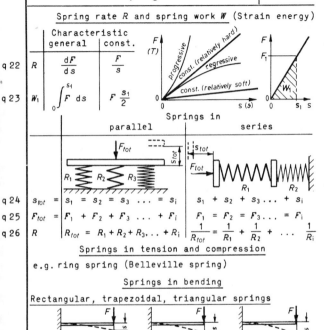

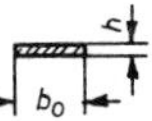

q 29 deflection $s = 4\,\psi\,\frac{l^3}{b_0\,h^3} \times \frac{F}{E}$

b_l/b_0	1 [1]	0·8	0·6	0·4	0·2	0 [2]
ψ	1·000	1·054	1·121	1·202	1·315	1·5

[1] Rectang. spring
[2] Triang. spring

continued on Q 7

continued from Q 6

Laminated leaf springs

Laminated leaf springs can be imagined as trapezoidal springs cut into strips and rearranged (spring in sketch can be replaced by two trapezoidal springs in parallel) of total spring width:

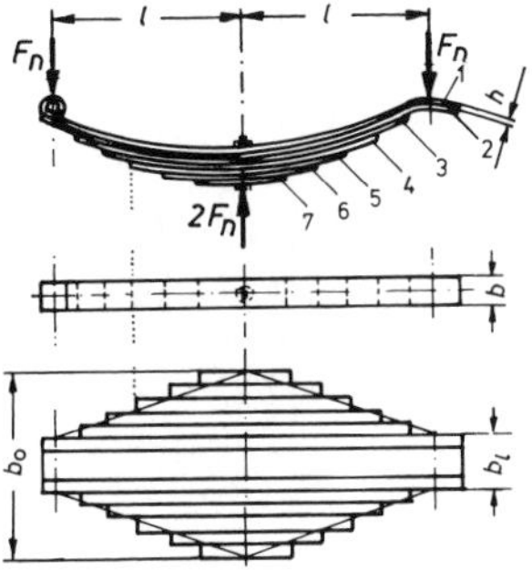

q 30 $$b_o = z\ b$$

z : no. of leaves.

Then (as q 28):

q 31 $$F_n \triangleq \frac{b\ h^2\ p_{bt}}{6\ l}$$

If leaves 1 and 2 are the same length (as in the sketch):

q 32 $$b_l = 2\ b$$

The calculation does not allow for friction. In practice, friction increases the carrying capacity by between 2...12 %.

Precise calculation according to Sheet 394, 1st edit. 1974 Beratungsstelle für Stahlverwendung, Düsseldorf

Disc springs (Ring springs)

Different characteristics can be obtained by combining n springs the same way and i springs the opposite way:

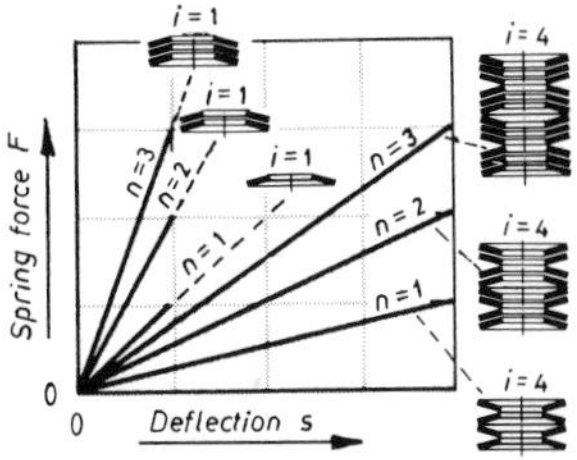

q 33 $$F_{tot} = n\ F_{single}$$

q 34 $$s_{tot} = i\ s_{single}$$

DIN 2092: Precise calculation of single disc springs
DIN 2093: Dimens. and charact. of standard disc springs

Material properties: Hot-worked steels for springs to BS 970; e.g. for leaf springs "250 A 53; 735 A 50".
Modulus of elasticity: $E = 200\ 000\ \text{N/mm}^2$.

p_{bt} : static 910 N/mm² [tempered
oscillating (500 ± 225) N/mm² scale removed and

continued on Q 8

continued from Q 7

Coiled torsion spring: The type shown in the sketch has both ends free and must be mounted on a guide post. Positively located arms are better.

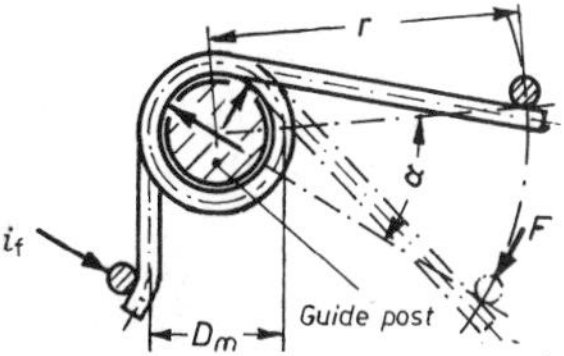

q 35 Spring force[1] $F_p \approx \dfrac{Z\, p_{bt}}{r}$

q 36 Angle of deflect. $\alpha \approx \dfrac{F\, r\, l}{I\, E}$

q 37 Spring coil length $l \approx D_m\, \pi\, i_f$

i_f : no. of coils.

(Additional corrections is needed for deflection of long arms).

For precise calculation, see DIN 2088.

Springs in torsion

Torsion bar spring

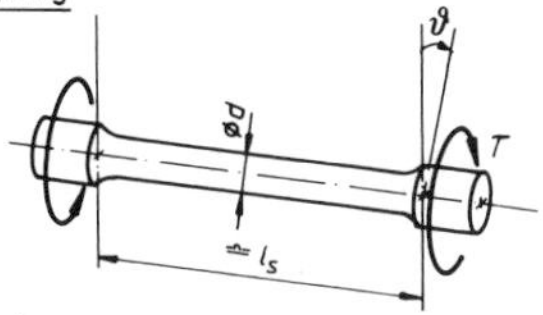

Shear stress	Torque	Angle of twist
q 38 $\tau = \dfrac{5\,T}{d^3}$	$T = \dfrac{d^3}{5} p_{qt}$	$\vartheta = \dfrac{T\, l_s}{G\, I_p} \approx \dfrac{10\, T\, l_s}{G\, d^4}$

l_s : spring length as shown in sketch.

Stress p_{qt} and fatigue strength τ_f in N/mm²

static			oscillating[2]		
q 39 p_{qt}	not preloaded	700	$\tau_f = \tau_m \pm \tau_A$	d = 20 mm	500 ± 350
	preloaded	1020		d = 60 mm	500 ± 240

τ_m : mean stress

τ_A : alternating stress amplitude of fatigue strength

Precise calculation see DIN 2091, especially for the spring length.

[1] Not allowing for the stress factor arising from the curvature of the wire.

[2] Surface ground and shot-blasted, preloaded.

continued on Q 9

continued from Q 8

Cylindrical helical spring (compression and tension)

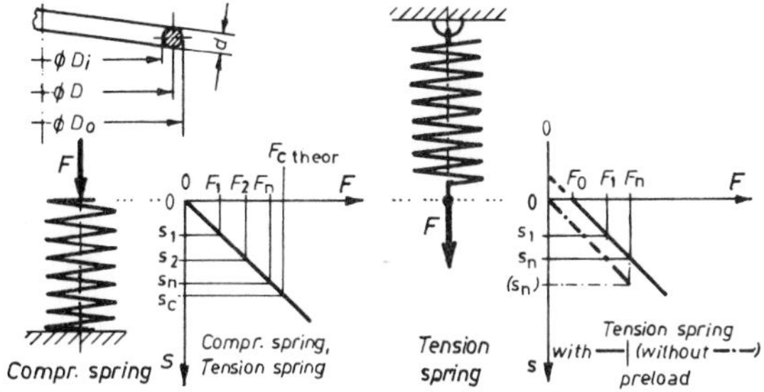

Normal coil ratio: $D/d = 5 \ldots 15;\ D = (D_o + D_i)/2$

Static stress

		compress. spring	tension spring
q 40	D known	$d = \sqrt[3]{\dfrac{8\, F_{c\,theor}\, D}{\pi\, p_{qt}}}$	$d = \sqrt[3]{\dfrac{8\, F_n\, D}{\pi\, p_{qt}}}$
q 41	D unknown, estimate D/d	$d = \sqrt{\dfrac{8\, F_{c\,theor}}{\pi\, p_{qt}} \times \dfrac{D}{d}}$	$d = \sqrt{\dfrac{8\, F_n}{\pi\, p_{qt}} \times \dfrac{D}{d}}$
q 42	max. perm. deflect.	$s_n = s_c - s_A$	
q 43	sum of min. distances between coils	$s_A = x\, d\, n$ with $x =$ 0·2 ... 0·7 at $D/d =$ 4 ... 20	solid length
q 44	no. of effective coils	$n = \dfrac{s}{F} \times \dfrac{G\, d^4}{8\, D^3}$	
q 45	permissible shear stress *)	$p_{qt} = 0{\cdot}56 \times R_m$ see diagram	$p_{qt} = 0{\cdot}45 \times R_m$ diagram $\times$ 0·8

*) For higher relaxation requirement see DIN 2089.

s_c : theoretical spring deflect. when just solid.

$F_{c\,theor}$: theoret. spring force when just solid.

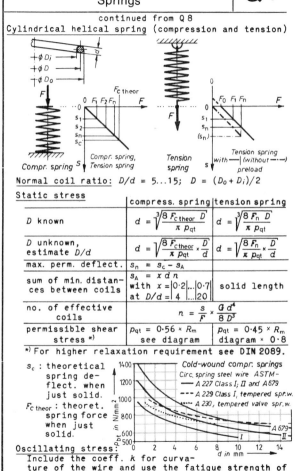

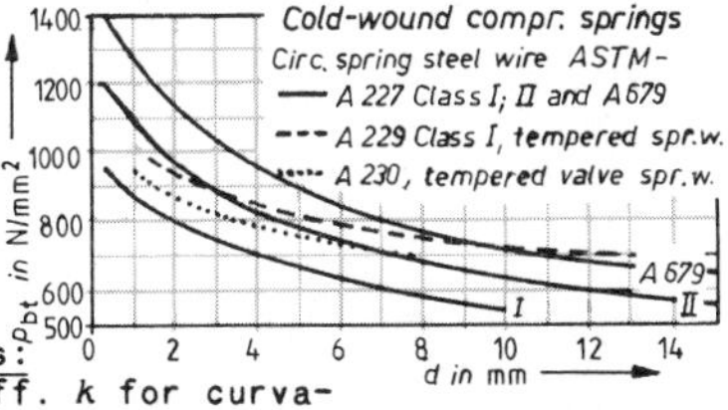

Oscillating stress:
Include the coeff. k for curvature of the wire and use the fatigue strength of spring steel (see DIN 2089) in the calculations.

MACHINE PARTS

Bearings

Rolling bearings

Use the formulae from the manufacturer's literature which gives load capacities and dimensions, e.g. S.K.F., Timken.

Journal bearings

Hydrodynamically-lubricated plain journal bearing

Bearings must be

running at proper temperatures and without excessive wear, i.e. separation of the journal and bearing by a film of lubricant.

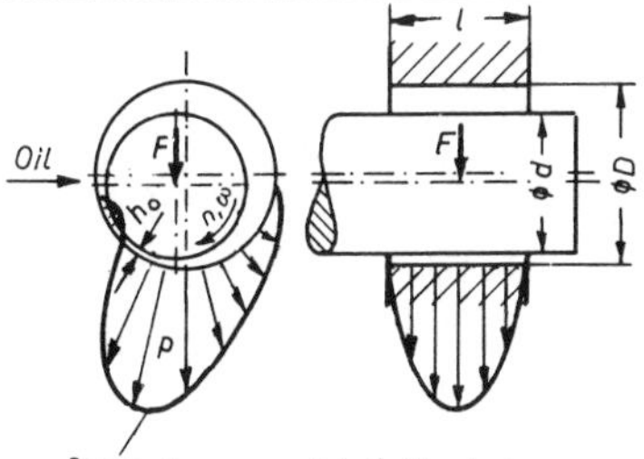

p_{max} *Pressure distribution in transverse and longitudinal sections*

Length/diameter ratio l/d

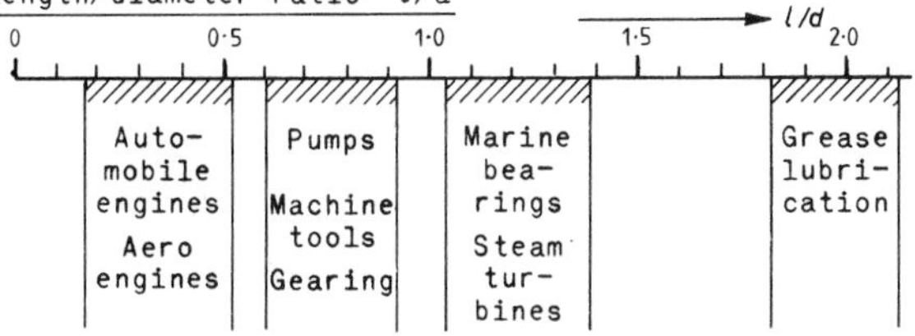

General properties

Short bearings	Long bearings
Large pressure drop at each end, therefore good cooling with adequate oil flow.	Small pressure drop at each and, therefore high load capacity.
Excellent for high rotational speeds.	Good at low rotational speeds.
Low load capacity at low rotational speeds.	Poor cooling facilities.
	Danger of edge loading.

continued on Q 11

continued from Q 10 (journal bearings)

Bearing pressure f_{bm}, f_{bmax}

q 46	mean	bearing pressure	$f_{bm} = \frac{F}{d\ b}$
q 47	max.		$f_{bmax} \leqslant \frac{2}{3}\ \sigma_{dF}$

f_{bmax} depends mainly on the relative thickness δ of the lubricating film (see Sommerfeld number q 56).

The adjacent diagram shows the ratio of maximum pressure to mean bearing pressure in relation to the relative thickness of the lubrication film. (According to Bauer, VDI 2204).

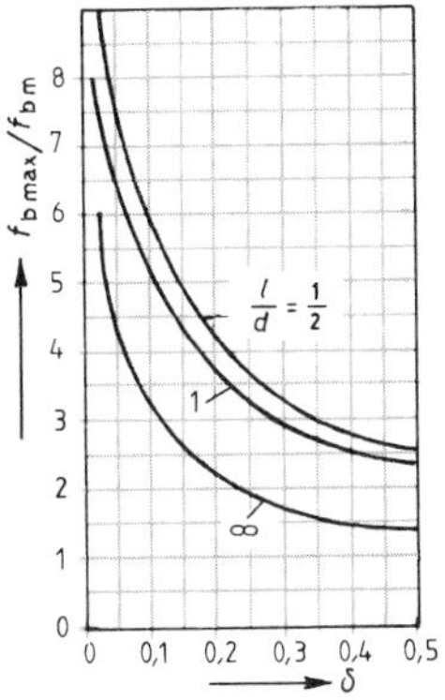

Bearing clearance s, relative bearing clearance ψ

q 48 $s = D - d$; $\psi = s/d$

ψ is basically the relative bearing clearance established during operation (including thermal expansion and elastic deformation).

q 49 Typical values $\psi = (0{\cdot}3 \ldots 1 \ldots 3)10^{-3}$ [1)]

Criteria for the choice of ψ:

	Lower value	Upper value
bearing material	soft (e.g. white metal)	hard (e.g. phosphor)
viscosity	relatively low	relatively high
peripheral speed	relatively low	relatively high
bearing pressure	relatively high	relatively low
length/dia. ratio	$l/d \leqslant 0{\cdot}8$	$l/d \geqslant 0{\cdot}8$
support	self-aligning	rigid

q 50 Minimum values for plastics $\psi \geqslant (3 \ldots 4)10^{-3}$

q 51 sintered metals $\psi \geqslant (1{\cdot}5 \ldots 2)10^{-3}$

q 52 1) Grease-lubricated plain bearings $\psi = (2 \ldots 3)10^{-3}$

continued on Q 12

continued from Q 11 (journal bearings)

Min. thickness of lubricating film h_{min} in µm

Theoretical	Actual
q 54 $h_o \geqslant h_{min}$ = shaft deflection + bearing distortion + sum of peak-to-valley heights $(R_{t1} + R_{t2})$	h_{min} = [(1)...3,5...10...(15)] µm (1)...3: special cases, e.g. some automobile crankshaft bearings 5...10: small shaft diameters 10...(15): large shaft diameters

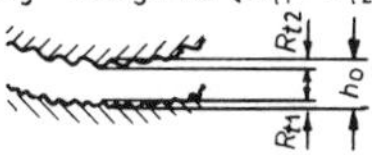

Relative thickness of lubricating film δ

q 55 $\delta = \dfrac{h_o}{s/2} = \dfrac{2\,h_o}{\psi\,d}$

Upper limit for bearings under steady conditions (rotational speed, load, constant load direction): $\delta \leqslant 0{\cdot}35$ to prevent instability and rough running.

Sommerfeld number S_o

(Dimensionless)

q 56 $S_o = \dfrac{p_m\,\psi^2}{\eta_m\,\omega}$

Inserting S_o in the adjacent diagram gives δ and therefore also h_o from q 55.

η_m viscosity based on mean temperature, t_m.

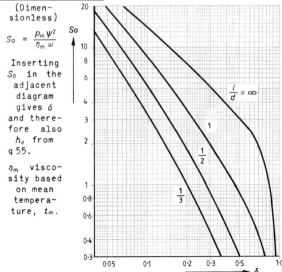

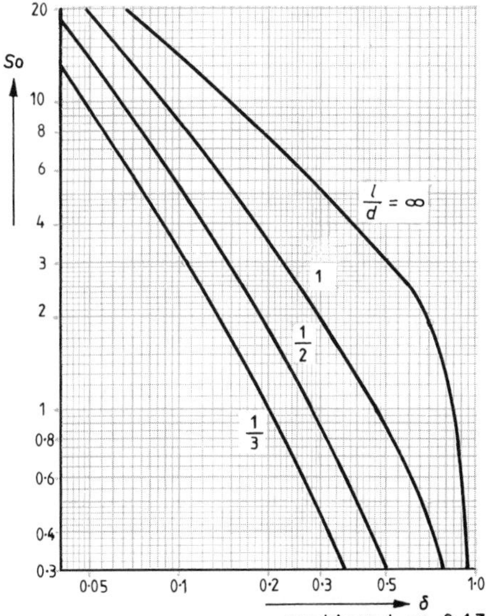

continued on Q 13

continued from Q 12 (journal bearings)

Lubricant flow rate $\dot{Q}$

The theoretical flow rate required to maintain hydrodynamic lubrication is:

q 57 $$\dot{Q} = 0{\cdot}5\, b\, \upsilon\, (s - 2\, h_0)$$

Rules: Oil inlet to the expanding part of the bearing.

q 58 In supply lines: $\upsilon \triangleq 2$ m/s ; $p_r = 0 \ldots 1 \ldots (5)$ bar

q 59 In return lines: $\upsilon \triangleq 0{\cdot}5$ m/s; $p_r = 0$.

Oil grooves (depth $\triangleq 2\,s$) must not be cut into the surfaces of the high pressure area (Q 10) and must not extend to the edges of the bearing. Only short oil grooves are needed with high pressures, but in some cases larger oil grooves are needed to remove heat.

No oil grooves in the load zone!

Heat removal

Requirement: Friction power ($P_F = \mu F \upsilon$) = rate of heat removal.

q 60 $$\mu = \frac{3\psi}{S_0} \quad \text{for } S_0 < 1 \text{ (high-speed range)}$$

q 61 $$\mu = \frac{3\psi}{\sqrt{S_0}} \quad \text{for } S_0 > 1 \text{ (heavy-load range)}$$

Natural cooling (determining oil type): Heat removal mainly from the housing surface A, therefore

q 62 $$P_F = h\, A(t_m - t_0) \qquad \text{by empirical formula}$$

for h:

q 63 $$h\left(\frac{\mathrm{W}}{\mathrm{m^2\,K}}\right) = 7 + \sqrt{w}\,, \qquad \text{where } w \text{ is in m/s.}$$

If effect of heat-transmitting surface A is unknown, guide values for pedestal bearings are:

q 64 $$A \triangleq (20 \ldots 30)\, d\, b \quad \text{for } d \leqslant 100 \text{ mm}$$

q 65 $$A \triangleq (15 \ldots 25)\, d\, b \quad \text{for } d > 100 \text{ mm.}$$

For a given design, load and required operating temperature, the necessary viscosity is:

q 66 $$\eta_m = \frac{p_m}{\omega} \times \frac{\psi\, h\, A(t_m - t_0)}{3\, F\, \upsilon} \qquad \text{(high-speed range)}$$

q 67 $$\eta_m = \frac{p_m}{\omega} \times \left[\frac{h\, A(t_m - t_0)}{3\, F\, \upsilon}\right]^2 \qquad \text{(heavy-load range)}$$

The required type of oil can be determined knowing t_m and η_m (see q 56).

continued on Q 14

Symbols see Q 14

continued from Q 13 (journal bearings)

Natural cooling (given oil type)

Since η is temperature-sensitive and the running oil temperature is initialy unknown, use iteration method with q 66 or q 67 for preliminary and successively improved estimates of t_m.

Graphical method according to VDI 2204.

Forced cooling

Heat removal through oil circuit, with oil cooler if necessary (convection, radiation and conduction from bearing are neglected).

q 68 $$P_F = \dot{Q}\, c\, \varrho\, (t_2 - t_1)$$

Guide values for simple calculations for mineral oils:

q 69 $$c \triangleq 1 \cdot 86 \text{ kJ/(kg K)}$$

q 70 $$\varrho \triangleq 900 \text{ kg/m}^3.$$

For more detailed calculations see VDI 2204.

---

c : specific heat (values see Z 5)
μ : coefficient of friction
F : radial bearing force
h : heat transfer coefficient
h_0 : min. thickness of lubricating film at running speed [(r.p.m.)
h_{min} : min. thickness of lubricating film at lowest running r.p.m. (min. thickness of oil lubricating film)
p_r : pressure (relative to atmospheric) of oil flow
p_m, p_{max} : mean, max. bearing pressure
s : radial clearance
v : peripheral speed of bearing journal or velocity [of oil flow
$\dot{Q}$: flow rate of lubricant
w : velocity of cooling air (m/s)
t : operating temperature of housing surface
t_0 : ambient temperature
t_1, t_2 : inlet, outlet temperature of oil
η : dynamic or absolute viscosity, see N 1
η_m: dynamic viscosity based on mean temperature of bearing t_m.
ϱ : density of lubricant (values see Z 5)
ψ : relative bearing clearance (values see Q 11)
ω : angular velocity

Crosshead guide

Crosshead guide will operate smoothly only when

q 71 $\tan \alpha < \dfrac{l}{(2h+l)\mu}$ or

the length ratio is

q 72 $\dfrac{l}{h} = \lambda > \dfrac{2\mu \tan \alpha}{1 - \mu \tan \alpha}$

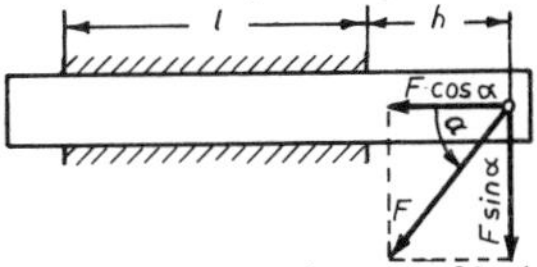

If the above conditions for $\tan \alpha$ are not satisfied there is a danger of tilting and jamming.

Friction clutches

Slip time and energy loss per operation

Drive side   *Clutch*   *Driven side*
                      T_C

$\boxed{I_1,\ T_M,\ \omega_1}$ —————||—————— $\boxed{I_2,\ T_L,\ \omega_2}$

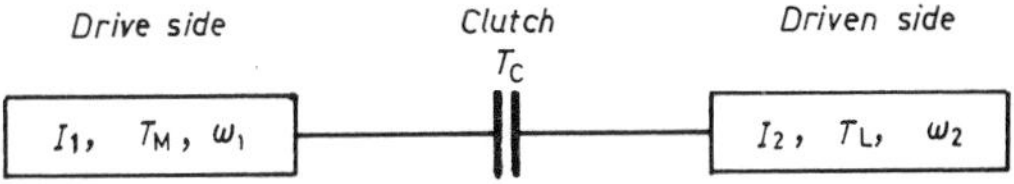

A simplified model with the following conditions is sufficient for approximate calculation:

Acceleration of driven side from $\omega_2 = 0$ to $\omega_2 = \omega_1$, $\omega_1 = $ const.; $T_L = $ const.; $T_C = $ const. $> T_L$. Then, per operation:

q 73   energy loss $W_l = I_2 \times \dfrac{\omega_1^2}{2} \times \left(1 + \dfrac{T_L}{T_C - T_L}\right)$

q 74   slip time $t_f = \dfrac{I_2\, \omega_1}{T_C - T_L}$

Calculating the area of the friction surface

single-plate	flat twin-plate	multi-plate	cone	cylindrical

clutches — clutches

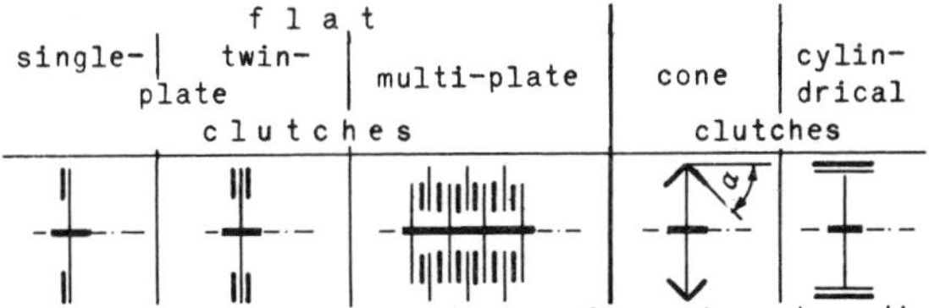

Number and size of friction surfaces depend on the permissible contact pressure p_b and the permissible thermal capacity per unit area q_p.

continued on Q 16

continued from Q 15 (friction clutches)

Calculation of contact pressure p_b

(for values see Z 19)

For all types of friction surfaces:

q 75 $$i\,A \geqslant \frac{T_C}{p_b\,\mu\,r_m}$$

q 76 where $$r_m = \frac{2}{3} \times \frac{r_a^3 - r_i^3}{r_a^2 - r_i^2} \approx \frac{r_a + r_i}{2}$$

	flat	conical	cylindrical
	friction surfaces		
q 77 operating force (axial)	$F_a = A\,p_b$	$F_a = A\,p_b \sin\alpha$	
q 78	for multi-plate clutches usually:	to prevent locking:	$r_a = r_i = r_m$
q 79	$\frac{r_i}{r_a} = 0{\cdot}6 \ldots 0{\cdot}8$	$\tan\alpha > \mu$	

Calculation permissible temperature rise

In HEAVY-LOAD STARTING the maximum temperature is reached in one operation. It depends on the energy loss, slip time, heat conduction, specific heat and cooling. These relationships cannot be incorporated in a general formula.

With CONTINUOUS OPERATION a constant temperature is only established after several operations. There are empirical values for permissible thermal capacity per unit area q_p with continuous operation (see Z 19).

q 80 Friction power $$P_F = W_l\,z$$

q 81 Condition $$i\,A \geqslant \frac{W_l\,z}{q_p}$$

Symbols see Q 17

Friction brakes

All friction clutches can also be used as brakes. But there are also:

Disc brakes

with caliper and pads.

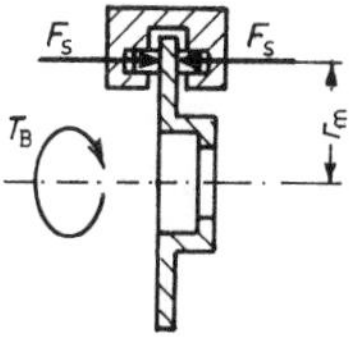

Braking torque T_B :

q 82 $T_B = 2 \mu F_s j r_m$

Expanding-shoe drum brakes

(Drawing of simplex brake showing, simplified, the forces acting on the shoes).

Leading shoes	Trailing shoes
3/84 $F_{n1} = \frac{F_s l}{a - \mu r}$ (Servo-action)	$F_{n2} = \frac{F_s l}{a + \mu r}$

Braking torque T_B :

q 85 $T_B = (F_{n1} + F_{n2}) \mu r$

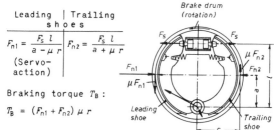

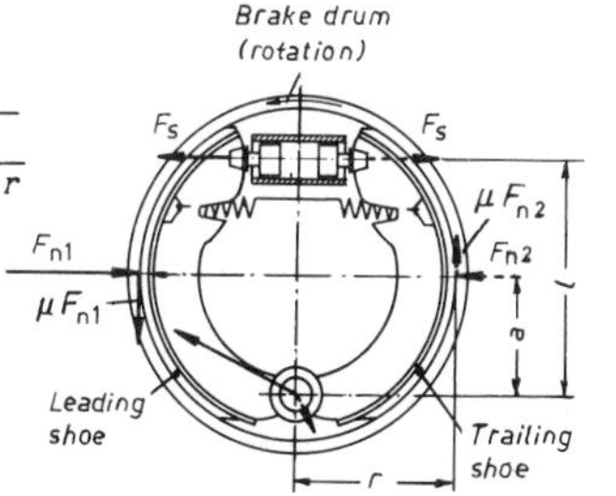

Band brakes see K 13

Notation for friction clutches and brakes (Q 15...Q 17)

A : area of friction surface
T_C : operating torque of clutch
T_L : load torque
T_M : motor torque
W_l : energy loss per operation
μ : coefficient of friction
i : no. of friction surfaces
j : no. of calipers on a disc brake
r : radius of friction surface
r_m , r_o , r_i : mean, outside, inside radius of friction [surface
z : operating frequency (EU: s^{-1}, h^{-1})
ω : angular velocity

(For properties of friction materials see Z 19)

Involute-tooth gears

Spur gears, geometry

q 86 Gear ratio $u = \frac{z_2}{z_1}$

q 87 Transmission ratio $i = \frac{\omega_a}{\omega_b} = \frac{n_a}{n_b} = -\frac{z_b}{z_a}$ [1)]

Transmission ratio of multi-stage gearing:

q 88 $i_{tot} = i_I \times i_{II} \times i_{III} \times \ldots \times i_n$

q 89 Involute function $\text{inv}\,\alpha = \tan\alpha + \alpha$

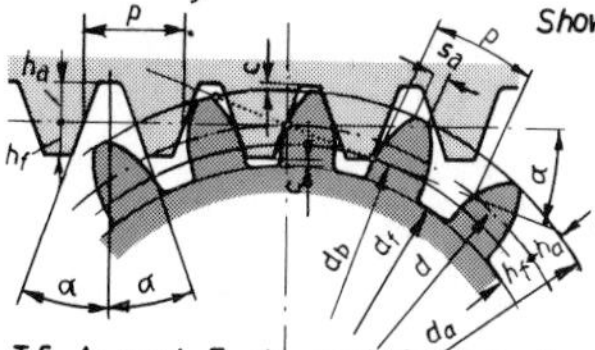

Showing the transverse path of contact
(See BS 2519 Part 1)

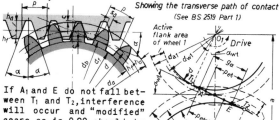

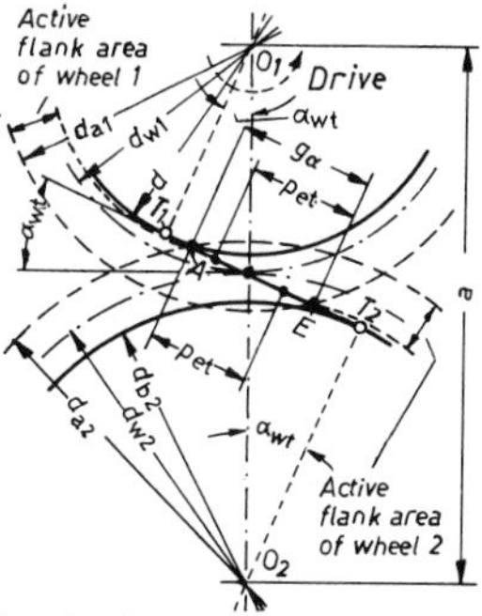

If A_1 and E do not fall between T_1 and T_2, interference will occur and "modified" gears as in Q 20 should be used.

1) Negative for external gears because rotation is opposite. Positive for internal gears. The sign can normally be disregarded.

		Standard gears	
		spur	helical
q 90, q 91	normal pitch	$p = \frac{\pi d}{z} = m\pi$	$p_n = m_n \pi$
q 92	circular pitch		$p_t = \frac{m_n \pi}{\cos\beta}$
q 93, q 94	normal module	$m = \frac{p}{\pi} = \frac{d}{z}$	$m_n = \frac{p_n}{\pi} = \frac{d}{z}\cos\beta$
q 95	circular module		$m_t = \frac{m_n}{\cos\beta} = \frac{d}{z}$
q 96	addendum	$h_a = h_{aP} = m$	
q 97	dedendum	$h_f = h_{fP} = m + c$	
q 98	bottom clearance	$c = (0\cdot1\ldots0\cdot3)\,m \simeq 0\cdot2\,m$	

continued on Q 19

For symbols see Q 29, suffixes see Q 23

continued from Q 18 (spur gears)

No.		Standard gears: spur	Standard gears: helical
9/100	reference diam.	$d = m\, z$	$d = \dfrac{m_n\, z}{\cos\beta} = m_t\, z$
q 101	tip diameter	$d_a = d + 2\, h_a$	
q 102	root diameter	$d_f = d - 2\, h_f$	
q 103	pressure angle	$\alpha = \alpha_n = \alpha_t = \alpha_P$	$\alpha_n = \alpha_P$
q 104			$\tan\alpha = \dfrac{\tan\alpha_n}{\cos\beta}$
5/106	base diameter	$d_b = d\cos\alpha$	$d_b = d\cos\alpha_t$
q 107	equivalent no of teeth		$z_{nx} = z\dfrac{1}{\cos^2\beta \times \cos\beta}$ table see DIN 3960
q 108			$\approx \dfrac{z}{\cos^3\beta}$
	min. no. of teeth to avoid interference		
q 109	theory	$z_{lim} = \dfrac{2}{\sin^2\alpha} \approx 17$ (for $\alpha_P = 20^\circ$)	
0/111	pract.	$z'_{lim} \approx 14$ (for $\alpha_P = 20^\circ$)	$z'_{lim\,h} \approx 14\cos^3\beta$
q 112	spread		$U = b\tan\lvert\beta\rvert$

No.		Standard gearing: spur	Standard gearing: helical
3/114	centre distance	$a = \dfrac{d_1+d_2}{2} = m\dfrac{z_1+z_2}{2}$	$a = \dfrac{d_1+d_2}{2} = m_n\dfrac{z_1+z_2}{2\cos\beta}$
q 115	length of path of contact (total length)	$g_\alpha = \dfrac{1}{2}\left[\sqrt{d_{a1}^2 - d_{b1}^2} + \sqrt{d_{a2}^2 - d_{b2}^2} - (d_{b1} + d_{b2})\tan\alpha_t\right]$	
6/117	transverse contact ratio	$\varepsilon_\alpha = \dfrac{g_\alpha}{p\cos\alpha}$	$\varepsilon_\alpha = \dfrac{g_\alpha}{p\cos\alpha}$
q 118	overlap ratio		$\varepsilon_\beta = \dfrac{b\sin\lvert\beta\rvert}{m_n\,\pi}$
q 119	contact ratio		$\varepsilon_\gamma = \varepsilon_\alpha + \varepsilon_\beta$

continued on Q 20

For symbols see Q 29, suffixes see Q 23

continued from Q 19 (gearing)

		modified gears spur	modified gears helical
	p, p_n, p_t, z, z_{nx}, m, m_n, m_t, d, d_b	see standard gears	
0/121	profile offset	$x\,m$	$x\,m_n$
2/123	profile offset fact. to avoid interference	$x_{min} = -\dfrac{z\,\sin^2\alpha}{2} + \dfrac{h_{a0} - \rho_{a0}(1-\sin\alpha)}{m}$	$x_{min} = -\dfrac{z\,\sin^2\alpha}{2\,\cos\beta} + \dfrac{h_{a0} - \rho_{a0}(1-\sin\alpha_n)}{m_n}$
		can be up to 0·17 mm	
4/125	profile offset fact. ditto[1]	$x \doteq \dfrac{14 - z}{17}$	$x \doteq \dfrac{14 - (z/\cos^3\beta)}{17}$
q 126	profile offset fact. to give a specific centre distance (tot.)	$x_1 + x_2 = \dfrac{(z_1 + z_2) \times (\mathrm{inv}\,\alpha_{wt} - \mathrm{inv}\,\alpha_t)}{2\,\tan\alpha_n}$	
q 127	α_{wt} calculated from	$\cos\alpha_{wt} = \dfrac{(z_1 + z_2)\,m_t}{2\,a}\cos\alpha_t$	
q 128	or	$\mathrm{inv}\,\alpha_{wt} = \mathrm{inv}\,\alpha_t + 2\,\dfrac{x_1 + x_2}{z_1 + z_2}\tan\alpha_n$	
q 129	centre distance	$a = a_d\,\dfrac{\cos\alpha_t}{\cos\alpha_{wt}}$	
q 130	addendum modification coeff.	$k\,m_n = a - a_d - m_n \times (x_1 + x_2)$ [2]	
q 131	addendum	$h_a = h_{aP} + x\,m_n + k\,m_n$	
q 132	dedendum	$h_f = h_{fP} - x\,m_n$	
q 133	outside diameter	$d_a = d + 2\,h_a$	
q 134	root diameter	$d_f = d - 2\,h_f$	
q 135	length of path of contact	$g_\alpha = \dfrac{1}{2}\left[\sqrt{d_{a1}^2 - d_{b1}^2} + \sqrt{d_{a2}^2 - d_{b2}^2} - (d_{b1} + d_{b2})\tan\alpha_{wt}\right]$	
/137	transverse contact ratio	$\varepsilon_\alpha = g_\alpha/(p\,\cos\alpha)$	$\varepsilon_\alpha = g_\alpha/(p_t\cos\alpha_t)$
q 138	overlap ratio		$\varepsilon_\beta = b\,\sin\lvert\beta\rvert/(m_n\,\pi)$
q 139	contact ratio		$\varepsilon_\gamma = \varepsilon_\alpha + \varepsilon_\beta$

[1] If tool data unknown take $\alpha_P = 20°$.

[2] Note the sign. With external gears $k \times m_n < 0$. When $k < 0{\cdot}1$ addendum modification can often be avoided

For symbols see Q 29, suffixes see Q 23

Spur gears, design

The dimensions are derived from

load-carrying capacity of the tooth root
load-carrying capacity of the tooth flank,

which must be maintained independently.

Gearing design is checked in accordance with DIN 3990. By conversion and rough grouping of various factors it is possible to derive some approximate formulae from DIN 3990.

Load capacity of tooth (approx. calculation)

Safety factor S_F against fatigue failure of tooth root:

q 140 $$S_F = \frac{\sigma_{F\,lim}}{\frac{F_t}{b\; m_n} \times Y_F \times Y_\varepsilon \times Y_\beta} \times \frac{Y_S \times K_{FX}}{K_I \times K_V \times K_{F\alpha} \times K_{F\beta}} \geqslant S_{F\,min}$$

Giving the approximate formulae:

q 141 $$m_n \geqslant \frac{F_t}{b} Y_F \times K_I \times K_V \times \underbrace{Y_\varepsilon \times Y_\beta \times K_{F\alpha}}_{\simeq 1} \times \underbrace{\frac{K_{F\beta}}{Y_S \times K_{FX}}}_{\simeq 1} \times \frac{S_{F\,min}}{\sigma_{F\,lim}}$$

Y_F : tooth from factor for external gearing (s. diagr.)

q 142 $K_I \times K_V$ = 1...3, rarely more, (allowing for external shock and irregularities exceeding the rated torque, additional internal dynamic forces arising from tooth errors and circumferential velocity).

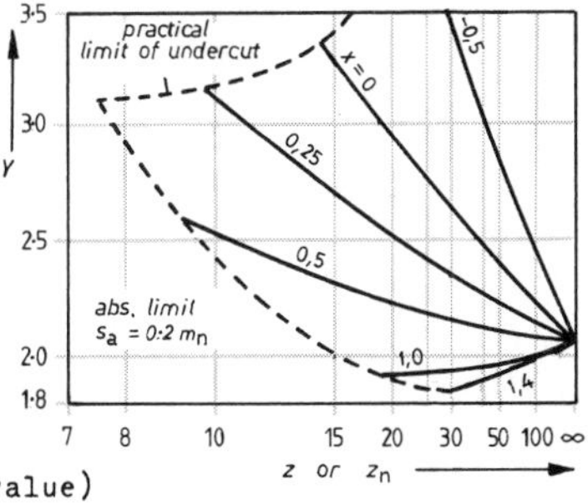

q 143 $S_{F\,min}$ = 1·7 (guide value)

q 144 $\sigma_{F\,lim}$: guide values see table on Q 22

continued on Q 22

For symbols see Q 29, suffixes see Q 23

continued from Q 21 (spur gears, design)

Load capacity of tooth flank (approx. calculation)

Safety factor S_H against pitting:

q 145 $$S_H = \frac{\sigma_{H\,lim}}{\sqrt{\frac{u+1}{u} \times \frac{F_t}{b\ d_1}} \times Z_H \times Z_M \times Z_\varepsilon} \times \frac{Z_V \times K_{HX} \times Z_R \times K_L}{\sqrt{K_I \times K_V \times K_{H\alpha} \times K_{H\beta}}} \geq S_{H\,min}$$

For metals the material factor Z_M is simplified to:

q 146 $$Z_M = \sqrt{0 \cdot 35\ E} \quad \text{where} \quad E = \frac{2\ E_1\ E_2}{E_1 + E_2}$$

Therefore, the approximate formula becomes:

q 147 $$d_1 \geq \sqrt{\frac{2\ T_1}{b} \times \frac{u+1}{u}\ 0 \cdot 35\ E} \times Z_H \times \underbrace{Z_\varepsilon \sqrt{K_{H\alpha}}}_{\simeq 1}\ \frac{\sqrt{K_1 \times K_V} \times \overbrace{\sqrt{K_{H\beta}}}^{\simeq 1}}{Z_V \times K_{HX} \times Z_R \times K_L} \times \frac{S_{H\,min}}{\sigma_{H\,lim}}$$

Mat.	Specificat. to Norm	Grade	$\sigma_{F\,lim}$ N/mm²	$\sigma_{H\,lim}$ N/mm²
CI	ASTM	A 48-50 B	80	360
CS	ASTM	A 536 120-90-02	230	560
CS	ASTM	A 572 Gr. 65	200	400
CS	SAE	1064	220	620
AS	SAE	4140	290	670
ASCH	SAE	3240	500	1630

CI: cast iron
CS: carbon steel
AS: alloy steel
ASCH: case harden. alloy steel

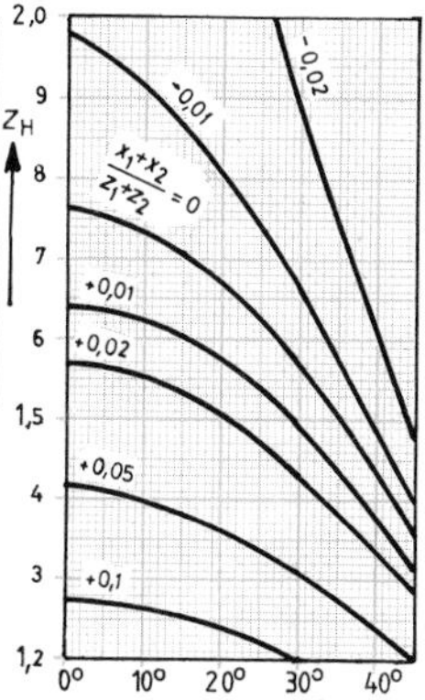

only valid for $\alpha_n = 20°$

continued on Q 23

Z_H : flank form factor (see diagram)
q 148 $K_I \times K_V$: see load capacity of tooth root (q 142)
q 149 $S_{H\,min} \simeq 1 \cdot 2$ (guide value)
q 150 $\sigma_{H\,lim}$: guide values see table
q 151 $Z_V \times K_{HX} \times Z_R \times K_L = 0 \cdot 5 \ldots 1$. Higher value for higher circumferential velocity, higher lubricating oil viscosity and lower roughness.

continued from Q 22, (spur gears, design)

In q 141, q 145 and q 147 b or b and d must be known. The following ratios are for estimating purposes and should be used for the initial calculation:

Pinion dimensions

	Either:	$\frac{d_1}{d_{shaft\,1}}$	Or: from gear ratio i and a specified centre distance a (see q113-114-129)
q 152	pinion integral with shaft	1·2 ... 1·5	
q 153	pinion free to turn on shaft	2	

Tooth width ratios

	Tooth- and bearing-quality	$\frac{b}{m}$	$\frac{b}{d_1}$
q 154	teeth smoothly cast or flame cut	6...10	
q 155	teeth machined; bearings supported each side on steel construction or pinion overhung	(6)...10...15	
q 156	teeth well machined; bearings supp. each side in gear casing	15...25	
q 157	teeth precision machined; good bearings each side and lubrication in gear casing: $n_1 \leqslant 50\ s^{-1}$.	20...40	
q 158	overhung gearwheel		$\leqslant 0{\cdot}7$
q 159	fully supported		$\leqslant 1{\cdot}2$

Suffixes for Q 18...25

- $_a$: driving wheel
- $_b$: driven wheel
- $_1$: small wheel or pinion
- $_2$: large wheel or wheel
- $_t$: tangential
- $_n$: normal
- $_m$: tooth middle for bevel gears
- $_v$: on back cone (or virtual cylindrical gear)

For symbols see Q 29

Bevel gears

Bevel gears, geometry

Equations q 86...q 88 are applicable and also:

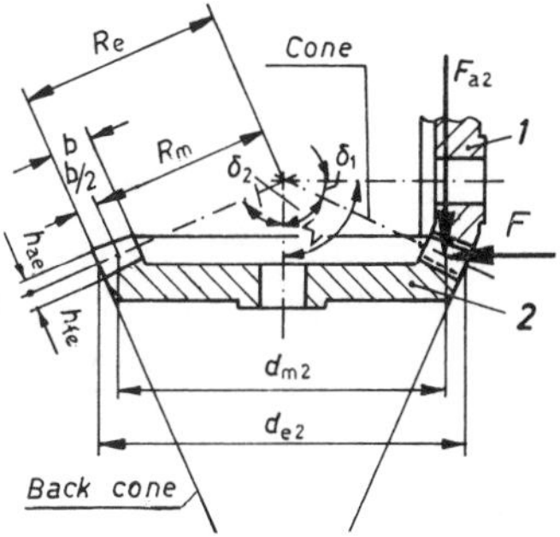

Only the axial and radial forces acting one mesh wheel 2 are shown

cone angle δ:

q 160 $\tan \delta_1 = \dfrac{\sin \Sigma}{\cos \Sigma + u}$;

q 161 $\left(\Sigma = 90^\circ \Rightarrow \tan \delta_1 = \dfrac{1}{u}\right)$

q 162 $\tan \delta_2 = \dfrac{\sin \Sigma}{\cos \Sigma + 1/u}$;

q 163 $(\Sigma = 90^\circ \Rightarrow \tan \delta_2 = u)$

q 164 angle between shafts $\Sigma = \delta_1 + \delta_2$

q 165 external pitch cone distance $R_e = \dfrac{d_e}{2 \sin \delta}$

Development of the back cone to examine the meshing conditions gives the virtual cylindrical gear (suffix "v" = virtual) with the values:

q 166	straight	bevel gears	$z_v = \dfrac{z}{\cos \delta}$	$u_v = \dfrac{z_{v2}}{z_{v1}}$
q 167				
q 168	spiral		$z_v \approx \dfrac{z}{\cos \delta \times \cos^3 \beta}$	

Formulae q 92, q 95...q 100 are also applicable to the surface of the back cone (suffix "e").

Bevel gears, design

The design is referred to the MID-POINT OF THE WIDTH b (suffix "m") with the values:

/170 $R_m = R_e - \dfrac{b}{2}$ | $m_m = \dfrac{d_m}{z}$

/172 $d_m = 2\, R_m \sin \delta$ | $F_{tm} = \dfrac{2\,T}{d_m}$

continued on Q 25

For symbols see Q 23, for suffixes see Q 23

continued from Q 24

Axial and radial forces in mesh

q 173 axial force $F_a = F_{tm} \tan\alpha_n \times \sin\delta$

q 174 radial force $F_r = F_{tm} \tan\alpha_n \times \cos\delta$

Load capacity of tooth root (approx. calculation)

Safety factor S_F against fatigue failure of tooth root:

q 175
$$S_F = \frac{\sigma_{F\,lim}}{\frac{F_{tm}}{b\, m_{nm}} \times Y_F \times Y_{\varepsilon V} \times Y_\beta} \times \frac{Y_S \times K_{FX}}{K_I \times K_V \times K_{F\alpha} \times K_{F\beta}} \geqslant S_{F\,min}$$

Giving the approximate formula:

q 176
$$m_{nm} \geqslant \frac{F_{tm}}{b} \times Y_F \times K_I \times K_V \times \underbrace{Y_{\varepsilon V} \times Y_\beta \times K_{F\alpha}}_{\approx 1} \times \underbrace{\frac{K_{F\beta}}{Y_S \times K_{FX}}}_{\approx 1} \times \frac{S_{F\,min}}{\sigma_{F\,lim}}$$

q 177 Y_F : substitute the number of teeth of the complementary spur gear z_v or, with spiral gears $z_{vn} \approx z_v/\cos^3\beta$. The graph for spur gears on page Q 21 is then also applicable to bevel gears.

For all other data see q 142, q 143 and q 144.

Load capacity of tooth flank (approx. calculation)

Safety factor S_H against pitting of tooth surface:

q 178
$$S_H = \frac{\sigma_{H\,lim}}{\sqrt{\frac{u+1}{u} \times \frac{F_{tm}}{b\, d_1} \times Z_H \times Z_M \times Z_{\varepsilon V}}} \times \frac{Z_V \times K_{HX} \times Z_R \times K_L}{\sqrt{K_I \times K_V \times K_{H\alpha} \times K_{H\beta}}} \geqslant S_{H\,min}$$

For metals the material factor Z_M is simplified to:

q 179
$$Z_M = \sqrt{0{\cdot}35\, E} \quad \text{with} \quad E = \frac{2\, E_1\, E_2}{E_1 + E_2}$$

Giving the approximate formula:

q 180
$$d_{vm1} \geqslant \sqrt{\frac{2\, T_1}{b} \times \frac{u_v+1}{u_v}\; 0{\cdot}35\, E} \times Z_{HV} \times \underbrace{Z_{\varepsilon V} \times \sqrt{K_{H\alpha}}}_{\approx 1} \times \frac{\sqrt{K_I \times K_V} \times \overbrace{\sqrt{K_{H\beta}}}^{\approx 1}}{Z_V \times K_{HX} \times Z_R \times K_L} \times \frac{S_{H\,min}}{\sigma_{H\,lim}}$$

q 181 Z_{Hv} : see diagram for Z_H (page Q 22), but only valid for $(x_1 + x_2)/(z_1 + z_2) = 0$ with $\beta = \beta_m$.

For all other data see q 148...q 151.

For symboles see Q 29, for suffixes see Q 23

MACHINE PARTS

Epicyclic gearing

Q 26

Velocity diagram and angular velocities (referred to fixed space)

		arm fixed	wheel 1 fixed	
q 182	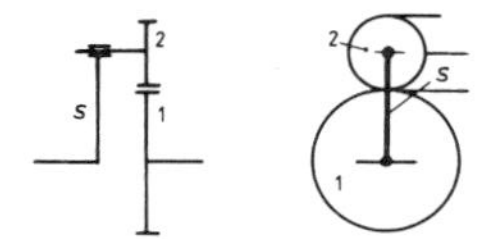	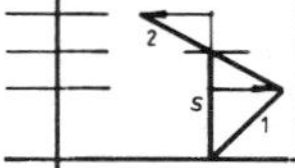 $\omega_2 = -\omega_1 \frac{r_1}{r_2}$	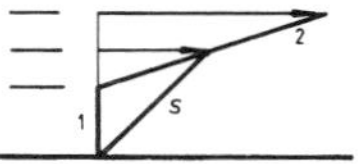 $\omega_2 = \omega_s \left(1 + \frac{r_1}{r_2}\right)$	
		arm fixed	**wheel 1 fixed**	**wheel 3 fixed**
q 183	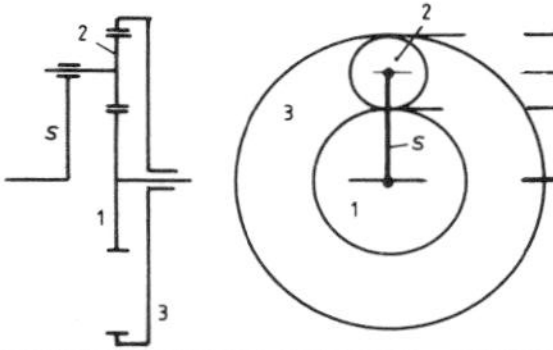	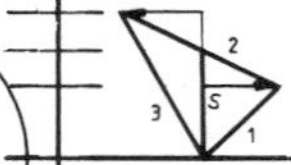 $\omega_3 = -\omega_1 \frac{r_1}{r_3}$	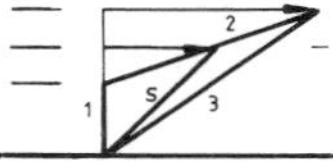 $\omega_3 = \omega_s \left(1 + \frac{r_1}{r_3}\right)$	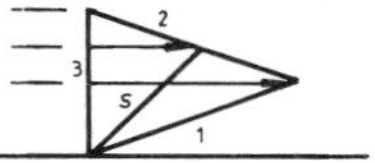 $\omega_1 = \omega_s \left(1 + \frac{r_3}{r_1}\right)$
		arm fixed	**wheel 1 fixed**	**wheel 4 fixed**
q 184	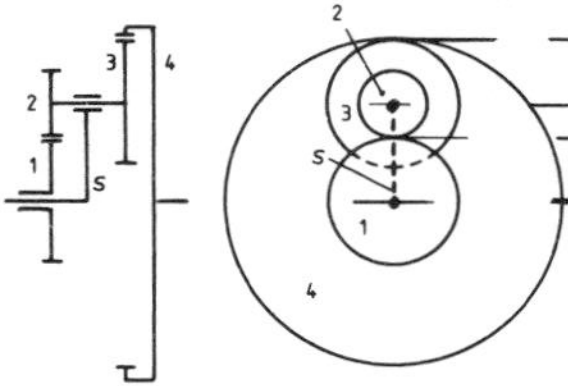	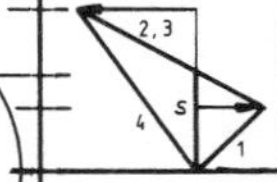 $\omega_4 = -\omega_1 \frac{r_1 \; r_3}{r_2 \; r_4}$	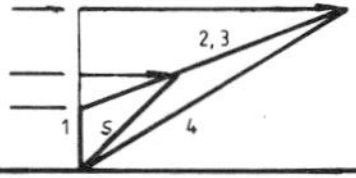 $\omega_4 = \omega_s \frac{(r_1+r_2)(r_2+r_3)}{r_2 \; r_4}$	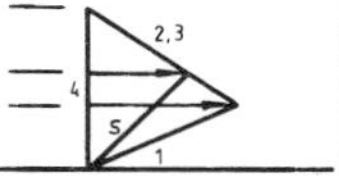 $\omega_1 = \omega_s \frac{(r_1+r_2)(r_2+r_3)}{r_1 \; r_3}$

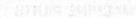

Worm gearing, geometry

(Cylindrical worm gearing, normal module in axial section, BS 2519, angle between shafts $\Sigma = 90^{\circ}$).

Drive worm

All the forces acting on the teeth in mesh are shown by the three arrows F_a F_t and F_r.

In the example: $z_1 = 2$, right-hand helix.

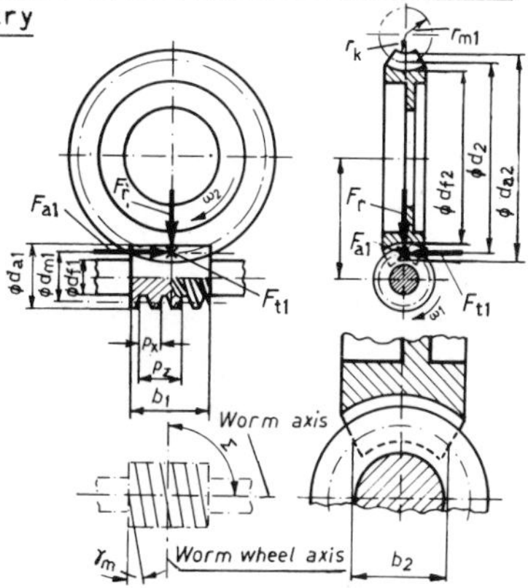

Gear tooth ratio and transmission ratio as q 86...88

		Worm, suffix 1	Worm wheel, suff. 2
q 185	module	$m_x = m = m_t$	
q 186	pitch	$p_x = m\,\pi = p_2 = d_2\,\pi / z_2$	
q 187	mean diameter (free to choose, for normal values see DIN 3976)	$d_{m1} = 2\,r_{m1}$	
q 188	form factor	$q = d_{m1}/m$	
q 189	centre helix angle	$\tan \gamma_m = \frac{m\,z_1}{d_{m1}} = \frac{z_1}{q}$	
q 190	pitch diameter		$d_2 = m\,z_2$
q 191	addendum	$h_{a1} = m$	$h_{a2} = m(1 + x)$ [1]
q 192	dedendum	$h_{f1} = m(1 + c_1^*)$	$h_{f2} = m(1 - x + c_2^*)$
q 193	tip clearance fact.	$c_1^* = (0{\cdot}167 \ldots \underline{0{\cdot}2} \ldots 0{\cdot}3) = c_2^*$	
q 194	outside diameter	$d_{a1} = d_{m1} + 2\,h_{a1}$	$d_{a2} = d_2 + 2\,h_{a2}$
q 195	tip groove radius		$r_k = a - d_{a2}/2$
q 196	tooth width	$b_1 \geq \sqrt{d_{a2}^{\,2} - d_2^{\,2}}$	$b_2 \approx 0{\cdot}9\,d_{m1} - 2\,m$
q 197	root diameter	$d_{f1} = d_{m1} - 2\,h_{f1}$	$d_{f2} = d_2 - h_{f2}$
q 198	centre distance	$a = (d_{m1} + d_2)/2 + x\,m$ [1]	

[1] Profile offset factor x for check of a pre-set centre distance; otherwise $x = 0$. continued on Q 28

For symbols see Q 29, for suffixes see Q 23

continued from Q 27

Worm gearing, design (worm driving)

No.		Worm	Worm wheel
199	peripheral force	$F_{t1} = \frac{2\ T_1}{d_{m1}} K_I \times K_V$	$F_{t2} = F_{a1}$
200	axial force	$F_{a1} = F_{t1} \times \frac{1}{\tan(\gamma + \varrho)}$	$F_{a2} = F_{t1}$
201	radial force	$F_{r1} = F_{t1} \times \frac{\cos \varrho \times \tan \alpha_n}{\sin(\gamma + \varrho)}$	$= F_r = F_{r2}$
202	rubbing speed	$\upsilon_g = \frac{d_{m1}}{2} \times \frac{\omega_1}{\cos \gamma_m}$	

Efficiency

No.	Worm driving	Worm wheel driving
203	$\eta = \tan \gamma_m / \tan(\gamma_m + \varrho)$	$\eta' = \tan(\gamma_m - \varrho)/\tan \gamma_m$

$(\gamma_m < \varrho) \Longrightarrow$ self-locking!

Coefficient of friction (typical values) $\mu = \tan \varrho$

	$\upsilon_g \approx 1$ m/s	$\upsilon_g \approx 10$ m/s
worm teeth hardened and ground	0·04	0·02
worm teeth tempered and macine cut	0·08	0·05

For calculation of worm shaft defelection see P 12

Calculation of module m

Load capacity of teeth root and flanks and temperature rise are combined in the approx. formula:

204 $F_{t2} = C\ b_2\ p_2$; where $b_2 \approx 0{\cdot}8\ d_{m1}$; $p_2 = m\,\pi$.

205 $m \approx \sqrt[3]{\frac{0{\cdot}8\ T_2}{C_{perm}\ q\ z_2}}$ | $F_{t2} = 2\ T_2/d_2 = 2\ T_2/(m\ z_2)$

206 $q \approx 10$ for $i = 10, 20, 40$

$q \approx 17$ for $i = 80$, self-locking

Assumed values for normal, naturally-cooled worm gears (worm hardened and ground steel, worm wheel of bronze):

υ_g	m s^{-1}	1	2	5	10	15	20
C_{perm}	N mm^{-2}	8	8	5	3·5	2·4	2·2

When cooling is adequate this value can be used for

207 all speeds: $C_{perm} \geqslant 8$ N mm^{-2}

For all symbols see Q 29, for suffixes see Q 23

Notation for Q 18...Q 28 (suffixes see Q 23)

a : standard centre distance
b : tooth width
h_{a0} : addendum of cutting tool
h_{aP} : addendum of reference profile (e.g. DIN 867)
h_{fP} : dedendum of reference profile
k : change of addendum factor
p_e : normal pitch ($p_e = p \cos \alpha$, $p_{et} = p_t \cos \alpha_t$)
z : no. of teeth
z_{nx} : equivalent no. of teeth
(C_{perm}), C : (permissible) load coefficient
F_t : peripheral force on pitch cylinder (plane section)
K_I : operating factor (external shock)
K_V : dynamic factor (internal shock)
$K_{F\alpha}$: end load distribution factor } for root stress
$K_{F\beta}$: face load distribution factor } for root stress
K_{FX} : size factor } for root stress
$K_{H\alpha}$: end load distribution factor } for flank stress
$K_{H\beta}$: face load distribution } for flank stress
R_e : total pitch cone length (bevel gears)
R_m : mean pitch cone length (bevel gears)
T : torque
$Y_{F'}$ (Y_S) : form factor, (stress concentration factor)
Y_β : skew factor
Y_ε : load proportion factor
Z_H : flank form factor
Z_ε : engagement factor
Z_R : roughness factor
Z_V : velocity factor
α_P : reference profile angle (DIN 867: $\alpha_P = 20°$)
α_W : operating angle
β / β_b : skew angle for helical gears — pitch cylinder / base cylinder
ρ : sliding friction angle ($\tan \rho = \mu$)
ρ_{a0} : tip edge radius of tool
$\sigma_{F\,lim}$: fatigue strength
$\sigma_{H\,lim}$: Hertz pressure (contact pressure)

Precise calculations for spur and bevel gears: DIN 3990
Terms and definitions for
spur gears and gearing : DIN 3960 } or BS 2519
bevel gears and gearing : DIN 3971 } or BS 2519
straight worm gearing : DIN 3975 } or BS 2519

Machine tool design: general considerations

Components of machine tools which are subjected to working stresses (frames with mating and guide surfaces, slides and tables, work spindles with bearings) are designed to give high accuracy over long periods of time. They are made with generous bearing areas and the means to readjust or replace worn surfaces should it become necessary.

The maximum permissible deflection at the cutting edge (point of chip formation) is approximately 0·03 mm. For spindle deflection refer to formula p 22 and for cutting forces see r 4.

Cutting drives (main drives) with v = const over the entire working range (max. and min. workpiece or tool-diameter) are obtainable with output speeds in geom. progression:

1 $$n_k = n_1\,\varphi^{k-1}$$

The progressive ratio φ for the speeds $n_1 \dots n_k$ with k number output speeds

2 are calculated by: $\varphi = \sqrt[k-1]{\dfrac{n_k}{n_1}}$

and the preferred series is selected.

Standardized progr. ratio φ: 1·12 – 1·25 – 1·4 – 1·6 – 2·0

Speed basic series R_{20} where $\varphi = \sqrt[20]{10} = 1{\cdot}12$:

...100-112-125-140-160-180-200-224-250-280-315-355-400-450-500-560-630-710-800-900-1000-... rpm.

Cutting gears are designated by the number of shafts and steps.

Example: A III/6 gear drive incorporates 3 shafts and provides 6 output speeds. Representation of gear unit as shown (for $k = 6$; $\varphi = 1{\cdot}4$; $n_1 = 180$; $n_k = 1000$):

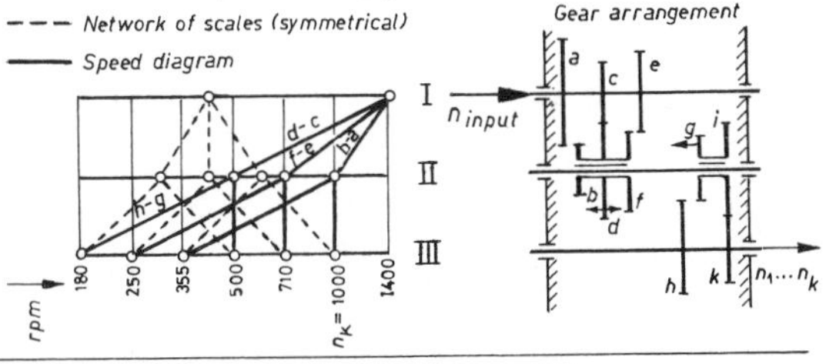

For explanation of symbols refer to R 4

PRODUCTION ENGINEERING

Machining

R 2

Cutting power P_s

r 3 Cutting power $P_s = \frac{F_s \times v}{\eta_{mech} \times \eta_{electr}}$

r 4 Cutting force $F_s = K \times k_{s\,1.1} \times b \times h^{1-z} \times z_e$

where h and b are in mm,
$k_{s\,1.1}$ is in N/mm² and F_s is in N.

Table of values for K, b, h, z_e ($k_{s\,1.1}$; $1-z$ see Z17)

No	Method	Sketch	K	b	h	z_e	Notes
r5	Turning external longitud.	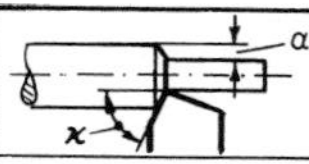	1 HM + HSS	$\frac{a}{\sin \varkappa}$	$s \times \sin \varkappa$	1	
r6	internal	analogous to r 5	1·2				
r7	Planing and shaping	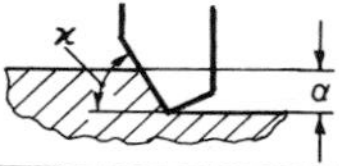	1·1 HM 1·2 HSS				
r8	Drilling and boring	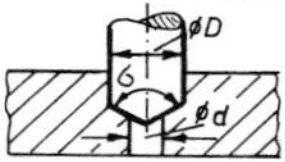	0·85 HM 1 HSS	$\frac{D-d}{2\sin\frac{\sigma}{2}}$	$s_z \times \sin\frac{\sigma}{2}$ $s_z = 0{,}5\ s$	2 for twist drill	$d=0$ when drilling $\sigma = 118°$ f. steel
r9	Plane milling down-cut and up-cut	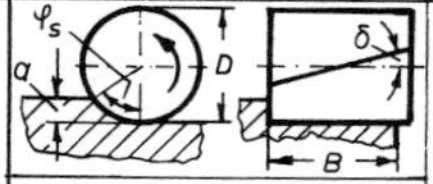	1·1 HM 1·2 HSS	$\frac{B}{\sin \delta}$	$\frac{2a}{D\,\varphi_s} s_z \times \sin \varkappa$	$\frac{\varphi_s\, z_s}{360°}$	$\cos \varphi_s = 1 - 2a/D$ $\varkappa = 90° - \delta$
r 10	End milling down-cut and up-cut	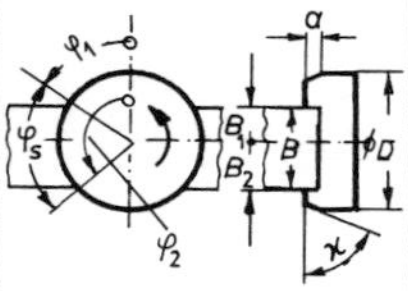		$\frac{a}{\sin \varkappa}$	$(\cos \varphi_1 - \cos \varphi_2) \times \frac{1}{\varphi_s} s_z \times \sin \varkappa$		$\varphi_s = \varphi_2 - \varphi_1$ $\cos \varphi_1 = \frac{2\,B_1}{D}$ $\cos \varphi_2 = \frac{2\,B_2}{D}$ calculate φ_1 and φ_2 in direction of rotation

For explanation of symbols refer to R 4

Feed drives

Feeds in geometrical progression with progressive ratio φ = 1·12 - 1·25 - 1·4 - 1·6 - 2·0.

Feed rate

	Method	Feed rate	Notes
r 11	Turning, longitudinal (external and internal)	$u = n \times s$	
r 12	Drilling	$u = n \times s_Z \times z_S$	for twist drills $z_e = z_S = 2$, $s_Z = 0{\cdot}5\, s$
r 13	Planing, shaping	$u = v$	
r 14	Milling, plane milling and end milling	$u = s_Z \times n \times z_S$	

Cutting times t_S

r 15 $$t_S = \frac{l_1}{u} \;; \quad \text{where} \quad l_1 = l + l'.$$

When calculating the cycle and machining times for each workpiece, the feed and infeed travels and also the lengths covered during non-cutting motions, divided by the corresp. speeds, must be taken into account.

Feed power P_V

r 16 Feed power $$P_V = \frac{u(F_R + F_V)}{\eta_{mech} \times \eta_{electr}}$$

r 17 Feed force $F_V \approx 0{\cdot}2\, F_S$; (F_S from r 4)

r 18 Friction force $F_R = m_b \times g \times \mu$

where m_b is the mass moved, e.g. in the case of milling machines the sum of the table and workpiece masses.

It must be determined whether the feed power as calculated under r 16 is sufficient to accelerate the moving components to rapid motion speed u_E within a given time t_b (in production machines $u_E \approx 0{\cdot}2$ m/s). Otherwise the following applies

r 19 $$P_V = u_E\, m_b \left(\mu\, g + \frac{u_E}{t_b}\right) \frac{1}{\eta_{mech} \times \eta_{electr}}$$

Explanation of symbols

used on pages R 1...R 3

a : infeed
b : width of chip
B : milling width
B_1, B_2: milling width measured from tool centre
d : diameter of pre-drilled hole
D : tool diameter
F_R : friction force
F_S : cutting force
F_V : feed force
g : gravitational acceleration
h : chip thickness
k : number of output speeds
$k_{s1.}$ basic cutting force related to area
K : method factor
l : cutting travel
l_1 : work travel
l' : overrun travel at both ends with feed rate u
n : speed
n_1 : minimum output speed
n_k : maximum output speed
s : feed
s_z : feed per cutting edge
t_b : acceleration time
t_s : cutting time
u : feed rate
u_E: rapid traverse speed
v : cutting speed
z_e : number of cutting edges in action
z_s : number of cutting edges per tool
δ : cutter helix angle
ε_s : slenderness ratio ($\varepsilon_s = a/s$)
η_{electr} : electrical efficiency
η_{mech} : mechanical efficiency
$\varkappa$: setting angle
μ : friction coefficient, see Z 7
σ : drill tip angle
φ : progressive ratio
φ_s : entrance angle of milling cutter

HM : carbide tip
HSS : high-speed tip

Cold working of sheet
Deep drawing

Initial blank diameter D

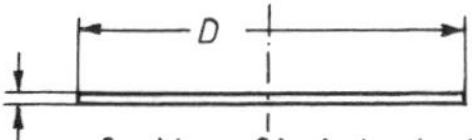

r 20 $D = \sqrt{\frac{4}{\pi} \times \Sigma A_{mi}}$

A_{mi} are the surface areas of the finished item which can be calculated from the following formulae b 30, c 12, c 16, c 21, c 25, c 27 or c 30. The surface areas at the transition radii for both drawing and stamping dies are calculated as follows:

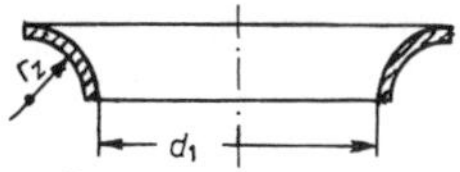

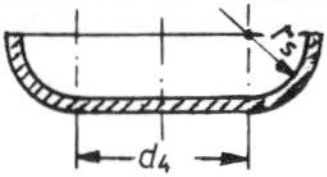

r 21 $A_m = \frac{\pi}{4}\left[2\pi d_1 r_z + 4(\pi - 2) r_z^2\right]$ | $A_m = \frac{\pi}{4}(2\pi d_4 + 8 r_s) r_s + \frac{\pi}{4} d_4^2$

Example: (assume $r_s = r_z = r$)

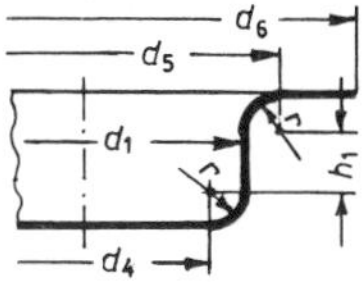

r 22 $D = \sqrt{d_4^2 + d_6^2 - d_5^2 + 4 d_1 h + 2\pi r (d_1 + d_4) + 4\pi r^2}$

1st and 2nd stages

	1st stage	2nd stage
	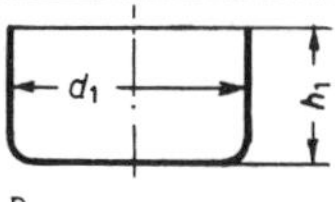	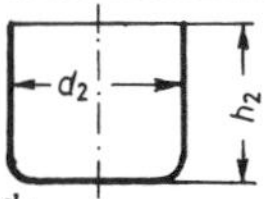
r 23	$\beta_1 = \frac{D}{d_1}$	$\beta_2 = \frac{d_1}{d_2}$
r 24	$\beta_{1\,max} = \beta_{100} + 0{\cdot}1 - \left(\frac{d_1}{s}\,0{\cdot}001\right)$	$\beta_{2\,max} = \beta_{100} + 0{\cdot}1 - \left(\frac{d_2}{s}\,0{\cdot}001\right)$
r 25	$F_{D1} = \pi\, d_1\, s\, k_{fm1}\, \varphi_1 \frac{1}{\eta_{E1}}$	$F_{D2} = \frac{F_{D1}}{2} + \pi\, d_2\, s\, k_{fm2}\, \varphi_2 \frac{1}{\eta_{E2}}$
r 26	$\varphi_1 = \left\lvert \ln\sqrt{0{\cdot}6\,\beta_1^2 - 0{\cdot}4} \right\rvert$	$\varphi_2 = \left\lvert \ln\sqrt{0{\cdot}6\,\beta_2^2 - 0{\cdot}4} \right\rvert$

r 27 r 28	$k_{fm1} = \frac{w}{\varphi_1}$	without	intermediate annealing	$k_{fm2} = \frac{k_{f1} + k_{f2}}{2}$
r 29		with		$k_{fm2} = \frac{w}{\varphi_2}$

continued on R 6

continued from R 5

The work w, related to the volume and the yield strength k_f is obtained from the deformation curves for the appropriate value of logarithmic deformation ratio φ (see Z 20).

Blank holding forces F_{B1} and F_{B2}

1st stage	2nd stage
r 30 $F_{B1}=(D^2-d_1^2)\frac{\pi}{4}\frac{R_m}{400}\left[(\beta_1^2-1)+\frac{d_1}{s}\right]$	$F_{B2}=(d_1^2-d_2^2)\frac{\pi}{4}\frac{R_m}{400}\left[(\beta_2^2-1)+\frac{d_2}{s}\right]$
Bottom tearing occurs if	
r 31 $R_m \leqslant \frac{F_{D1}+0{\cdot}1\,F_{B1}}{\pi\, d_1\, s}$	$R_m \leqslant \frac{F_{D2}+0{\cdot}1\,F_{B2}}{\pi\, d_2\, s}$

r 32 Maximum drawing conditions β and R_m

Material Sheet metal Description	Specificat. to		β_{100}	without intermediate annealing β_{2max}	with intermediate annealing β_{2max}	R_m N/mm^2
Steel	ASTM	A 366-79	1·7	1·2	1·5	390
		A 619-75	1·2	1·2	1·6	360
		A 620-75	2·0	1·3	1·7	340
		A 283 Gr. C	1·7	–	–	410
Stainless Steel	SAE	3310	2·0	1·2	1·8	600
Aluminum-manganese-silic.-alloy	Alu-Association	AA 6004 soft	2·05	1·4	1·9	150

Notation for R 5 and R 6

A_{mi} : surface area
F_{D1} , F_{D2} : drawing force in 1st and 2nd stage
k_{fm1} or k_{fm2} : mean yield strength, 1st or 2nd stage
k_{f1} , k_{f2} : yield strength for φ_1 and φ_2
r : radius
r_s : radius of stamping die
r_d : radius of drawing die
w : work per unit volume = $\frac{\text{work of deformation}}{\text{forminged volume}}$
β_1, β_2 : drawing ratio, 1st and 2nd stage
β_{100} : max. drawing ratio for s = 1 mm and d = 100 mm
β_{1max}, β_{2max} : max. drawing ratio, 1st and 2nd stage
η_{E1} , η_{E2} : process efficiency, 1st and 2nd stage
φ_1, φ_2 : logarithm. deformation ratio, 1st and 2nd stage

The most important electrical quantities and their units. – Basic rules

1 Note regarding capital and small letters used as symbols

Electrical engineering quantities that are independent of time are mainly denoted by capital letters. Quantities that vary with time are denoted by small letters or by capital letters provided with the subscript t.

Examples: formulae s 8, s 9, s 13

Exceptions: f, ω, $\hat{\imath}$, $\hat{u}$, p_{Fe10}

Electrical work W

Electrical work W is equivalent to mechanical work W as explained on M 1. Energy conversion, however, is subject to losses.

Units: J; W s (wattsecond); kW h; MW h

$$1\ \text{W s} = 1\ \text{Joule} = 1\ \text{J} = 1\ \text{N m}$$

Further the following relation applies, using quantities explained on S1 and S2:

2 $$W = I\,V\,t = \frac{V^2}{R}\,t = I^2 R\,t$$

Electrical power P

Electrical power P is equivalent to mechanical power P, as explained on M 1. Energy conversion, however, is subject to losses.

Units: W (watt); kW; MW

$$1\ \text{W} = 1\ \frac{\text{J}}{\text{s}} = 1\ \frac{\text{N m}}{\text{s}}$$

Further the following relation applies, using quantities explained on S1 and S2:

3 $$P = \frac{V^2}{R} = I^2 R$$

Frequency f see L 1

Period T see L 1

Angular frequency ω, angular velocity ω see L 1

Current I

Is a base quantity (see preface and instructions)

Units: A (ampere); mA; kA

The current of 1 A has been defined by means of the attracting force which two parallel current-carrying conductors exert on each other.

continued on S 2

continued from S 1

Current density J

s 4 $$J = \frac{I}{A}$$

Applicable only, where distribution of current I is uniform over cross section A.

Units: A/m^2; A/mm^2

Potential difference V

s 5 $$V = \frac{P}{I}$$

Units: V (volt); mV; kV

Where a direct current of 1 A through a conductor converts energy at a rate of 1 W, the voltage across this conductor is 1 V.

$$1\,V = \frac{W}{A} = 1\frac{J}{s\,A} = 1\,A\,\Omega = 1\frac{N\,m}{s\,A}$$

Resistance R

s 6 $$R = \frac{V}{I} \qquad \text{(Ohm's law)}$$

Units: Ω (ohm); kΩ; MΩ

Where a voltage of 1 V across a conductor causes a current of 1 A to the flow through it, resistance is 1 Ω.

$$1\,\Omega = \frac{1\,V}{1\,A} = 1\frac{W}{A^2} = 1\frac{W}{s\,A^2} = 1\frac{N\,m}{s\,A^2}$$

Conductance G

Conductance G is the reciprocal of resistance R.

s 7 $$G = 1/R$$

Unit: 1/Ω

$$1/\Omega = [1\,\text{Mho}]$$

Quantity of electricity, charge Q

s 8 $$q = \int i\,dt \qquad \text{(see s 1)}$$

For direct current:

s 9 $$Q = I\,t$$

Unit: C (coulomb)

$$1\,C = 1\,A\,s$$

continued on S 3

continued from S 2

Capacitance C

The capacitance C of a capacitor is the ratio of quantity of electricity Q stored in it and voltage V across it:

S 10
$$C = \frac{Q}{V}$$

Units: F (farad); µF; nF; pF

Where a capacitor requires a charge of 1 C to be charged to a voltage of 1 V, its capacitance is 1 F.

$$1\,\text{F} = 1\,\frac{\text{C}}{\text{V}} = 1\,\frac{\text{A s}}{\text{V}} = 1\,\frac{\text{A}^2\text{s}}{\text{W}} = 1\,\frac{\text{A}^2\text{s}^2}{\text{J}} = 1\,\frac{\text{A}^2\text{s}^2}{\text{N m}}$$

Magnetic flux Φ

S 11
$$\Phi = \frac{1}{N}\int \upsilon \, dt \qquad \text{(see s 1)}$$

Here N is the number of turns of a coil and υ the voltage induced, when the magnetic flux Φ linked with the coil varies with time.

Units: Wb (weber) = V s = 10^8 M (maxwell)

1 Wb is the magnetic flux which, linking a circuit of 1 turn, induces in it a voltage of 1 V as it is reduced to zero at a uniform rate in 1 s.

Magnetic induction (flux density) B

The magnetic induction in a cross section A is:

S 12
$$B = \frac{\Phi}{A}$$

Here A is the cross-sectional area traversed perpendicularly by the homogeneous magnetic flux Φ.

Units: T (tesla); µT; nT; V s/m^2; G (gauss)

$$1\,\text{T} = 1\,\frac{\text{V s}}{\text{m}^2} = 10^{-4}\frac{\text{V s}}{\text{cm}^2} = \left[10^4\,\text{G} = 10^4\,\frac{\text{M}}{\text{cm}^2}\right]$$

Where a homogeneous magnetic flux of 1 Wb perpendicularly traverses an area of 1 m^2, its magnetic induction is 1 T.

continued on S 4

continued from S 3

Inductance L

S 13 $$L = N\frac{\Phi}{I} = N\frac{\Phi_t}{i} \qquad \text{(see s 1)}$$

Here I is the current flowing through a coil of N turns and Φ the magnetic flux linked with this coil.

Units: H (henry); mH

1 H is the inductance of a closed loop of one turn which, positioned in vacuum and passed through by a current of 1 A, encloses a magnetic flux of 1 Wb.

$$1\ \text{H} = 1\ \frac{\text{Wb}}{\text{A}} = 1\ \frac{\text{V s}}{\text{A}}$$

Magnetic field strength H

S 14 $$H = \frac{B}{\mu_0\ \mu_r}$$

Units: A/m; A/cm; A/mm; (Ampere Turn/m)

Magnetomotive force F

S 15 $$F = N I$$

Units: A; kA; mA; (Ampere Turn)

Magnetomotive force F_i in the i-th section of a magnetic circuit:

S 16 $$F_i = H_i\ l_i$$

here l_i is the length of this section.

S 17 $$\sum_{i=1}^{n} F_i = F$$

Reluctance S of a homogeneous section of a magnetic circuit:

S 18 $$S = \frac{F}{\Phi} \qquad \left(\begin{array}{c}\text{equivalent to}\\ \text{Ohm's law for}\\ \text{magnetic circuits}\end{array}\right)$$

Units: 1/H;= A/V s; (Ampere Turn/Wb)

for symbols see S 16

ELECTRICAL ENGINEERING

Electric circuits

S 5

Basic properties of electric circuits

Directions of currents, voltages, arrows representg. them

20	Direction of the current and of arrows	generator $- \rightarrow +$
21	representing positive currents in	load $+ \rightarrow -$
22	Direction of the potential difference and of arrows representing positive voltages always	$+ \rightarrow -$

Directions of arrows representg. currents or voltages

	where funct. of element (generator or load) as well as polarity is	determine directions of arrows	where calcul. results in a positive value, direction with respect to arrow of current or voltage is	
			positive	negative
23	known	as stated above	—	—
24	unknown	at random	equal	opposite

Special rule

Arrows representing voltage drop across a resistor and current causing it, should always be determined in same direction (as $R > 0$).

Ohm's Law

Current through a resistor:

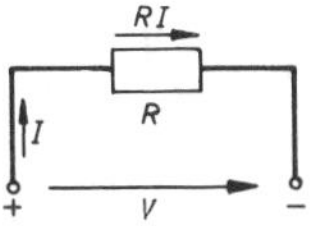

25 $$I = \frac{V}{R}$$ (see also S 6)

Resistance R of conductor

26 $$R = \frac{\varrho\, l}{A} = \frac{l}{\gamma\, A}$$

Resistance R of conductor at temperature ϑ (in degrees centigrade)

27 $$R = R_{20}\left[1 + \alpha(\vartheta - 20^{\circ}\mathrm{C})\right]$$

Electric heating of a mass m

28 $$V\, I\, t\, \eta = c\, m\, \Delta\vartheta$$

α : temperature coeff. (see Z 21)
γ : conductivity (see Z 21)
ϱ : resistivity (see Z 21)
c : specific heat (see Z 1...4)
η : efficiency [and O 2]

$\Delta\vartheta$: temper. change
t : time
R_{20} : resistance at $\vartheta = 20^{\circ}\mathrm{C}$

continued on S 6

continued from S 5

1st Kirchhoff Law

The algebraic sum of all currents entering a branch point (node) is zero.

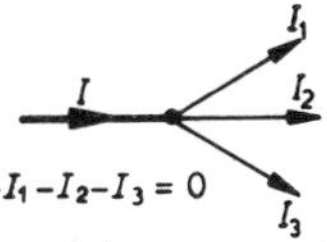

29 $\Sigma I = 0$ $\qquad I - I_1 - I_2 - I_3 = 0$

Here currents	into	the node	positive
	out of	are considered	negative

Ratio of currents

Where several resistors are connected in parallel, total current and partial currents are inversely proportional to their respective resistances.

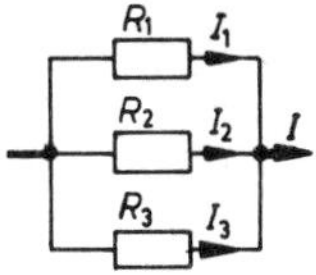

30 $I : I_1 : I_2 : I_3 = \frac{1}{R} : \frac{1}{R_1} : \frac{1}{R_2} : \frac{1}{R_3}$

Current division

Partial currents of 2 resistors connected in parallel:

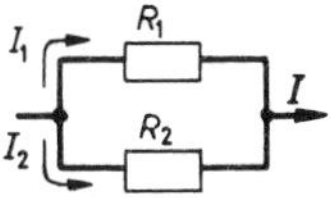

31 $I_1 = I \frac{G_1}{G_1 + G_2} = I \frac{R_2}{R_1 + R_2}$

2nd Kirchhoff Law

The algebraic sum of all voltages around a closed mesh (loop) is zero.

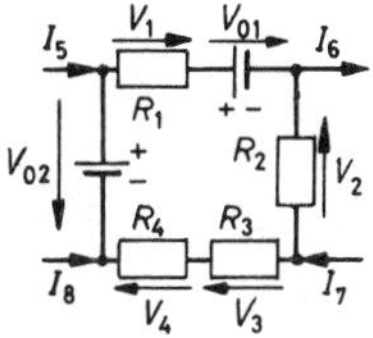

$\Sigma V = 0$

32 Here voltages traversed in accordance with (opposite to) direction of arrow are considered positive (negative).

$V_1 + V_{01} - V_2 + V_3 + V_4 - V_{02} = 0$

Ratio of voltages

Where several resistors are connected in series, the ratio of partial voltages is equal to the ratio of the respect. resistances.

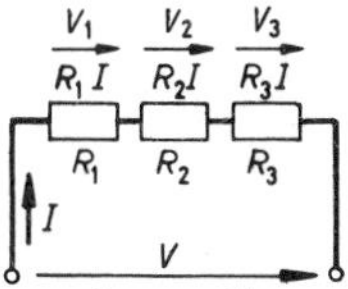

33 $V_1 : V_2 : V_3 = R_1 : R_2 : R_3$

Voltage divider

Partial voltages across 2 resistors connected in series:

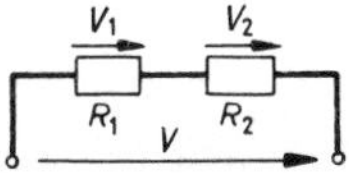

34 $V_1 = V \frac{G_2}{G_1 + G_2} = V \frac{R_1}{R_1 + R_2}$

Series connection

Total resistance R (according to s 26)

generally:

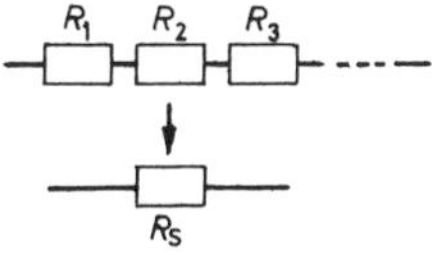

s 35 $$R_S = R_1 + R_2 + R_3 + \ldots$$

for n equal resistances R:

s 36 $$R_S = n R$$

Parallel connection

Total resistance R (according to s 30)

generally:

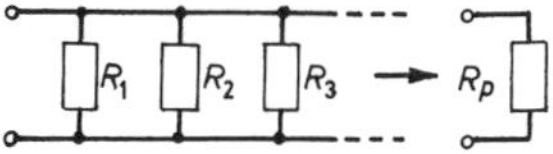

s 37 $$\frac{1}{R_p} = \frac{1}{R_1} + \frac{1}{R_2} + \frac{1}{R_3} + \ldots$$

s 38 $$G_p = G_1 + G_2 + G_3 + \ldots$$

	for 2 several resistances	for 3 several resistances	for n equal resistances
s 39	$R_p = \dfrac{R_1 R_2}{R_1 + R_2}$	$R_p = \dfrac{R_1 R_2 R_3}{R_1 R_2 + R_2 R_3 + R_1 R_3}$	$R_p = \dfrac{R}{n}$
s 40	$= \dfrac{1}{G_1 + G_2}$	$= \dfrac{1}{G_1 + G_2 + G_3}$	$= \dfrac{1}{n G}$

Multiple connection

A multiple connection of several known resistances is subdivided into parallel and series connections, proceeding outwards. These are separately converted as to be conveniently combined again, e.g.:

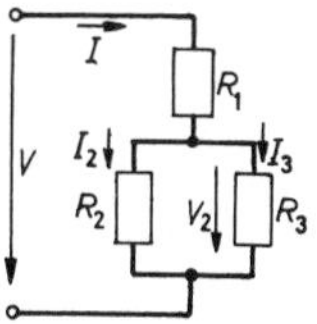

s 41 $$I = \frac{R_2 + R_3}{R_1 R_2 + R_1 R_3 + R_2 R_3} V = \frac{G_1 (G_2 + G_3)}{G_1 + G_2 + G_3} V$$

s 42 $$I_3 = \frac{R_2}{R_1 R_2 + R_1 R_3 + R_2 R_3} V = \frac{G_1 G_3}{G_1 + G_2 + G_3} V$$

s 43 $$V_2 = \frac{R_2 R_3}{R_1 R_2 + R_1 R_3 + R_2 R_3} V = \frac{G_1}{G_1 + G_2 + G_3} V$$

Transformation of a delta to a star-circuit and vice versa

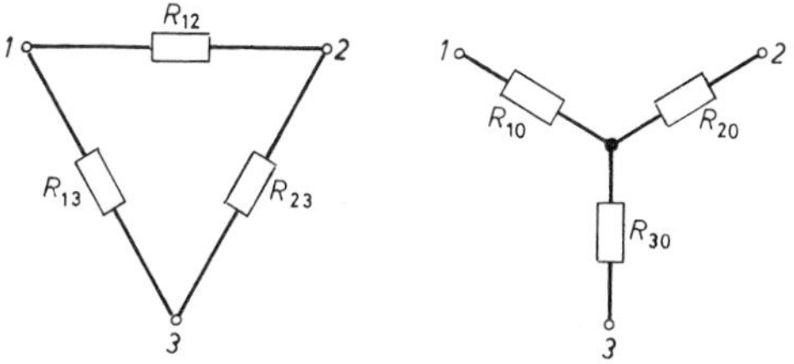

s 44 $$R_{12} = \frac{R_{10} \times R_{20} + R_{10} \times R_{30} + R_{20} \times R_{30}}{R_{30}} \qquad R_{10} = \frac{R_{12} \times R_{13}}{R_{23} + R_{12} + R_{13}}$$

s 45 $$R_{13} = \frac{R_{10} \times R_{20} + R_{10} \times R_{30} + R_{20} \times R_{30}}{R_{20}} \qquad R_{20} = \frac{R_{23} \times R_{12}}{R_{23} + R_{12} + R_{13}}$$

s 46 $$R_{23} = \frac{R_{10} \times R_{20} + R_{10} \times R_{30} + R_{20} \times R_{30}}{R_{10}} \qquad R_{30} = \frac{R_{23} \times R_{13}}{R_{23} + R_{12} + R_{13}}$$

Potential divider

Potential dividers are used to provide reduced voltages.

s 47 $$V_V = \frac{R_2 R_V}{R_1 R_2 + R_1 R_V + R_2 R_V} V$$

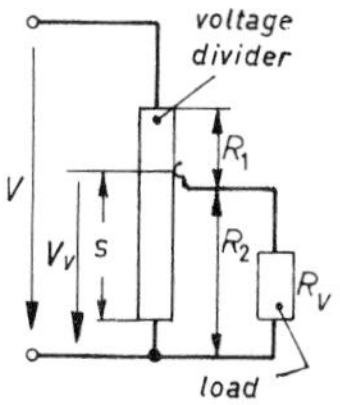

For applications, where V_V has to be approximately proportional

s 48 to s, the condition $R_V \geqq 10(R_1 + R_2)$ has to be satisfied.

s : distance of sliding contact from zero position

Applications in electrical measurements

Extending the range of a voltmeter

s 49 $$R_V = R_M\left(\frac{V_{max}}{V_{M\,max}} - 1\right)$$

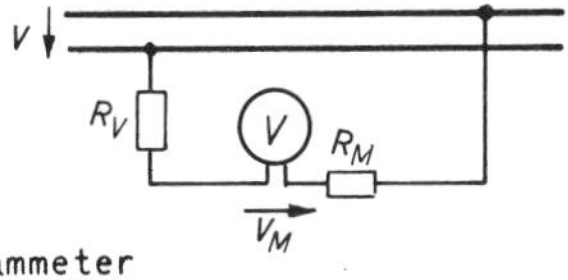

Extending the range of an ammeter

s 50 $$R_N = R_M\,\frac{I_{M\,max}}{I_{max} - I_{M\,max}}$$

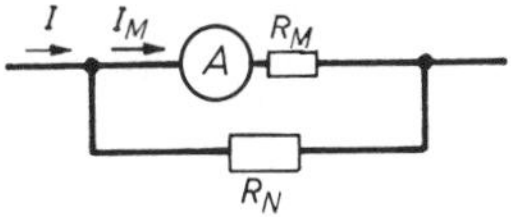

Wheatstone bridge for measuring an unknown resistance R_X

A slide-wire Wheatstone bridge may be used for measuring resistances of between 0·1 and 10^6 ohms. The calibrated slide wire is provided with a scale reading $a/(l-a)$. The sliding contact is adjusted, until the detector current I_B is zero. Then

s 51 $$\frac{R_X}{R} = \frac{a}{l - a}$$

and hence

s 52 $$R_X = R\,\frac{a}{l - a}$$

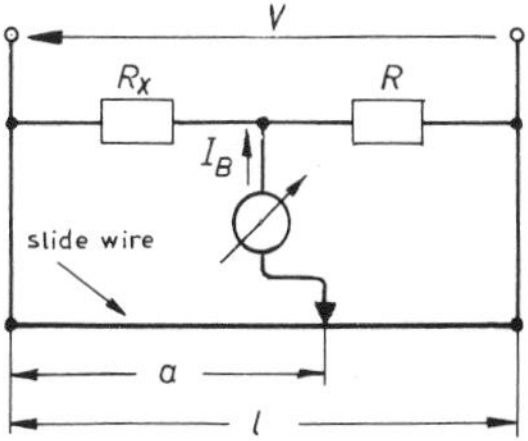

Wheatstone bridge used as a primary element

In many types of measuring equipment Wheatstone bridges serve as comparators for evaluating voltage differences.

R_1 : sensor resistor, the variation of which is proportional to the quantity x to be measured (e.g. temp., distance, angle etc.)

R_2 : zero value of R_1

Approx. the relat. applies

s 53 $$V_M \sim \Delta R \sim x$$

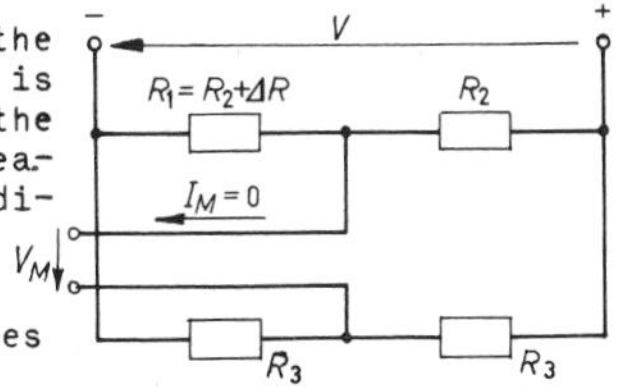

R_M : internal resistance of the measurement

Capacitance C of a capacitor

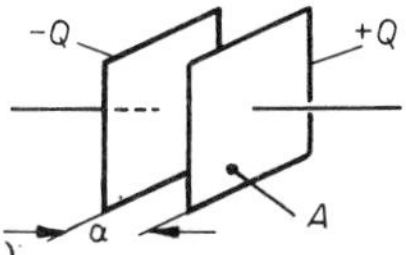

s 54 $$C = \frac{\varepsilon_0 \varepsilon_r A}{a}$$

Quantity of electricity Q (see s 8)

Electrical work W_C stored in an electric field

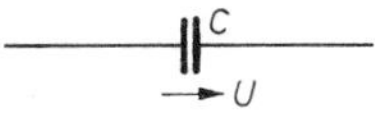

s 55 $$W_C = \frac{1}{2} C V^2$$

Capacitors connected in parallel

Where capacitors are added in parallel, the total capacitance C increases.

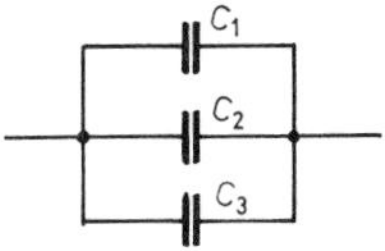

s 56 $$C = C_1 + C_2 + C_3$$

Capacitors connected in series

Where capacitors are added in series, the total capacitance C decreases.

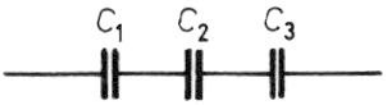

s 57 $$\frac{1}{C} = \frac{1}{C_1} + \frac{1}{C_2} + \frac{1}{C_3}$$

Capacitance of two coaxial cylinders

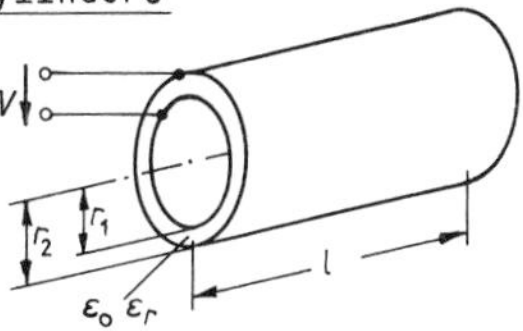

s 58 $$C = 2\pi \varepsilon_0 \varepsilon_r \frac{l}{\ln \frac{r_2}{r_1}}$$

ε_r : relative permittivity (siehe Z 22)
s 59 ε_0 : absolute permittivity $\varepsilon_0 = 8 \cdot 85 \times 10^{-12}$ A s/(V m)
A : plate area (one side)
a : thickness of dielectric
r_1 : radius of inner cylinder
r_2 : radius of outer cylinder
l : length of cylinders

s 60 Deflection of a magnetic needle

The N-pole of a magnetic needle is attracted by a magnetic S-pole and repelled by a magnetic N-pole.

Fixed conductors and coils

s 61 Magn. flux about a current-carrying conductor

Assuming a corkscrew were screwed in the direction of the current, its direction of rotation would indicate the direction of the lines of magnetic flux.

s 62 Magnetic flux within a current-carrying coil

Assuming a corkscrew were rotated in the direction of the current through the coil, the direction of its axial motion would indicate the direction of the lines of magnetic flux through the coil.

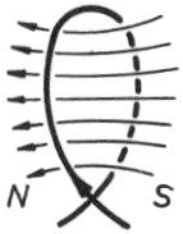

Movable conductors and coils

s 63 Parallel conductors

Two parallel conductors carrying currents of the same direction attract. Two carrying currents of opposite direction repel each other.

s 64 Two coils facing each other

Where two coils positioned face to face carry currents of the same direction, they attract, where they carry currents of opposite directions, they repel each other.

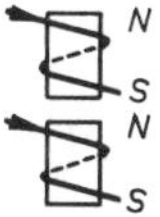

Machines

s 65 Right Hand Rule (generator)

Where the thumb points in the direction of the magnetic flux and the middle finger in the direction of motion, the index finger indicates the direction of current flow.

s 66 Left Hand Rule (motor)

Where the thumb points in the direction of the magnetic flux and the index finger in the direction of current flow, the middle finger indicates the direction of motion.

Quantities of magnetic circuits

Magnetic flux Φ

s 67 $$\Phi = \frac{F}{S} = \frac{N I}{R_m}$$ (see also s 11)

Magnetic induction (flux density) B

s 68 $$B = \frac{\Phi}{A} = \mu_r \mu_0 H$$ (see also s 12)

Inductance L

s 69 $$L = N\frac{\Phi}{I} = N^2\Lambda = \frac{N^2}{R_m}$$ (see also s 13)

For calculation of L see also s 140 through s 146

Magnetic field strength H (Magnetising force)

s 70 $$H = \frac{B}{\mu_r \mu_0} = \frac{F_i}{l_i}$$ (see also s 14)

Magnetomotive force F

s 71 $$F = N I = \sum_{i=1}^{n} F_i$$ (see also s 15)

Magnetomotive force F_i

s 72 $$F_i = H_i \, l_i$$ (see also s 16)

Reluctance S

s 73 $$S = \frac{F}{\Phi} = \frac{l}{\mu_r \mu_0 A}$$ (see also s 18)

Energy W_m stored in a magnetic field

s 74 $$W_m = \frac{1}{2} N I \Phi = \frac{1}{2} L I^2$$

Leakage flux Φ_L

Part of the total magnetic flux Φ leaks through the air and is thus lost for the desired effect. Φ_L is related to the useful flux Φ_u. Hence the leakage

s 75 coefficient is: $$\sigma = \frac{\Phi}{\Phi_u} = \frac{\text{total flux}}{\text{useful flux}} \quad (1{\cdot}15 \ldots 1{\cdot}25)$$

For symbols see S 16

The magnetic field and its forces

Force F_m acting between magnetic poles

In the direction of the magnetic flux a tensile force F_m occurs:

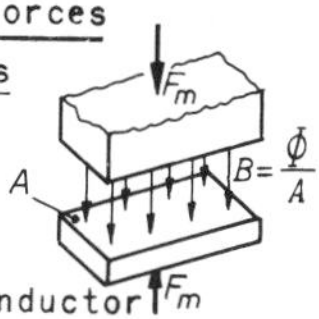

s 78 $F_m = \frac{1}{2} \times \frac{B^2 A}{\mu_0}$

Force F_l acting on a current-carrying conductor

A conductor carrying a current I encounters a transverse force F_l over its length l perpendicular to the lines of magnetic flux:

s 79 $F_l = B\, l\, I$

When applied to the armature of a DC-machine, the moment is:

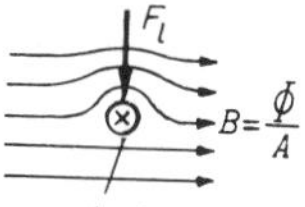

s 80 $M_i = \frac{1}{2\pi} \Phi I \frac{p}{a} z$

s 81 Φ : flux per pole

Induced voltage V_i (induction law)

Where a coil of N turns and resistance R_i is threaded by a magnetic flux Φ that varies with time, an open-circuit voltage

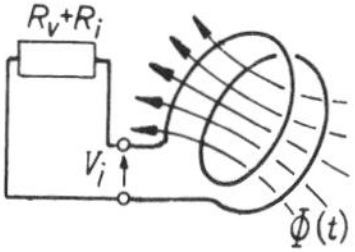

s 82 $V_i = N \frac{d\Phi}{dt}$ (see also s 11)

is induc. across its terminals. This voltage causes a current through an external load resistor R_v.

voltage induced by

	motion of conduct. perpendicular to flux	rotation of conductor loop in magnetic field	rotat. of generator armature
s 83	$V_i = B\, l\, v$	$V_i = \omega \Phi_{max} \times \sin(\omega t)$	$V_i = \Phi\, n\, z \frac{p}{a}$
s 84		$\Phi_{max} = l\, d\, B$	$= l\, d\, B \frac{z\, p}{2\pi a} \omega$
	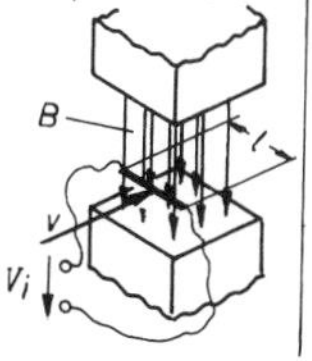	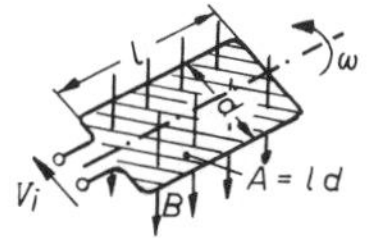	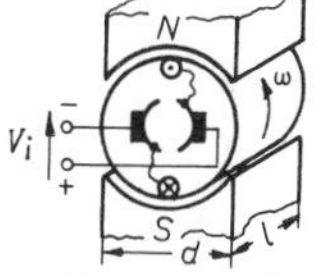

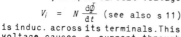

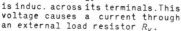

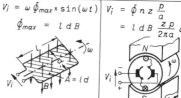

s 85 Voltage V_i due to self-induction: $V_i = L \times di/dt$

Continued on page S 16

For symbols see S 16

General terms relating to alternating-current circuit

Sense of phase angles

In vector diagrams arrows are sometimes used to represent phase angles. Here counterclockwise arrows are taken positive, clockwise arrows negative.

Example

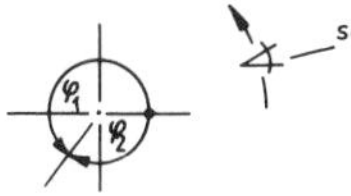

s 86 $$\varphi_1 - \varphi_2 = 360^\circ = 0$$

s 87 $$\varphi_1 = \varphi_2$$

Peak values (see also s 1)

Current i and voltage υ of an alternating current vary periodically with time t, usually sinusoidally. The maximum values $\hat{i}$ and $\hat{\upsilon}$ are called peak values. At an angular frequency $\omega = 2\pi f$ the angle covered in time t is:

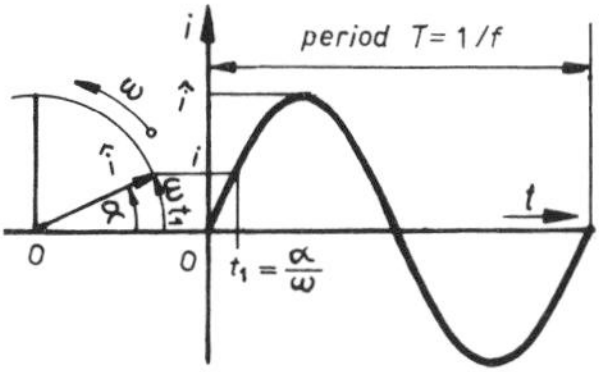

s 88 $$\alpha = \omega t = 2\pi f t$$

Hence at this time

s 89 the current is $i = \hat{i} \sin(\omega t) = \hat{i} \sin\alpha$

s 90 the voltage is $\upsilon = \hat{\upsilon} \sin(\omega t) = \hat{\upsilon} \sin\alpha$

Root-mean-square (rms) values

These are used for practical calculations and are usually indicated by meters.

	generally	for sine waves
s 91	$I = I_{eff} = \sqrt{\frac{1}{T}\int_0^T i^2\,dt}$	$I = I_{eff} = \frac{\hat{i}}{\sqrt{2}}$
s 92	$V = V_{eff} = \sqrt{\frac{1}{T}\int_0^T \upsilon^2\,dt}$	$V = V_{eff} = \frac{\hat{\upsilon}}{\sqrt{2}}$

s 93 With these values the relation $P = V I$ also applies for alternating current, if $\cos\varphi = 1$ (see s 105).

continued on S 15

continued from S 14

Phase shift, phase angle φ

Where different kinds of load (resistance, inductance and/or capacitance) are present in an alternating-current circuit, a phase shift between current and voltage occurs.

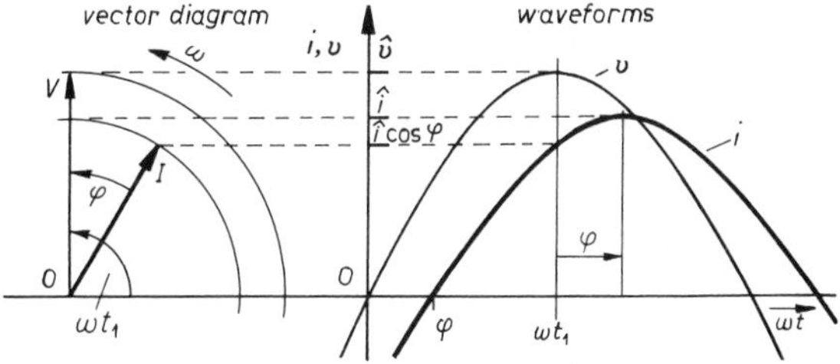

s 94 $\upsilon = \hat{\upsilon} \sin(\omega t)$ $\quad\quad$ $i = \hat{i} \sin(\omega t - \varphi)$

Q factor, damping factor $\tan\delta$, loss angle δ

The Q factor of a circuit has been defined by:

s 95 $$Q = \frac{2\pi\hat{w}}{W_{VP}}$$

Here $\hat{w}$ is the peak value of the energy stored in the circuit and W_{VP} the loss energy dissipated in one period.

The reciprocal of Q factor is called damping factor

s 96 $\tan\delta = 1/Q$ $\quad$ (δ is the loss angle)

For a choke (s 115 and s 118) and for a capacitor-resistor combination (s 116 and s 119) this definition results in the simple relations:

s 97 $Q = \tan\varphi$ | $\tan\delta = 1/Q = 1/\tan\varphi$
s 98 $\delta = 90^\circ - \varphi$ | $= V_w/V_b$ (for series connection)
s 99 $= I_w/I_b$ (for parallel connection)

For formulae regarding $\tan\varphi$ see S 17 and S 18. Formulae s 128 and s 129 applicable for resonant circuits are not so simple.

Basic equations for single phase alternating current

	Impedance	Z see S 17 and S 18
s 100	Admittance	$Y = 1/Z$
s 101	Voltage across impedance	$V = I\,Z$
s 102	Current through impedance	$I = \dfrac{V}{Z}$
s 103	Reactance	$X = Z \sin\varphi$
s 104	Active power	$P = V\,I \cos\varphi = I^2 R$
s 105	Reactive power	$P_q = V\,I \sin\varphi = I^2 X$
s 106	Apparent power	$P_s = V\,I = \sqrt{P^2 + P_q^2} = I^2 Z$
s 107	Power factor	$\cos\varphi = \dfrac{P}{V\,I} = \dfrac{P}{P_s}$
s 108	Alternating magnetic flux in a coil	$\hat{\Phi} = \dfrac{V_L}{4 \cdot 44\,N f}$

s 85 contd. (V_i due to self-induction)

Where the current i flowing through a coil changes with time, the magnetic field caused by this current also changes. Thereby a voltage V_i is induced in the coil. Its direction is such that it counteracts the instantaneous change of current (Lenz's law).

Symbols used on page S 13:

s 109 μ_0 : absolute permeability ($\mu_0 = 4\pi \times 10^{-7}$ V s/A m)

μ_r : relative permeability
for vacuum, gases, fluids, and most solids: $\mu_r = 1$,
for magnetic materials take μ_r from Z 23

a : number of parallel paths through winding
l : length of magnetic circuit
N : number of turns of coil
p : number of pole pairs
z : number of conductors

R_R	resistance in	series	equivalent circuit of choke
R_P	resistance in	parallel	equivalent circuit of choke
L_R	inductance in	series	equivalent circuit of choke
L_P	inductance in	parallel	equivalent circuit of choke

Components, series- and parallel connections carrying alternating current

	component		symbol	vector diagram	phase relat.	phase angle	impedance	$\tan\varphi =$
s 110	resistive lamp, coil with bi-filar winding		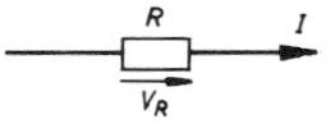		I and V in phase	$\varphi = 0°$	$Z = R$	0
s 111	inductive ideal inductance		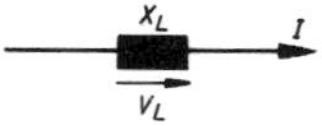	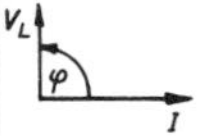	I lags V by 90°	$\varphi = 90°$	$Z = X_L = \omega L$	∞
s 112	capacitive capacitor		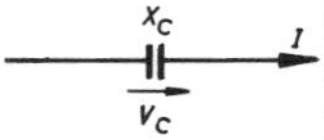	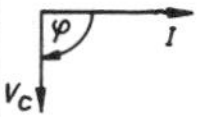	I leads V by 90°	$\varphi = -90°$	$Z = X_C = \frac{1}{\omega C}$	∞
s 113	resist.+ induct.+ capacit. conn. in series	$\omega L_R < \frac{1}{\omega C}$	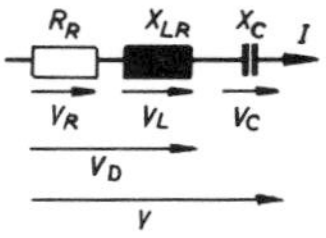	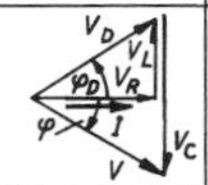	I leads V	$-90° < \varphi < 0°$	$Z = \sqrt{R_R^2 + \left(\omega L_R - \frac{1}{\omega C}\right)^2}$	$\frac{\omega L_R - \frac{1}{\omega C}}{R_R}$
s 114	choke in series with capacit.	$\omega L_R > \frac{1}{\omega C}$		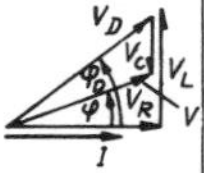	I lags V	$0° < \varphi < 90°$		
s 115	resistive and inductive in series choke		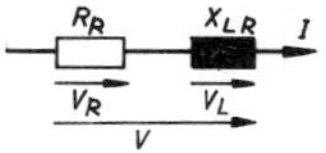	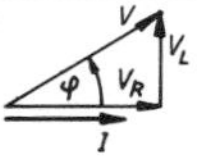	I lags V by less than 90°	$0° < \varphi < 90°$	$Z = \sqrt{R_R^2 + (\omega L_R)^2}$	$\frac{\omega L_R}{R_R}$

continued S 18

continued from S 17

	kind of load	symbol	vector diagram	phase relat.	phase angle	impedance	$\tan\varphi =$
s 116	resist. + capac. in series resistor in series with capacitor	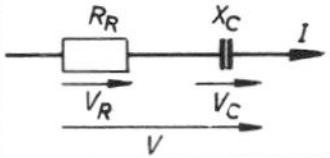	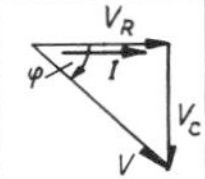	I leads V by less than 90°	$-90° < \varphi < 0°$	$Z = \sqrt{R_R^2 + \left(\frac{1}{\omega C}\right)^2}$	$-\frac{1}{R_R \omega C}$
s 117	resist.+induct. and capacit. in parallel choke + capacit. in parallel	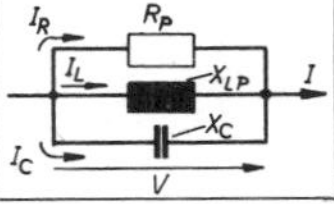	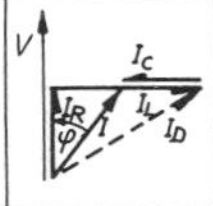	I leads or lags V depending on values	$-90° < \varphi < 90°$	$Z = \frac{1}{\sqrt{\left(\frac{1}{R_P}\right)^2 + \left(\frac{1}{\omega L_P} - \omega C\right)^2}}$	$R_P\left(\frac{1}{\omega L_P} - \omega C\right)$
s 118	resistive and inductive in parallel choke	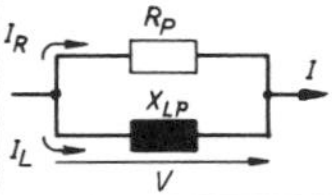	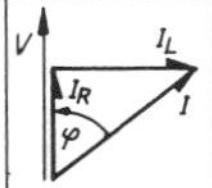	I lags V	$0° < \varphi < 90°$	$Z = \frac{1}{\sqrt{\left(\frac{1}{R_P}\right)^2 + \left(\frac{1}{\omega L_P}\right)^2}}$	$\frac{R_P}{\omega L_P}$
s 119	resist.+ capac. in parallel resistor + capacit. in parall.	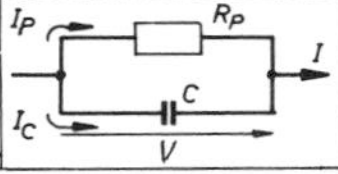	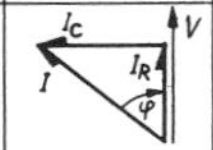	I leads V	$-90° < \varphi < 0°$	$Z = \frac{1}{\sqrt{\left(\frac{1}{R_P}\right)^2 + (\omega C)^2}}$	$-R_P \omega C$

s 120 The given values R and L of a choke always are the values R_R and L_R of the series equivalent circuit (see s 115). However, where a choke is connected in parallel with a capacitor, it is desirable to use the parallel equivalent circuit of a choke (see s 118).

s 121 The values R_P and L_P contained therein may be calculated by:

$$R_P = R_R + \frac{(\omega L_R)^2}{R_R}$$

$$L_P = L_R + \frac{R_R^2}{\omega^2 L_R}$$

Resonant circuits

		series-	parallel-
		resonant circuit	
	symbol and general vector diagram	see s 113	see s 117
	vector diagram at resonance	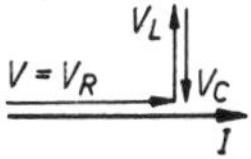	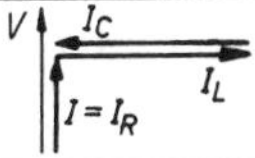
s 122	resonance condition	$V_L = V_C$	$I_L = I_C$
s 123		$\omega_r L_R - \frac{1}{\omega_r C} = 0$	$\frac{1}{\omega_r L_P} - \omega_r C = 0$
s 124		$\omega_r^2 L_R C = 1$	$\omega_r^2 L_P C = 1$
s 125	resonant frequency	$f_r = \frac{1}{2\pi\sqrt{L_R C}}$	$f_r = \frac{1}{2\pi\sqrt{L_P C}}$
		where line frequ. $f = f_r$, resonance occurs	
s 126	current at resonance	$I_r = \frac{V}{R_R}$	$I_r = \frac{V}{R_P} = \frac{R_R C V}{L_R}$
s 127		at $V_b = V_L - V_C = 0$ $\varphi = 0$	at $I_b = I_L - I_C = 0$ $\varphi = 0$
s 128	Q factor	$Q_R = \frac{\omega_r L_R}{R_R} = \frac{1}{\omega_r C R_R}$	$Q_P = \omega_r C R_P = \frac{R_P}{\omega_r L_P}$
s 129	loss angle δ from	$\tan\delta_R = \frac{1}{Q_R} = \frac{R_R}{\omega_r L_R}$	$\tan\delta_P = \frac{1}{Q_P} = \frac{1}{\omega_r C R_P}$
s 130	wavelength	$\lambda = \frac{c}{f_r} = \frac{300 \times 10^6\ \mathrm{m}}{f_r\,\mathrm{s}}$	
s 131	resonant period	$T_r = 2\pi\sqrt{L_R C}$	$T_r = 2\pi\sqrt{L_P C}$

Wave trap

A parallel resonant circuit has its maximum impedance Z_{max} at its resonant frequency. Therefore it acts as a rejector for currents of this frequency.

s 132 $$Z_{max} = R_P = \frac{L_R}{R_R C} \quad \text{and current} \quad I = \frac{V}{Z_{max}}$$

for symbols see S 16

Alternating-current bridge

AC bridges are used to determine capacitances and inductances. For balancing the bridge variable capacitor C_2 and resistor R_2 are adjusted until the sound in the low resistance headphone K reaches its minimum or vanishes. The following circuits are independent of frequency.

measurement of

	capacitance	inductance
	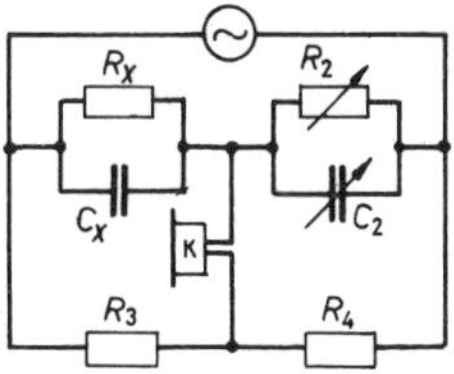	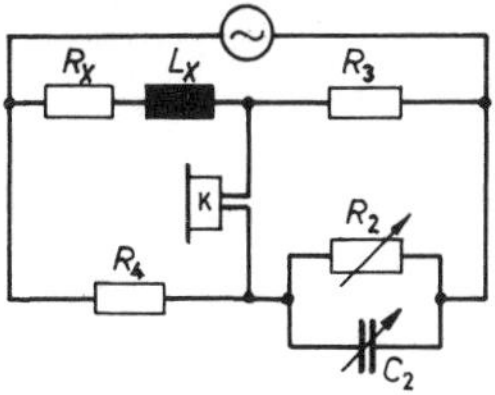
s 133	$C_X = \frac{R_4}{R_3} C_2$	$L_X = C_2 R_3 R_4$
s 134	$R_X = \frac{R_3}{R_4} R_2$	$R_X = \frac{R_3 R_4}{R_2}$
s 135	$\tan \delta_X = \frac{1}{R_X \omega C_X}$	$\tan \delta_X = \frac{R_X}{\omega L_X}$

Determination of an unknown impedance by measuring the voltages across this impedance and an auxiliary resistor:

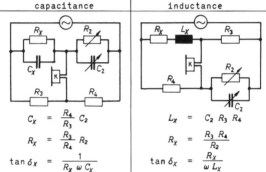

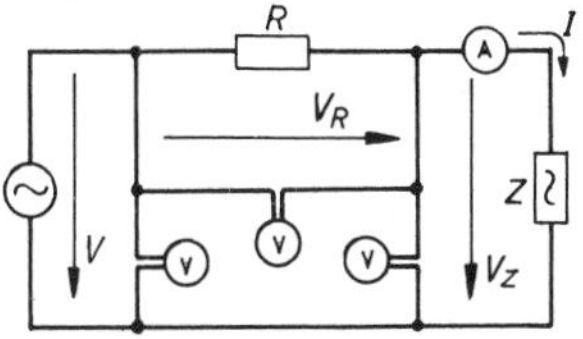

s 136 $$P_{WZ} = \frac{V^2 - V_R^2 - V_Z^2}{2R}$$

s 137 $$\cos \varphi_Z = \frac{P_{WZ}}{V_Z I}$$

s 138 $$Z = \frac{V_Z}{I}$$

s 139 select auxiliary resistor R such that $V_R \approx |V_Z|$

C_X : unknown capacitance
L_X : unknown inductance
δ_X : loss angle, see S 15
$R_{2 \dots 4}$: known resistances
R_X : unknown restistance of coil or capacitor
C_2 ,: C_4 : calibrated adjustable capacitances
Z : unknown impedance (inductive or capacitive)

Inductance L from impedance and resistance

Calculating L from impedance and resistance

140 Pass an alternating current ($J = I/A \approx 3\ \text{A/mm}^2$) through a coil and measure the terminal voltage V, current I, active power P:

141 impedance $Z = \frac{V}{I}$; resistance $R = \frac{P}{I^2}$

142 $$L = \frac{1}{\omega}\sqrt{Z^2 - R^2}$$

Calculating L for a toroidal coil

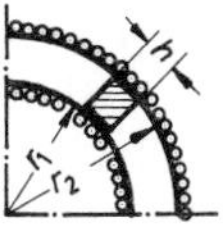

143 $$L = \frac{\mu_0 h N^2}{2\pi} \ln\frac{r_2}{r_1}$$

Calculating L for a square coil

armatures must be circular

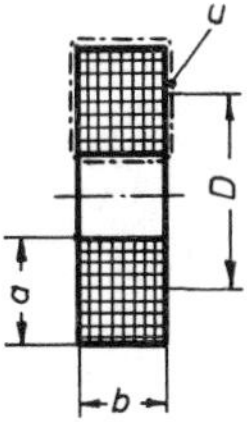

	$\frac{D}{u}$	inductance
144	< 1	$L = 1{\cdot}05\,\frac{D}{m}\,N^2\sqrt[4]{\left(\frac{D}{u}\right)^3}\ \mu\text{H}$
145	> 1	$L = 1{\cdot}05\,\frac{D}{m}\,N^2\sqrt{\frac{D}{u}}\ \mu\text{H}$
146	$\geqq 3$	values become unreliable

$$1\ \mu\text{H} = 10^{-6}\ \frac{\text{V s}}{\text{A}}$$

a : thickness of winding
A : cross section of wire
b : width of coil
d_a : external diameter of wire and insulation
D : mean diameter of coil
l_o : internal length of armature winding
147 l_m : mean length of armature winding ($l_m = l_o + \pi a$)
N : number of turns
u : circumference of coil cross section
α : ratio $a : b$
148 β : degree of loosening of turns $\left(\beta = \frac{a\,b}{N\,d_a^2}\right)$

Non-magnetic coils with specified inductance L

High frequency coils

	$\frac{D}{u}$	formula
149	< 1	$\left(\frac{D}{m}\right)^{3.5} N^{3.25} \approx \frac{1}{39}\left(\frac{d_o}{m}\right)^{1.5}\left(\frac{L}{H}\right)^2 \times 10^{-22}$
150	> 1	$\left(\frac{D}{m}\right)^{3} N^{3.5} \approx \frac{1}{55}\left(\frac{d_o}{m}\right)\left(\frac{L}{H}\right)^2 \times 10^{-22}$

here:

$$d_o = \frac{u}{2\sqrt{N}} = d_a(1+a)\sqrt{\frac{\beta}{a}}$$

Low frequency coils

Assuming that

151 $\beta = 1$ and $D = u$, then

152 $$N \approx 975 \times 10^3 \sqrt{\frac{L}{H}\,\frac{m}{D}}$$

153 $$a = \frac{1}{4}\left(u \pm \sqrt{u^2 - 16\,N\,d_a^{\,2}}\right); \qquad b = \frac{u}{2} - a$$

Calculation of number of turns N of a coil

From cross section

154 $$N \approx \frac{a\,b}{d_a^{\,2}}$$

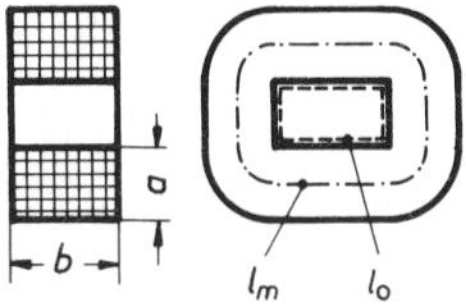

From resistance

155 $$N \approx \frac{R\,A}{\varrho\, l_m}$$

Using reference coil

Position unknown coil of N_x turns and reference coil of N_o turns at short distance on closed iron core. Magnetize core by alternating voltage V_e applied to magnetizing coil N_e. Measure voltages V_x and V_o using high impedance voltmeter. Then

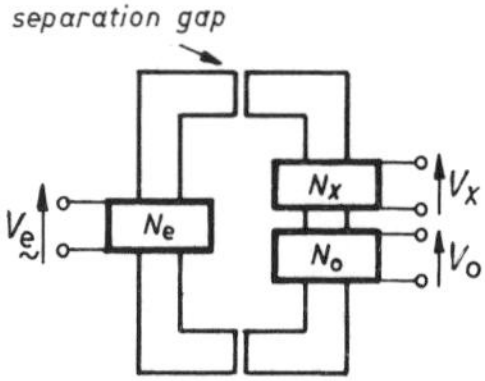

156 $$N_x = N_o \frac{V_x}{V_o}$$

For explanation of symbols see S 21

Hysteresis

Remanent-flux density B_r

A residual magnetism of flux density B_r remains in the iron core, after the external magnetic field strength H has been removed.

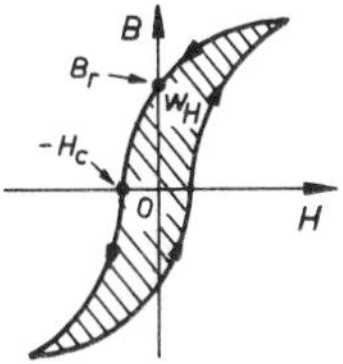

Coercive force H_c

The coercive force H_c has to be applied to reduce the flux density B to zero.

Hysteresis work W_H

The energy W_H dissipated during a single cycle of the hysteresis loop is equal to the product of area of the hysteresis loop w_H and core volume V_{Fe}:

157 $$W_H = w_H V_{Fe}$$

Hysteresis power P_{VH}

158 $$P_{VH} = W_H f = w_H V_{Fe} f$$

Eddy currents

According to the induction law alternating voltages are also induced inside an iron core. Depending on the resistivity of the core iron these voltages cause induction currents called eddy currents. They are kept small by lamination (making up the core of thin metal sheets, which are insulated from each other).

Core losses (iron losses)

Core losses per unit mass p_{Fe}

They are the combined hysteresis and eddy-current losses per unit mass. They are measured at a peak induction $\hat{B}$ = 1 T = 10 kG or 1·5 T = 15 kG and at a frequency f = 50 Hz and are then denoted p_{Fe10} or p_{Fe15} respectively. For values see Z 4.

Total core losses P_{Fe}

159 $$P_{Fe} = p_{Fe10}\left(\frac{\hat{B}}{\mathrm{T}} \times \frac{f}{50\,\mathrm{Hz}}\right)^2 m_{Fe}(1 + \varkappa)$$

m_{Fe}: mass of core | $\varkappa$: addit. for punchg. ridges etc. (0·1...1·0)

Choke coil

Choke coil used as a dropping impedance

It is used in an ac circuit to reduce the line voltage V down to a value V_V for a restitive load with minimum losses.

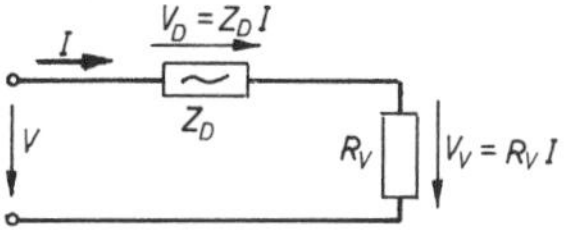

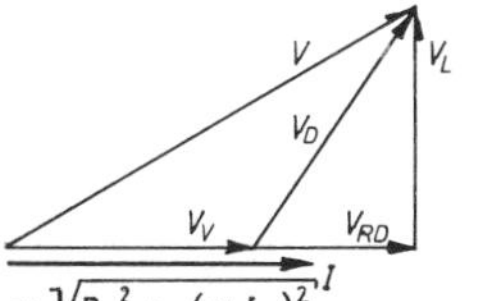

s 160 impedance of | choke $\quad Z_D = \sqrt{R_R^{\,2} + (\omega L_R)^2}$

s 161 impedance of | total circuit $\quad Z = \sqrt{(R_R + R_V)^2 + (\omega L_R)^2}$

s 162 inductance $\quad L_R = \frac{1}{\omega}\sqrt{\left(\frac{V R_V}{V_V}\right)^2 - (R_V + R_R)^2}$

In a rough calculation of L_R neglect the unknown resistance R_R of the choke. After dimensioning the choke R_R is known, and Z may be determined exactly.

s 163 Check V_V by $\quad V_V = \frac{V R_R}{Z}$

and repeat procedure, if necessary.

Choke of constant inductance without core

Dimension according to S 21. Make preliminary assumptions regarding values r_2/r_1 (toroid coil) or D/u (straight coil). In case of unfavourable results repeat procedure. Determine resistance of choke according to s 26.

Choke coil of constant inductance with iron core

The iron core essentially serves for guiding the magnetic flux and should incorporate as many as possible single air gaps δ_1. These should be filled with insulating layers and should not exceed 1 cm in length. The m.m.f. required to magnetize the core is neglected. Peak values of H and B are used for calculations. The variation of inductance L_R

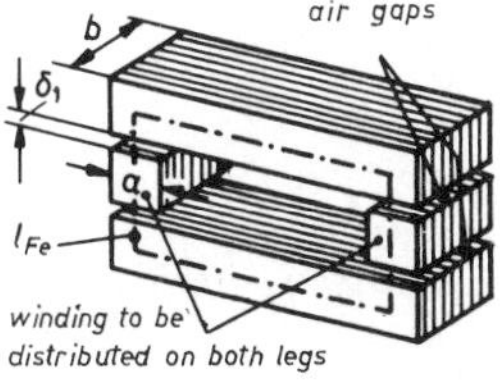

continued on S 25

continued from S 24

may be expressed in terms of the maximum relative current-depending variation of inductance

s 164 $$g_L = \frac{|L_{R\,tot} - L_R|}{L_R}\ ; \qquad \frac{1}{g_L} = \frac{A_{Fe}\,\hat{B}_{Fe}\,\delta}{\hat{H}_{Fe}\,l_{Fe}\,\mu_0\,A_L} + 1$$

If $g_L > g_{L\,requ}$, repeat dimensioning with greater A_{Fe} and smaller $\hat{B}_{Fe}$ at unchanged product $A_{Fe} \times \hat{B}_{Fe}$.

Dimensioning. Given: L_R, f, $g_{L\,requ}$, $V_{L\,eff}$ or I_{eff}, then the preliminary / final dimensions are

	then the		preliminary	final
s 165, s 166	effective cross section of core		$A_{Fe}' = \sqrt{K\,I_{eff}\,V_{L\,eff}}$ with $I_{eff} = \frac{V_{L\,eff}}{2\pi f L_R}$	take A_{Fe} from stand. or determine a and b by $A_{Fe} = 0{\cdot}9\,a\,b \approx A_{Fe}'$
s 167	number of turns		$N = \frac{V_{L\,eff}}{4{\cdot}44\,f\,\hat{B}_{Fe}\,A_{Fe}}$	
s 168	cross section of air gap		$A_L' = ab + 5\text{cm}(a+b)$	$[A_L = a\,b + 5(a+b)\delta_1]$
s 169	total	length of air gap	$\delta' = \frac{N^2\mu_0 A_L'}{L_R}$	$\delta = \frac{a\,b\,n\,N^2\mu_0}{n\,L_R - 5\,N^2\mu_0(a+b)}$
s 170	single	length of air gap	$\delta_1' = \delta'/n < 1$ cm	$\delta_1 = \delta/n < 1$ cm
s 171	diameter of wire		$d' = 2\sqrt{\frac{I}{J'\,\pi}}$	use next stand. values f. d; d_a including insulation
s 172	cross section of winding		$A_W = 1{\cdot}12\,d_a^2\,N$	
	length of limb		l_S to be determined from dimensions of core section and A_W	

Choke coil of current-depending inductance

This type of choke employs an iron core without an air gap. It is used only for special purposes, e.g. as a magnetic amplifier.

K : power coefficient of choke
$\approx 0{\cdot}24\ \text{cm}^4/\text{VA}$ for air-cooled chokes } core section
$\approx 0{\cdot}15\ \text{cm}^4/\text{VA}$ for oil chokes } see S 24
for ▭▭ core section increase values by 75%

J' : preliminary current density for air-cooled choke $J' = 2\ \text{A/mm}^2$, for oil choke $J' \approx 3 \ldots 4\ \text{A/mm}^2$

$\hat{B}_{Fe}$: core induction (take approx. $1 \ldots 1{\cdot}2$ T)

$\hat{H}_{Fe}$: field strength in core corr. to $\hat{B}_{Fe}$ to be taken from Z 23 according to material employed

n : number of single air gaps, increase reduces stray [flux

R_{Cu}: resistance of winding according to s 26

R_R : resistance of choke includ. core losses ($R_R \approx 1{\cdot}3\,R_{Cu}$)

l_{Fe}: mean length of magnetic path through iron

Transformer

Designation of windings

distinction by			
nominal voltages		function in circuit (direction of power transfer)	
winding with higher nominal voltage	winding with lower nominal voltage	input winding	output winding
high-end winding	low-end winding	primary (index 1) winding	secondary (index 2) winding

Nominal values (index N)

s 173 rated power $P_{SN} = V_{1N} \times I_{1N} = V_{2N} \times I_{2N}$

s 174 nominal transformation ratio $u = V_{1N} / V_{20} = I_{2N} / I_{1N}$

By the rated secondary voltage V_{2N} we mean the open-circuit secondary voltage ($V_{2N} = V_{20}$), not the one at nominal load.

Core losses P_{Fe} and open-circuit measurements

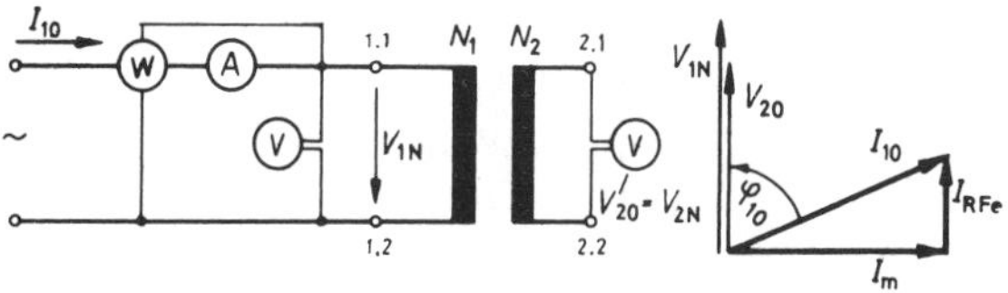

The core losses P_{Fe} only depend on primary voltage V_1 and frequency f, not on the kind of load.

s 175 $$P_{10} = P_{Fe}$$

Core losses P_{Fe} and nominal transformation ratio ü are determined by open-circuit measurements (see circuit diagram: secondary open, values provided with index 0). The primary current's resistive component I_{RFe} covers the core losses, its reactive component is the magnetizing current I_m. The copper losses are negligibly small. The core losses P_{Fe} are required for calculating operational power dissipation and efficiency.

continued on S 27

continued from S 26

Copper losses P_{Cu} and short-circuit measurements

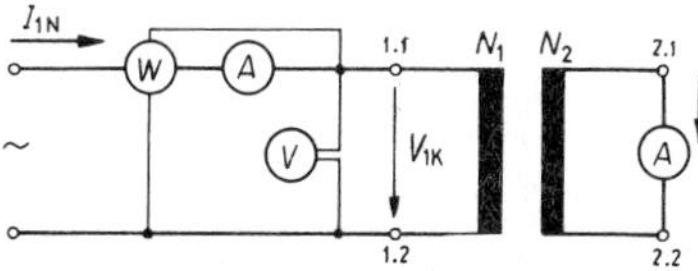

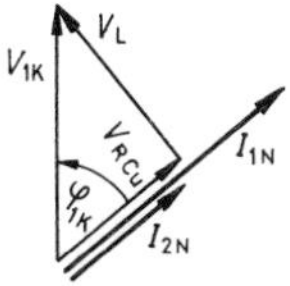

P_{Cu} depends only on the primary current I_1 and is determined by short-circuit measurements (see circuit diagram, values provided with index K). With the secondary shorted, the primary voltage is adjusted to a value V_{1K}, which causes the rated currents to flow. V_{1K} is so small that $I_{R\,Fe}$ and I_m are negligible. Then the short-circuit primary power P_{1K} is equal to the rated copper losses $P_{Cu\,N}$ of the total transformer at rated currents. P_{1K} is required for calculating operational power dissipation

s 176 and efficiency. $P_{1K} = P_{Cu\,N}$

The values measured are used for calculating the relative short-circuit voltage v_K, which, for bigger transf., is always indicated on the name plate:

s 177 $v_K = 100(V_{1K}/V_{1N})\%$

The following quantities may be determined using the vector diagram:

s 178 $R_{Cu} = V_R/I_{1N}$; $L = V_L/\omega I_{1N}$; $\cos\varphi_{1K} = V_R/V_{1K} = \dfrac{P_{Cu\,N}}{v_K\,P_{SN}}$

Operating conditions

simplified equivalent circuit

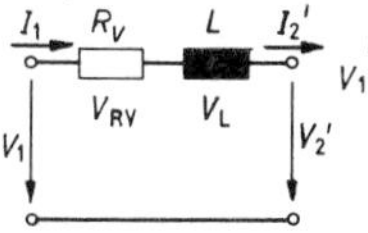

simplified vector diagram

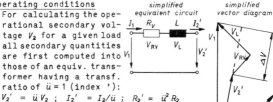

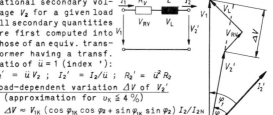

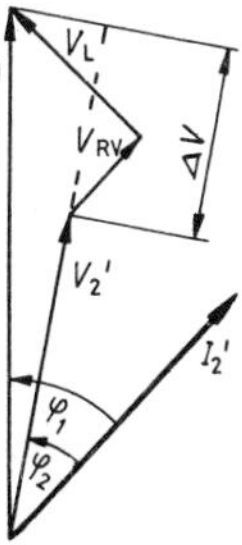

For calculating the operational secondary voltage V_2 for a given load all secondary quantities are first computed into those of an equiv. transformer having a transf. ratio of $\ddot{u} = 1$ (index '):

s 179 $V_2' = \ddot{u}\,V_2$; $I_2' = I_2/\ddot{u}$; $R_2' = \ddot{u}^2 R_2$

Load-dependent variation ΔV of V_2'

(approximation for $v_K \leqq 4\,\%$)

s 180 $\Delta V \approx V_{1K}\,(\cos\varphi_{1K}\cos\varphi_2 + \sin\varphi_{1K}\sin\varphi_2)\,I_2/I_{2N}$

Secondary voltage V_2

s 181 $V_2' \approx V_1 - \Delta V$; $V_2 = V_2'/\ddot{u}$

Basic connections

Star

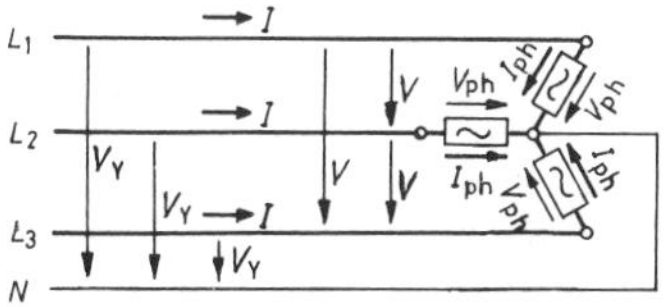

s 182 $V = V_{ph}\sqrt{3}$

s 183 $I = I_{ph}$

Delta

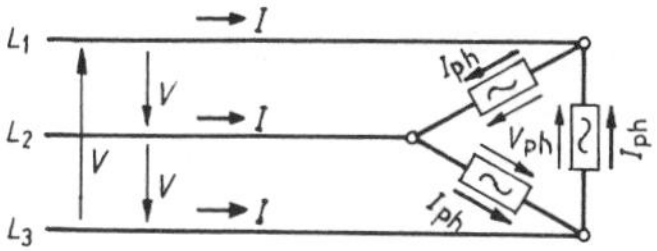

s 184 $V = V_{ph}$

s 185 $I = I_{ph}\sqrt{3}$

Measuring threephase power

Load balanced

connection

with neutral point (star connected)	without neutral point (delta connected)
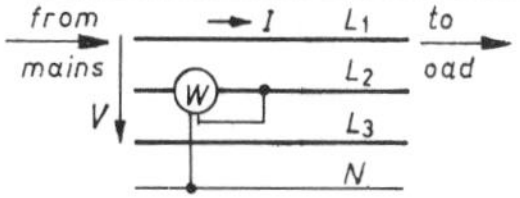	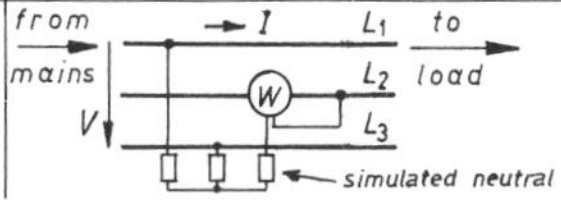

s.186 total power $P = 3P_{w\,ph} = \sqrt{3}\,V\,I\cos\varphi$

Load unbalanced (Two wattmeter method)

For delta connected without neutral point. (Also for balanced load without neutral point).

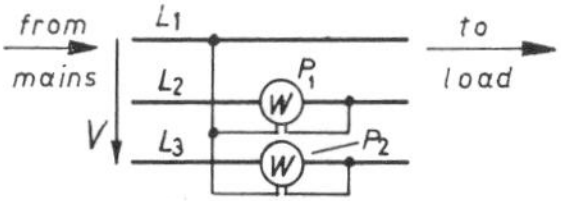

s 187 total power $P = P_1 + P_2$

I_{ph} : phase current | V_{ph} : phase voltage
I : line current | V : line voltage
L_1, L_2, L_3 : outer conductor
N : neutral conductor
$P_{w\,ph}$: active power of one phase

Reactive and active power, power factor
(for symmetrical load)

s 188 reactive power $P_q = \sqrt{3}\, V I \sin\varphi$

s 189 active power $P = \sqrt{3}\, V I \cos\varphi$

s 190 power factor $\cos\varphi = \dfrac{P}{\sqrt{3}\, V I}$

Power factor correction
(for inductive consumers)

General

Adjust to power factor according to current rate, usually $\cos\varphi = 0{\cdot}8 \ldots 0{\cdot}9$. Adjust large consumers separately and directly and small consumers centrally to main or subdistributors.

Calculating the required capacitor power

Calculate power factor $\cos\varphi$ as above, use wattmeter (see connection in S 28) or a current meter to determine P.

s 191 capacitor power $P_q = (\tan\varphi_1 - \tan\varphi_2) P_w$

s 192 inherent consumption of condenser $P_c \approx 0{\cdot}003\, P_q$

Table (numerical)

cos φ	tan φ	cos φ	tan φ	cos φ	tan φ	cos φ	tan φ
0·42	2·161	0·62	1·265	0·81	0·724	0·91	0·456
0·44	2·041	0·64	1·201	0·82	0·698	0·92	0·426
0·46	1·930	0·66	1·138	0·83	0·672	0·93	0·395
0·48	1·828	0·68	1·078	0·84	0·646	0·94	0·363
0·50	1·732	0·70	1·020	0·85	0·620	0·95	0·329
0·52	1·643	0·72	0·964	0·86	0·593	0·96	0·292
0·54	1·559	0·74	0·909	0·87	0·567	0·97	0·251
0·56	1·479	0·76	0·855	0·88	0·540	0·98	0·203
0·58	1·405	0·78	0·802	0·89	0·512	0·99	0·142
0·60	1·333	0·80	0·750	0·90	0·484	1	0·000

$\tan\varphi_1$ or $\tan\varphi_2$ can be calculated from the above table, $\cos\varphi_1$ representing the required power factor and $\cos\varphi_2$ the consumer power factor.

Direct-current machine

(motor and generator)

General

s 193	moment constant	$C_M = \frac{p\,z}{2\,\pi\,a}$
s 194	rotational source voltage	$V_q = C_M \Phi\,\omega = 2\,\pi\,C_M\,\Phi\,n$
s 195	torque	$M = C_M \Phi\,I_a$
s 196	armature current	$I_a = \frac{\pm(V - V_q)}{R_a}$ *)
s 197	terminal voltage	$V = V_q \pm I_a R_a$ *)
s 198	speed	$n = \frac{V \mp I_a R_a}{2\,\pi\,C_M \Phi}$ **)
s 199	internal power	$P_i = M_i\,\omega = V_q\,I_a$
s 200	mechanical power supplied to generator	$P_G = \frac{1}{\eta}\,V\,I_{tot}$
s 201	mechanical power supplied by motor	$P_M = \eta\,V\,I_{tot}$

Shunt motor (for circuit diagram see S 31)
Easy starting, speed is fairly independent of load and, within certain limits, easy to regulate.

Series motor (for circuit diagram see S 31)
Easy starting with powerful starting torque. Speed depends greatly on load. When running free may run away.

Compound wound motor (for circuit diagram see S 31)
Operates almost like a shunt motor. Main circuit winding ensures a powerful starting torque.

a : number of armature pairs
p : number of pole pairs
Φ : magnetic flux
z : number of conductors
R_a : armature resistance

*) + motor
− generator

**) − motor
+ generator

		motors		generators	
rotation	clockwise	counter-clockwise	clockwise	counter-clockwise

s 202	with shunt winding
s 203	with series winding
s 204	with compound winding

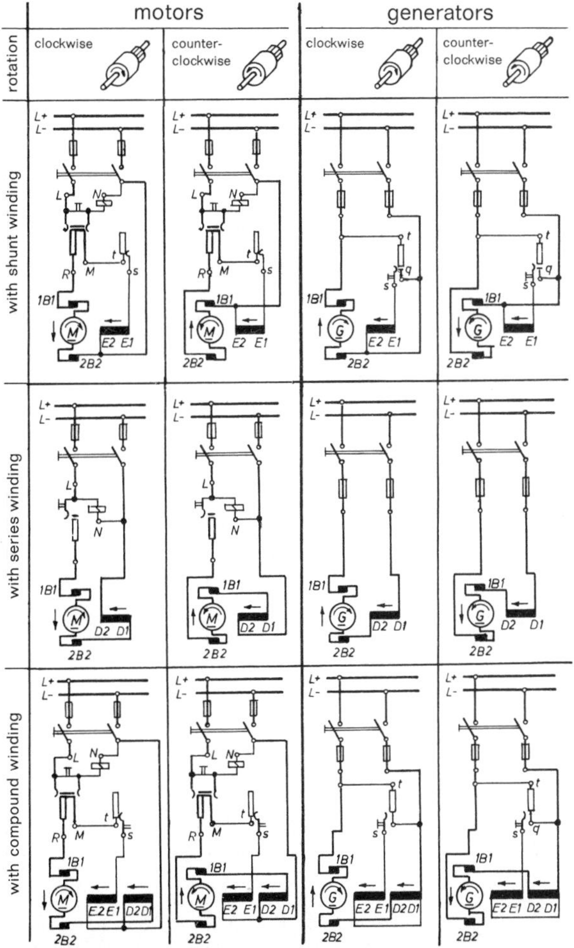

Three-phase motor

Speed

At a given frequency f the speed is determined by the number of pole pairs p.

s 205 speed $n = \frac{f}{p} = \frac{60\, f\, \mathrm{s}}{p} \times \frac{1}{\mathrm{min}}$

Switching

Where both terminals of each winding are accessible on the switchboard, the three-phase motor can be connected either in star or in delta.

phase voltage	
in star	in delta
$V_{ph} = \frac{V}{\sqrt{3}}$	$V_{ph} = V$

s 206

A 400/230 volt motor operates with its nominal values of current, torque and power, when connected to

s 207 $V = 230\,\mathrm{V}$ in delta, meaning $V_{ph} = V = 230\,\mathrm{V}$

s 208 $V = 400\,\mathrm{V}$ in star, meaning $V_{ph} = \frac{V}{\sqrt{3}} = \frac{400\,\mathrm{V}}{\sqrt{3}} = 230\,\mathrm{V}$

Star-delta connection

Higher powered motors usually operate in delta. To avoid excessive inrush currents, particularly in relatively low current networks, the motor is started in star and then switched over into delta. If, for instance, a 400/230 volt motor is connected in star to a 230/135 volt network, it is supplied with only $1/\sqrt{3}$ times its nominal voltage.

Induction motor

The rotating field of the stator causes voltage and current to be induced in the armature winding. Due to slip the rotational speed of the armature is about 3 to 5% lower than that of the rotating field; it remains almost constant under load.

Synchronous motor

Requires direct current for excitation and is synchronized with the speed of the rotating field by means of an auxiliary squirrel-cage armature. Can be used directly as a generator.

ELECTRICAL ENGINEERING — S 33

Transformer switch groups

Switch groups generally used for transformers

	type: key-numb.	type: switch group	sign: PV	sign: SV	switch diagram: PV	switch diagram: SV	ratio $V_1 : V_2$
			Threephase-output transformers				
s 209	0	D d 0	1V 1U 1W	2V 2U 2W	1U 1V 1W	2U 2V 2W	$\frac{N_1}{N_2}$
s 210	0	[Y y 0]	1V 1U 1W	2V 2U 2W	1U 1V 1W	2U 2V 2W	$\frac{N_1}{N_2}$
s 211	0	D z 0	1V 1U 1W	2V 2U 2W	1U 1V 1W	2U 2V 2W	$\frac{2\,N_1}{3\,N_2}$
s 212	5	[D y 5]	1V 1U 1W	2U 2W 2V	1U 1V 1W	2U 2V 2W	$\frac{N_1}{\sqrt{3}\,N_2}$
s 213	5	[Y d 5]	1V 1U 1W	2U 2W 2V	1U 1V 1W	2U 2V 2W	$\frac{\sqrt{3}\,N_1}{N_2}$
s 214	5	[Y z 5]	1V 1U 1W	2U 2W 2V	1U 1V 1W	2U 2V 2W	$\frac{2\,N_1}{\sqrt{3}\,N_2}$
s 215	6	D d 6	1V 1U 1W	2W 2U 2V	1U 1V 1W	2U 2V 2W	$\frac{N_1}{N_2}$
s 216	6	Y y 6	1V 1U 1W	2W 2U 2V	1U 1V 1W	2U 2V 2W	$\frac{N_1}{N_2}$
s 217	6	D z 6	1V 1U 1W	2W 2U 2V	1U 1V 1W	2U 2V 2W	$\frac{2\,N_1}{3\,N_2}$
s 218	11	D y 11	1V 1U 1W	2V 2U 2W	1U 1V 1W	2U 2V 2W	$\frac{N_1}{\sqrt{3}\,N_2}$
s 219	11	Y d 11	1V 1U 1W	2V 2U 2W	1U 1V 1W	2U 2V 2W	$\frac{\sqrt{3}\,N_1}{N_2}$
s 220	11	Y z 11	1V 1U 1W	2V 2U 2W	1U 1V 1W	2U 2V 2W	$\frac{2\,N_1}{\sqrt{3}\,N_2}$
			Single phase-output transformers				
s 221	0	I i 0	1.1 1.2	2.1 2.2	1.1 1.2	2.1 2.2	$\frac{N_1}{N_2}$

PV: primary voltage; SV: secondary voltage; D / d: delta; Y / y: star; – / z: zig-zag

Key numbers are used to calculate the phase angle (= key number × 30°) between the primary and secondary voltage, e.g. for Dy5 the phase angle is 5×30 = 150°

Note: Use the framed switch groups for preference.

The most important measuring instruments

Symbol	Type of instrument	Construction	Basically measured quantity	Scale	Used for measuring	
	moving coil	moving coil in uniform radial field of a permanent magnet. 2 spiral springs or tension bands serve for counter moment and applicat. of curr.	dc value (arithm. mean)	linear	—	I and V
	moving coil with rectifier		arithm. mean of rectified value	~linear	~ [1]	I and V
	cross coil	2 coils fixed to each other moving in non-uniform field of a permanent magnet 2 current leads without counter moment	$\frac{I_1}{I_2}$	almost square-law	— ~ [2]	$\frac{I_1}{I_2}$
	moving coil with thermo-couple	Thermocouple in close thermal contact with heater. Thermocouple feeds moving coil instruments	root-mean-square value	almost square-law	— ~ [2] [3]	I and V
	soft iron	1 moving and 1 fixed piece of soft iron, fixed coil, spiral spring as counter moment	root-mean-square value	non-linear	— ~ [2] [4]	I and V
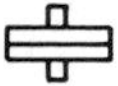	electro-dynamic	moving coil in uniform field of fixed coil, 2 spiral springs or tension bands as counter moment and current leads, magnetic screen	$I_1 \times I_2 \times \cos\varphi$	square-law for I and V, linear for P	— ~ [2]	I, V, P and $\cos\varphi$
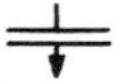	electro-static	1 fixed and 1 moving capacitor plate	root-mean-square value	non-linear	— ~ [2]	V 100 V

[1] for sinusoidal waveforms only
[2] also for non-sinusoidal waveform
[3] also for rf currents and voltages
[4] $f < 500$ Hz

RADIATION PHYSICS

Photometry and Optics

T 1

General

For every photometric quantity there is a corresponding radiation-physical quantity and the same relationships apply to both. They are differentiated by different suffixes, v for visual and e for energy.

	Photometry			Radiation physics		
	Quantity	Symbol	Units	Quantity	Symbol	Units
t 1	luminous intensity	I_v	candela cd	radiant intensity	I_e	$\frac{W}{sr}$
t 2	luminous flux	$\Phi_v = \Omega I_v$	lumen lm = cd sr	radiant power	$\Phi_e = \Omega I_e$	W = J/s
t 3	quantity of light, luminous energy	$Q_v = \Phi_v t$	lumen-second lm s, also lm h	radiant energy, quantity of radiat.	$Q_e = \Phi_e t$	J = W s
t 4	luminance	$L_v = \frac{I_v}{A_1 \cos \varepsilon_1}$	$\frac{cd}{m^2}$	radiance	$L_e = \frac{I_e}{A_1 \cos \varepsilon_1}$	$\frac{W}{sr\ m^2}$
t 5	illuminance	$E_v = \frac{\Phi_v}{A_2}$	lux $lx = \frac{lm}{m^2}$	irradiance	$E_e = \frac{\Phi_e}{A_2}$	$\frac{W}{m^2}$
t 6	light exposure	$H_v = E_v\ t$	lx s	radiant exposure	$H_e = E_e\ t$	$\frac{W\ s}{m^2}$

Definition of the base unit "candela" (cd)

The luminous intensity of a surface of $1/600\,000\ m^2$ ($= 1\tfrac{2}{3}\ mm^2$) of a black body at a temperature of 2042 K.

Photometric radiation equivalent

t 7 1 watt = 680 lm for wavelength 555 nm.

Luminous flux consumption for lighting (val. see Z 25)

A surface A lit to an illumination E_v will require a luminous flux of

t 8 $$\Phi_v = \frac{A\ E_v}{\eta}$$

For symbols see T 2

RADIATION PHYSICS
Distance law, Refraction of light

T 2

Optical distance law

The illumination of a surface is inversely proportional to the square of its distance from the light source:

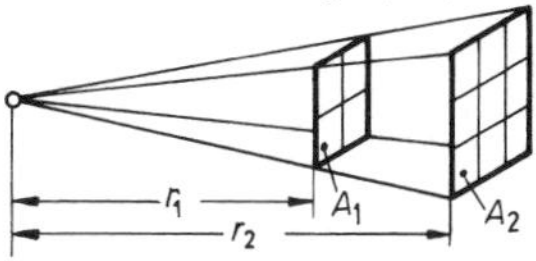

t 9 $$\frac{E_{v1}}{E_{v2}} = \frac{r_2^2}{r_1^2} = \frac{A_2}{A_1}$$

Where two light sources produce equal illumination of a surface, the ratio of the squares of their distances from the surface is equal to the ratio of their luminous intensities:

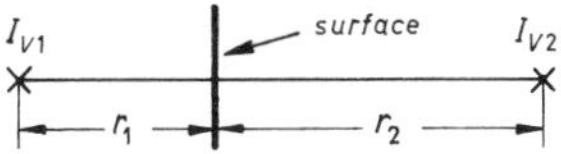

t10 $$\frac{I_{v1}}{I_{v2}} = \frac{r_1^2}{r_2^2}$$

Light refraction

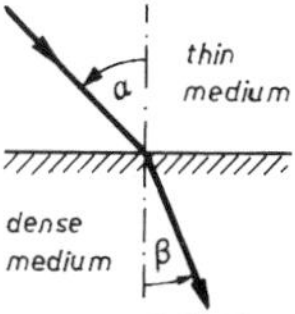

t11 $$\frac{n_b}{n_a} = \frac{\sin\alpha}{\sin\beta} = \text{const. for all angles.}$$

t12 Where $\sin\beta \geqslant \frac{n_a}{n_b}$ { total reflection occurs

t13 Refractive index for yellow sodium lighting $\lambda = 589{\cdot}3$ nm

solid matter		fluid matter		gasous matter	
in relation to		atmosphere		in relation to vacuum	
plexiglas	1·49	water	1·33	hydrogen	1·000 292
quartz	1·54	alcohol	1·36	oxygen	1·000 271
crown glass	1·56	glycerine	1·47	atmosphere	1·000 292
diamond	2·41	benzol	1·50	nitrogen	1·000 297

A_1 : area of radiating surface

A_2 : area of illuminated or irradiated surface

$A_1 \cos\varepsilon_1$: projection of the radiating surface A_1 perpendicular to the direction of radiation

n_a, (n_b): refractive index of thin (dense) medium

ε_1 : angle between emergent beam and normal to radiating surface A_1

Ω : solid angle Ω is the ratio of the area A_k intercepted on a sphere of radius r_k to the square
t14 of the radius: $\Omega = A_k/r_k^2$; unit sr = m^2/m^2.
t15 The solid angle of a point is $\Omega = 4\pi$ sr = 12·56 sr.

η : luminous efficacy (see table Z 25)

RADIATION PHYSICS

Wavelengths, Mirror

T 3

Wavelengths (in atmosphere)

t 16

Type of radiation		Wavelength $\lambda = c/f$
X-rays	hard	0·0057 nm... 0·08 nm
	soft	0·08 nm... 2·0 nm
	ultra-soft	2·0 nm...37·5 nm
optical radiation	UV-C ... IR-C	100 nm... 1 mm
ultra-violet radiation	UV-C	100 nm...280 nm
	UV-B	280 nm...315 nm
	UV-A	315 nm...380 nm
visible radiation, light	violet	380 nm...420 nm
	blue	420 nm...490 nm
	green	490 nm...530 nm
	yellow	530 nm...650 nm
	red	650 nm...780 nm
infra-red radiation	IR-A	780 nm... 1·4 µm
	IR-B	1·4 µm... 3·0 µm
	IR-C	3·0 µm... 1 mm

Mirrors

Plane mirrors

The image is at the same distance behind the mirror as the object is in front of it:

t 17 $u = -v$

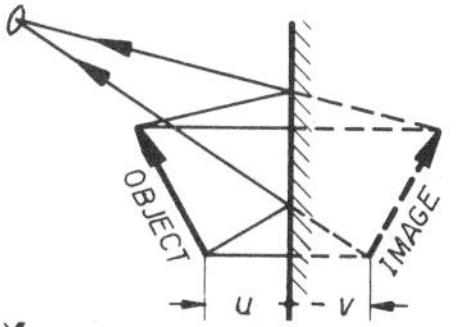

Concave mirrors

t 18 $$\frac{1}{f} = \frac{1}{u} + \frac{1}{v}$$

Depending upon the position of object, the image will be real or virtual:

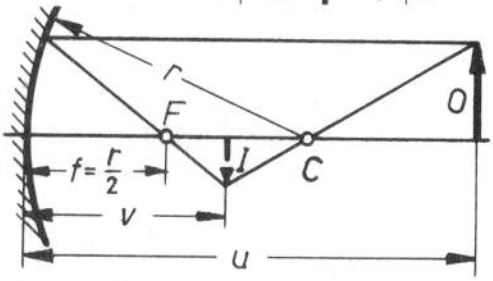

u	v	image
∞	f	at focal point
$> 2f$	$f < v < 2f$	real, inverted, smaller
$2f$	$2f$	real, inverted, of equal size
$2f > u > f$	$> 2f$	real, inverted, larger
f	∞	no image
$< f$	negative	virtual, larger

Convex mirrors

Produce only virtual and smaller images. Similar to concave mirror where:

$f = -r/2$

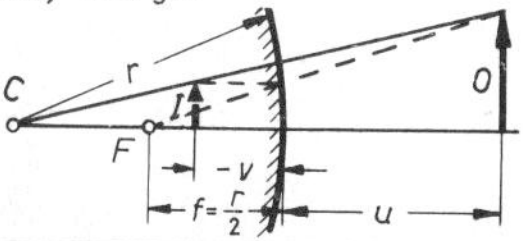

t 19 $c \hat{=} 0{\cdot}3 \times 10^9$ m/s (velocity of light)

RADIATION PHYSICS

Lenses

T 4

Lenses

Refraction D of a lens

t 20 $D = \frac{1}{f}$; Unit: 1 dpt = 1 dioptrics = $\frac{1}{\text{m}}$

Lens equation (thin lenses only)

t 21 $\frac{1}{f} = \frac{1}{v} + \frac{1}{u}$

t 22 $= (n-1)\left(\frac{1}{r_1} + \frac{1}{r_2}\right)$

t 23 $m = \frac{B}{G} = \frac{v}{u}$

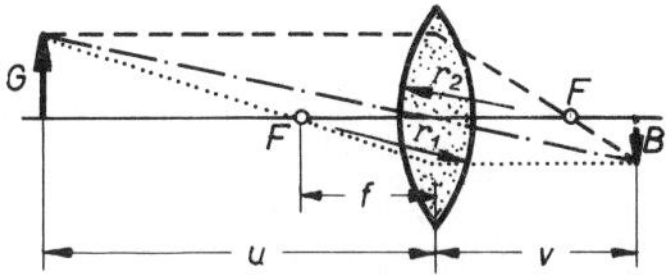

Where two lenses with focal depths f_1 and f_2 are placed immediately one behind the other, the equivalent focal length f, is given by

t 24 $\frac{1}{f} = \frac{1}{f_1} + \frac{1}{f_2}$

Magnifying lens

general	where object is in focus
t 25 $m = \frac{s}{f} + 1$	$m = \frac{s}{f}$

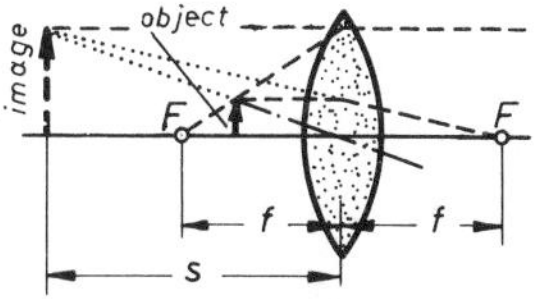

Microscope

total magnification

t 26 $m = \frac{t}{f_1}\,\frac{s}{f_2}$

t 27 $= m_1 \times m_2$

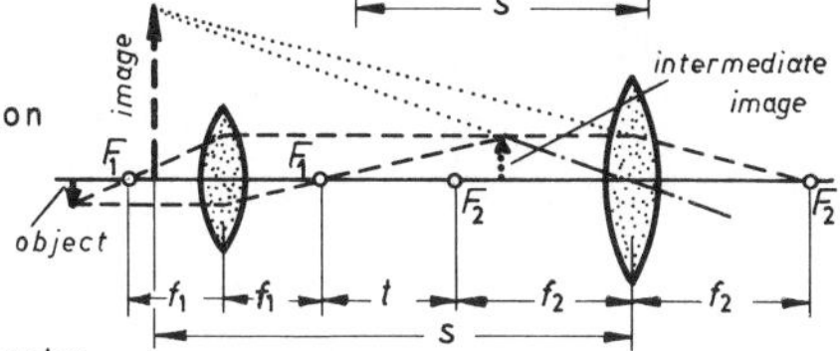

Macro photography

t 28 camera extension $a = f(m + 1)$

t 29 distance of object $c = \frac{a}{m} = f(1 + \frac{1}{m})$

B : size of image	n : refractive index (see T2)
F : focus	r : radius of curvature
f : focal length	t : optical length of tube
G : size of object	m : magnification factor

s : range of vision (= 25 cm for normal vision)

Ionizing radiation

Ionizing radiation is any radiation of charged particles which causes direct or indirect ionization or excitation of a permanent gas.

	Accumulated values	Units	time rate values	Units
t 31	amount of absorbed energy (measured value) $J = \frac{Q}{m}$	$1\,\frac{A\,s}{kg} = 1\,\frac{C}{kg}$ [1 Röntgen = 1 R = $258\,\frac{\mu C}{kg}$]	rate of absorbing energy $\dot{J} = \frac{J}{t} = \frac{I}{m}$	$1\,\frac{A}{kg}$ [$1\frac{R}{s} = 258\frac{\mu A}{kg}$ $1\frac{R}{a} = 8{\cdot}2\frac{pA}{kg}$]
t 32 t 33	absorbed dose $D = f\,J$ $= \frac{W}{m}$	1 gray = 1 gy $= 1\frac{V\,A\,s}{kg} = 1\frac{W\,s}{kg}$ $= 1\frac{J}{kg}$ [1 rad = 1 rd $= \frac{cJ}{kg} = 0{\cdot}01$ gy $= 6{\cdot}242 \times 10^{16}\frac{eV}{kg}$]	absorbed dose, -rate $\dot{D} = \frac{D}{t} = \frac{P}{m}$	$1\frac{gy}{s} = 1\frac{W}{kg}$ $= 31{\cdot}56 \times 10^{6}\frac{J}{kg\,a}$ [$1\frac{rd}{s} = 10\frac{mW}{kg}$ $= 0{\cdot}01\frac{gy}{s}$]
t 34 t 35	dose equivalent (theoretical value) $H = D_q = q\,D$ $= q\,f\,J$	$1\frac{V\,A\,s}{kg} = 1\frac{W\,s}{kg}$ $= 1\frac{J}{kg}$ [1 REM = 1 rem $= 1\frac{cJ}{kg}$]	dose equivalent, -rate $\dot{H} = \dot{D}_q = \frac{D_q}{t}$ $= q\,\dot{D}$	$1\frac{W}{kg} = 1\frac{gy}{s}$ [$1\frac{rem}{s} = 10\frac{mW}{kg}$ $1\frac{rem}{a} = 317\frac{pW}{kg}$]

t 36	Activity $A = -\,dN/dt = \lambda\,N$	1 Becquerel = 1 Bq = 1/s [1 Curie = 1 Ci = 37×10^{9} Bq]
t 37	Decay $\lambda = \ln 2/T_{1/2}$	(s, min, h, d, a)$^{-1}$

Units in [ ] are valid until 31.12.85

Explanation of terms, symbols and units see T 6

RADIATION PHYSICS

Ionizing radiation

T 6

Ionization current I: When air molecules are ionized by radiation and a voltage is applied, an ionization current I flows. (Instrument: the ionization chamber).

Charge Q: When an ionization current I flows for a time t it produces a charge

t 38 $$Q = I\,t$$

Radiation energy W: W is the radiation energy necessary for ionization. Each pair of ions in the air molecule requires the energy

t 39 $$W_L = 33{\cdot}7\text{ eV}.$$

t 40 (Charge of one electron: $1\text{ e} = 1{\cdot}602 \times 10^{-19}\text{As}$)

t 41 (1 electron volt: $1\text{ eV} = 1{\cdot}602 \times 10^{-19}\text{ As} \times 1\text{ V} = 1{\cdot}602 \times 10^{-19}\text{ J}$)

t 42 Dose: Dose is a value related to mass m, e.g. $J = Q/m$.

Activity A: The activity A is the number of atoms of a radioactive substance that disintegrates per unit time.

Notation of symbols used

	N : no. of radioactive atoms		
	$T_{1/2}$: half-life		
t 43	f_L : ionization constant for air	$(f_L = \frac{W_L}{e} = 33{\cdot}7\text{ V})$
	f : ionization constant for	
t 44	tissue	$f = f_L$
t 45	bone	$f = (1...4)\,f_L$
	m : mass (base unit)	
	q : quality factor for	
t 46	β, γ and X-rays	$q = 1$
t 47	other radiation	$q = 1...20$

Notation of units used

A : ampere | C : coulomb | J : joule

t 48 a : annum ($1\text{ annum} = 1\text{ a} = 31{\cdot}56 \times 10^6\text{ s}$ or approx. 1 year)

Exposure to radiation (dose equivalent): In 1975 the average person in the Federal Republic of Germany would have been exposed to the following radiation:

Type	H in mJ/kg	[mrem]
from natural sources	1·1	110
for medical reasons	0·5	50
other artificial radiation*)	< 0·1	< 10
*) permitted by law	≤ 0·3	≤ 30

CHEMISTRY

Elements

U 1

element	symbol	atomic mass in u	element	symbol	atomic mass in u
aluminum	Al	26·9815	neodymium	Nd	144·240
antimony	Sb	121·75	neon	Ne	20·183
argon	Ar	39·948	nickel	Ni	58·71
arsenic	As	74·9216	niobium	Nb	92·906
barium	Ba	137·34	nitrogen	N	14·0067
beryllium	Be	9·0122	osmium	Os	190·2
bismuth	Bi	208·980	oxygen	O	15·9994
boron	B	10·811	palladium	Pd	106·4
bromine	Br	79·909	phosphorus	P	30·9738
cadmium	Cd	112·40	platinum	Pt	195·09
caesium	Cs	132·905	potassium	K	39·102
calcium	Ca	40·08	praseodymium	Pr	140·907
carbon	C	12·0112	radium	Ra	226·04
cerium	Ce	140·12	rhodium	Rh	102·905
chlorine	Cl	35·453	rubidium	Rb	85·47
chromium	Cr	51·996	ruthenium	Ru	101·07
cobalt	Co	58·9332	samarium	Sm	150·35
copper	Cu	63·54	scandium	Sc	44·956
erbium	Er	167·26	selenium	Se	78·96
fluorine	F	18·9984	silicon	Si	28·086
gadolinium	Gd	157·25	silver	Ag	107·870
gallium	Ga	69·72	sodium	Na	22·9898
germanium	Ge	72·59	strontium	Sr	87·62
gold	Au	196·967	sulfur	S	32·064
helium	He	4·0026	tantalum	Ta	180·948
hydrogen	H	1·008	tellurium	Te	127·6
indium	In	114·82	thallium	Tl	204·37
iodine	I	126·9044	thorium	Th	232·038
iridium	Ir	192·2	thulium	Tm	168·934
iron	Fe	55·847	tin	Sn	118·69
krypton	Kr	83·80	titanium	Ti	47·90
lanthanum	La	138·91	tungsten	W	183·85
lead	Pb	207·19	uranium	U	238·03
lithium	Li	6·939	vanadium	V	50·942
magnesium	Mg	24·312	xenon	Xe	131·30
manganese	Mn	54·9381	yttrium	Y	88·905
mercury	Hg	200·59	zinc	Zn	65·37
molybdenum	Mo	95·94	zirconium	Zr	91·22

u : atomic mass unit ($1\ u = 1{\cdot}66 \times 10^{-27}\ kg$)

CHEMISTRY

Chemicals

Chemical terms

trade name	chemical name	chemical formula
acetone	acetone	$(CH_3)_2 \cdot CO$
acetylene	acetylene	C_2H_2
ammonia	ammonia	NH_3
ammonium (hydrox. of)	ammonium hydroxide	NH_4OH
aniline	aniline	$C_6H_5 \cdot NH_2$
bauxite	hydrated aluminum [oxides	$Al_2O_3 \cdot 2\,H_2O$
bleaching powder	calcium hypochlorite	$CaCl(OCl)$
blue vitriol	copper sulphate	$CuSO_4 \cdot 5\,H_2O$
borax	sodium tetraborate	$Na_2B_4O_7 \cdot 10\,H_2O$
butter of zinc	zinc chloride	$ZnCl_2 \cdot 3\,H_2O$
cadmium sulphate	cadmium sulphate	$CdSO_4$
calcium chloride	calcium chloride	$CaCl_2$
carbide	calcium carbide	CaC_2
carbolic acid	phenol	C_6H_5OH
carbon dioxide	carbon dioxide	CO_2
carborundum	silicon carbide	SiC
caustic potash	potassium hydroxide	KOH
caustic soda	sodium hydroxide	$NaOH$
chalk	calcium carbonate	$CaCO_3$
cinnabar	mercuric sulphide	HgS
ether	di-ethyl ether	$(C_2H_5)_2O$
fixing salt or hypo	sodium thiosulphate	$Na_2S_2O_3 \cdot 5\,H_2O$
glauber's salt	sodium sulphate	$Na_2SO_4 \cdot 10\,H_2O$
glycerine or glycerol	glycerine	$C_3H_5(OH)_3$
graphite	crystaline carbon	C
green vitriol	ferrous sulphate	$FeSO_4 \cdot 7\,H_2O$
gypsum	calcium sulphate	$CaSO_4 \cdot 2\,H_2O$
heating gas	propane	C_3H_8
hydrochloric acid	hydrochloric acid	HCl
hydrofluoric acid	hydrofluoric acid	HF
hydrogen sulphide	hydrogen sulphide	H_2S
iron chloride	ferrous chloride	$FeCl_2 \cdot 4\,H_2O$
iron sulphide	ferrous sulphide	FeS
laughing gas	nitrous oxide	N_2O
lead sulphide	lead sulphide	PbS
limestone	calcium carbonate	$CaCO_3$

continued on U 3

continued from U 2

trade name	chemical name	chemical formula
limestone	calcium carbonate	$CaCO_3$
magnesia	magnesium oxide	MgO
marsh gas	methane	CH_4
minium or red lead	plumbate	$2\,PbO \cdot PbO_2$
nitric acid	nitric acid	HNO_3
phosphoric acid	ortho phosphoric acid	H_3PO_4
potash	potassium carbonate	K_2CO_3
potassium bromide	potassium bromide	KBr
potassium chlorate	potassium chlorate	$KClO_3$
potassium chloride	potassium chloride	KCl
potassium chromate	potassium chromate	K_2CrO_4
potassium cyanide	potassium cyanide	KCN
potassium dichromate	potassium dichromate	$K_2Cr_2O_7$
potassium iodide	potassium iodide	KI
prussic acid	hydrogen cyanide	HCN
pyrolusite	manganese dioxide	MnO_2
quicklime	calcium monoxide	CaO
red prussiate of pot.	potassium ferrocyan.	$K_3Fe(CN)_6$
salammoniac	ammonium chloride	NH_4Cl
silver bromide	silver bromide	$AgBr$
silver nitrate	silver nitrate	$AgNO_3$
slaked lime	calcium hydroxide	$Ca(OH)_2$
soda ash	hydrated sodium carb.	$Na_2CO_3 \cdot 10\,H_2O$
sodium monoxide	sodium oxide	Na_2O
soot	amorphous carbon	C
stannous chloride	stannous chloride	$SnCl_2 \cdot 2\,H_2O$
sulphuric acid	sulphuric acid	H_2SO_4
table salt	sodium chloride	$NaCl$
tinstone, tin putty	stannic oxide	SnO_2
trilene	trichlorethylene	C_2HCl_3
urea	urea	$CO(NH_2)_2$
water	water	H_2O
white lead	basic lead carbonate	$2\,PbCO_3 \cdot Pb(OH)_2$
white vitriol	zinc sulphate	$ZnSO_4 \cdot 7\,H_2O$
yellow pruss. of pot.	potass. ferrocyanide	$K_4Fe(CN)_6 \cdot 3H_2O$
zinc blende	zinc sulphide	ZnS
zinc or chinese white	zinc oxide	ZnO

pH values

The negative log of the hydrogen-ion-concentration c_{H^+} indicates its *pH* value:

$$pH = -\log c_{H^+}$$ u 1

c_{H^+}	1	10^{-1}	10^{-2}	…	10^{-7}	…	10^{-12}	10^{-13}	10^{-14}
pH-value	0	1	2		7		12	13	14
	← acid			→	neu-tral	← alkaline			→

Establishing *pH* values by using suitable indicators.

Acid-base-indicators

Indicator	*pH*-Range	Colour change from	to
thymol blue	1·2–2·8	red	yellow
p-dimethylamino-azo-[benz.	2·9–4·0	red	orange-yellow
bromophenolblue	3·0–4·6	yellow	red-violet
congo red	3·0–4·2	blue-violet	red-orange
methyl orange	3·1–4·4	red	yellow-(orange)
brom cresol green	3·8–5·4	yellow	blue
methyl red	4·4–6·2	red	(orange)-yellow
litmus	5·0–8·0	red	blue
bromocresol purple	5·2–6·8	yellow	purple
brom phenol red	5·2–6·8	orange yell.	purple
bromothymol blue	6·0–7·6	yellow	blue
phenol red	6·4–8·2	yellow	red
neutral red	6·4–8·0	(blue)-red	orange-yellow
cresol red	7·0–8·8	yellow	purple
meta cresol purple	7·4–9·0	yellow	purple
thymol blue	8·0–9·6	yellow	blue
phenolphtalein	8·2–9·8	colourless	red-violet
alizarin yellow 66	10·0–12·1	light-yell.	light brown-yell.

CHEMISTRY

Reagents, Equations, Freezing mixtures

U5

Reagents

	reagent	indicator	colouration
u 2	acids	blue litmus paper	red
u 3		red phenolphthalein	colourless
u 4		yellow methylorange	red
u 5	bases	red litmus paper	blue
u 6		colourless phenolphthalein	red
u 7		red methylorange	yellow
u 8	ozone	potassium-iodide starch paper	blue-black
u 9	H_2S	lead iodide paper	brown-black
u 10	ammonia solution	hydrochloric acid	white fumes
u 11	carbonic acid	calcium hydroxide	sediment

Preparation of chemicals

	to prepare	use reaction
u 12	ammonia	$CO(NH_2)_2 + H_2O \rightarrow 2\,NH_3 + CO_2$
u 13	ammonium chloride	$NH_4OH + HCl \rightarrow NH_4Cl + H_2O$
u 14	ammonium hydroxide	$NH_3 + H_2O \rightarrow NH_4OH$
u 15	cadmium sulphide	$CdSO_4 + H_2S \rightarrow CdS + H_2SO_4$
u 16	carbon dioxide	$CaCO_3 + 2\,HCl \rightarrow CO_2 + CaCl_2 + H_2O$
u 17	chlorine	$CaOCl_2 + 2\,HCl \rightarrow Cl_2 + CaCl_2 + H_2O$
u 18	hydrogen	$H_2SO_4 + Zn \rightarrow H_2 + ZnSO_4$
u 19	hydrogen sulphide	$FeS + 2\,HCl \rightarrow H_2S + FeCl_2$
u 20	lead sulphide	$Pb(NO_3)_2 + H_2S \rightarrow PbS + 2\,HNO_3$
u 21	oxygen	$2\,KClO_3 \rightarrow 3\,O_2 + 2\,KCl$
u 22	sodium hydroxide	$Na_2O + H_2O \rightarrow 2\,NaOH$
u 23	zinc sulphide	$ZnSO_4 + H_2S \rightarrow ZnS + H_2SO_4$

Freezing mixtures

	Drop in temperature from °C	Drop in temperature to °C	Mixture (The figures stand for proportions by mass)
u 24	+10	−12	4 H_2O + 1 KCl
u 25	+10	−15	1 H_2O + 1 NH_4NO_3
u 26	+ 8	−24	1 H_2O + 1 $NaNO_3$ + 1 NH_4Cl
u 27	0	−21	3,0 ice (crushed) + 1 $NaCl$
u 28	0	−39	1,2 ice (crushed) + 2 $CaCl_2 \cdot 6\,H_2O$
u 29	0	−55	1,4 ice (crushed) + 2 $CaCl_2 \cdot 6\,H_2O$
u 30	+15	−78	1 methyl alcohol + 1 CO_2 solid

U 6 CHEMISTRY

Moisture, Drying agents, Water-hardness

Atmospheric relative humidity in closed containers

Relative humidity above the solution (%) 20°C = 65°F	Supersaturated aqueous solution	
92	$Na_2CO_3 \cdot 10\ H_2O$	u 31
86	$K\,Cl$	u 32
80	$(NH_4)_2\,SO_4$	u 33
76	$Na\,Cl$	u 34
63	$NH_4\,NO_3$	u 35
55	$Ca(NO_3)_2 \cdot 4\ H_2O$	u 36
45	$K_2CO_3 \cdot 2\ H_2O$	u 37
35	$Ca\,Cl_2 \cdot 6\ H_2O$	u 38

Drying agents (desiccants) for desiccators

Water remaining after drying at 25°C (77°F), g/m³ air	desiccant		
	name	formula	
1·4	copper sulphate, dehydr.	$Cu\,SO_4$	u 39
0·8	zinc chloride	$Zn\,Cl_2$	u 40
0·14...0·25	calcium chloride	$Ca\,Cl_2$	u 41
0·16	sodium hydroxide	$Na\,OH$	u 42
0·008	magnesium oxide	$Mg\,O$	u 43
0·005	calcium sulphate, dehydr.	$Ca\,SO_4$	u 44
0·003	hydrated aluminium	$Al_2\,O_3$	u 45
0·002	potassium hydroxide	$K\,OH$	u 46
0·001	silica gel	$(Si\,O_2)_x$	u 47
0·000 025	phosphorus pentoxide	$P_2\,O_5$	u 48

Hardness of a water

$1°$German hardness $\mathrel{\hat{=}} 1°d \mathrel{\hat{=}} \dfrac{10\ \text{mg CaO}}{1\ \text{l water}} \mathrel{\hat{=}} \dfrac{7 \cdot 19\ \text{mg MgO}}{1\ \text{l water}}$ u 49

$1°d = 1 \cdot 25°$ English hardness $= 1 \cdot 78°$ French hardness u 50

$= 17 \cdot 8°$ American hardness (1·00 ppm $CaCO_3$) u 51

Classification of hardness

0... 4°d	very soft	12...18°d	rather hard	u 52
4... 8°d	soft	18...30°d	hard	u 53
8...12°d	slightly hard	above 30°d	very hard	u 54

Mixture rule for fluids (mixture cross)

a	capa-	starting	fluid	in weight-%	$a \rightarrow c \rightarrow x = \lvert b - c \rvert$	u 55
c	city	mixed				
b	of the	admixture			$b \rightarrow c \rightarrow y = \lvert c - a \rvert$	u 56

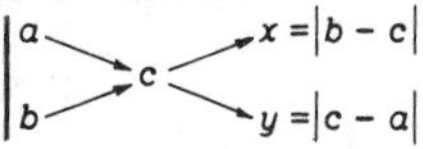

for water is $b = 0$.

Example: $a = 54\%$; $b = 92\%$; c shall become 62%. – One should mix thus 30 weight-sharings of a with 8 of b.

TABLES

Properties of solids

Z 1

Reference conditions

Density ρ at t = 20°C

Boiling point t: The values in brackets refer to sublimation, i.e. direct transition from the solid to the gaseous state.

Thermal conductivity λ at t = 20°C

Specific heat c for the temperature range 0 < t < 100°C

Substance	density ρ	melting point t	boiling point t	thermal conductivity λ	specific heat c
	kg/dm³	°C	°C	W/(m K) [1]	kJ/(kg K) [2]
agate	2·6	1600	2600	10·89	0·80
alumin. bronze	7·7	1040	2300	127·9	0·436
aluminum cast.	2·6	658	2200	209·4	0·904
aluminum roll.	2·7	658	2200	209·4	0·904
amber	1·0	300	·	·	·
antimony	6·67	630	1440	22·53	0·209
arsenic	5·72	815	·	·	0·348
artific. wool	1·5	·	·	·	1·357
asbestos	2·5	1300	·	0·17	0·816
barium	3·59	704	1700	·	0·29
barytes	4·5	1580	·	·	0·46
beryllium	1·85	1280	2970	165	1·02
bismuth	9·8	271	1560	8·1	0·13
boiler scale	2·5	1200	2800	1·2...3·5	0·80
borax	1·72	740	·	·	0·996
brass, cast	8·4	900	1100	113	0·385
brass, rolled	8·5	900	1100	113	0·385
brick	1·8	·	·	1·0	0·92
bromine	3·14	− 7·3	63	·	·
bronce(Cu Sn 6)	8·83	910	2300	64	0·37
brown iron ore	5·1	1570	·	0·58	0·67
cadmium	8·64	321	765	92·1	0·234
calcium	1·55	850	1439	·	0·63
carbon	3·51	3600	·	8·9	0·854
cast iron	7·25	1200	2500	58	0·532
cerium	6·77	630	·	·	·
chalk	1·8	·	·	0·92	0·84

[1] 1 W/(m K) = 0·8589 kcal/(h m K)

[2] 1 kJ/(kg K) = 0·2388 kcal/(kg K)

Z 2 TABLES

Properties of solids

Substance	density ρ	melting point t	boiling point t	thermal conductivity λ	specific heat c
	kg/dm³	°C	°C	W/(m K)[1]	kJ/(kg K)[2]
charcoal	0·4	·	·	0·084	0·84
chromium	7·1	1800	2700	69	0·452
clay	1·8...2·1	1600	2980	1	0·88
cobalt	8·8	1490	3100	69·4	0·435
coke	1·4	·	·	0·184	0·84
concrete reinf	2·4	·	·	0·8...1·7	0·88
constantan	8·89	1600	2400	23·3	0·41
copper, cast	8·8	1083	2500	384	0·394
copper, rolled	8·9	1083	2500	384	0·394
cork	0·2...0·3	·	·	0·05	2·0
diamond	3·5	·	[3540]	·	0·52
dripping, beef	0·9...1·0	40...50	350	·	0·88
duralium	2·8	650	2000	129·1	0·92
ebonite	1·2...1·8	·	·	0·17	·
electron	1·8	650	1500	162·8	1·00
emery	4·0	2200	3000	11·6	0·96
fire brick	1·8...2·2	2000	2900	0·47	0·88
glass, window	2·5	700	·	0·81	0·84
glass-wool	0·15	·	·	0·04	0·84
gold	19·29	1063	2700	310	0·130
graphite	2·24	3800	4200	168	0·71
ice	0·92	0	100	2·3	2·09
ingot iron	7·9	1460	2500	47...58	0·49
iodine	4·95	113·5	184	0·44	0·218
iridium	22·5	2450	4800	59·3	0·134
iron, cast	7·25	1200	2500	58	0·532
iron, forged	7·8	1200	·	46...58	0·461
iron-oxide	5·1	1570	·	0·58	0·67
lead	11·3	327·4	1740	34·7	0·130
leather	0·9...1·0	·	·	0·15	1·5
limestone	2·6	·	·	2·2	0·909
lithium	0·53	179	1372	301·2	0·36
magnesia	3·2...3·6	·	·	·	·
magnesium	1·74	657	1110	157	1·05
" , alloy	1·8	650	1500	70...145	1·01

[1] 1 W/(m K) = 0·8598 kcal/(h m K)
[2] 1 kJ/(kg K) = 0·2388 kcal/(kg K)

TABLES
Properties of solids

Z 3

Substance	density ρ	melting point t	boiling point t	thermal conductivity λ	specific heat c
	kg/dm³	°C	°C	W/(m K) [1]	kJ/(kg K) [2]
manganese	7·43	1221	2150	·	0·46
marble	2·0...2·8	·	·	2·8	0·84
mica	2·8	·	·	0·35	0·87
molybdenum	10·2	2600	5500	145	0·272
nickel	8·9	1452	2730	59	0·461
osmium	22·48	2500	5300	·	0·130
oxide of chrom	5·21	2300	·	0·42	0·75
palladium	12·0	1552	2930	70·9	0·24
paper	0·7...1·1	·	·	0·14	1·336
paraffin	0·9	52	300	0·26	3·26
peat	0·2	·	·	0·08	1·9
phosphorbronce	8·8	900	·	110	0·36
phosphorus	1·82	44	280	·	0·80
pig iron, white	7·0...7·8	1560	2500	52·3	0·54
pinchbeck	8·65	1000	1300	159	0·38
pitch	1·25	·	·	0·13	·
pit coal	1·35	·	·	0·24	1·02
platinum	21·5	1770	4400	70	0·13
porcelain	2·2...2·5	1650	·	0·8 ... 1·0	0·92
potassium	0·86	63	762·2	1	1
quartz	2·5	1470	2230	9·9	0·80
radium	5	960	1140	·	·
red lead	8·6...9·1	·	·	0·7	0·092
red metal	8·8	950	2300	127·9	0·381
rhenium	21·4	3175	5500	71	0·14
rhodium	12·3	1960	2500	88	0·24
rosin	1·07	100...300	·	0·32	1·30
rubber, raw	0·95	125	·	0·2 ... 0·35	·
rubidium	1·52	39	700	58	0·33
sand, dry	1·4...1·6	1550	2230	0·58	0·80
sandstone	2·1...2·5	1500	·	2·3	0·71
selenium	4·4	220	688	0·20	0·33
silicon	2·33	1420	2600	83	0·75
" , carbide	3·12	·	·	15·2	0·67
silver	10·5	960	2170	407	0·234

[1] 1 W/(m K) 0·8598 kcal/(h m K)

[2] 1 kJ/(kg K) = 0·2388 kcal/(kg K)

TABLES

Properties of solids

Substance	density ϱ	melting point t	boiling point t	thermal conductivity λ	specific heat c
	kg/dm³	°C	°C	W/(m K)[1]	kJ/(kg K)[2]
slate	2·6...2·7	2000	·	0·5	0·76
snow	0·1	0	100	·	4·187
sodium	0·98	97·5	880	126	1·26
soot	1·6...1·7	·	·	0·07	0·84
steatite	2·6...2·7	1600	·	2	0·83
steel	7·85	1460	2500	47...58	0·49
sulfur, cryst.	2·0	115	445	0·20	0·70
tantalum	16·6	2990	4100	54	0·138
tar	1·2	− 15	300	0·19	·
tellurium	6·25	455	1300	4·9	0·201
thorium	11·7	1800	4000	38	0·14
timber, alder	0·55	·	·	0·17	1·4
" , ash	0·75	·	·	0·16	1·6
" , birch	0·65	·	·	0·14	1·9
" , larch	0·75	·	·	0·12	1·4
" , longl.p.	0·85	·	·	·	1·4
" , maple	0·75	·	·	0·16	1·6
" , oak	0·85	·	·	0·17	2·4
" , pitchpine	0·75	·	·	0·14	1·3
" , pockwood	1·28	·	·	0·19	2·5
" , red bch.	0·8	·	·	0·14	1·3
" , red pine	0·65	·	·	0·15	1·5
" , white "	0·75	·	·	0·15	1·5
tin, cast	7·2	232	2500	64	0·24
tin, rolled	7·28	232	2500	64	0·24
titanium	4·5	1670	3200	15·5	0·47
tungsten	19·2	3410	5900	130	0·13
uranium	19·1	1133	3800	28	0·117
vanadium	6·1	1890	3300	31·4	0·50
wax	0·96	60	·	0·084	3·43
welding iron	7·8	1600	2500	54·7	0·515
white metal	7·5...10	300...400	2100	35...70	0·147
zinc, cast	6·86	419	906	110	0·38
zinc, die-cast	6·8	393	1000	140	0·38
zinc, rolled	7·15	419	906	110	0·38

[1] 1 W/(m K) = 0·8598 kcal/(h m K)
[2] 1 kJ/(kg K) = 0·2388 kcal/(kg K)

Reference conditions

Density ρ at t = 20°C and p = 1·0132 bar.

Melting point and boiling point t at p = 1·0132 bar.

Thermal conductivity λ at t = 20°C. For other temperatures see Z 15.

Specific heat c for the temperature range 0 < t < 100°C.

Substance	density ρ	melting point t	boiling point t	thermal conductivity λ	specific heat c
	kg/dm³	°C	°C	W/(m K) [1]	kJ/(kg K) [2]
acetic acid	1·08	16·8	118	·	·
acetone	0·79	− 95	56·1	·	·
alcohol	0·79	− 130	78·4	0·17...0·2	2·43
benzene	0·89	5·4	80	0·137	1·80
benzine	0·7	− 150	50 ... 200	0·16	2·1
chloroform	1·53	− 70	61	·	·
diesel oil	0·88	− 5	175	0·13	·
ether	0·73	117	35	0·14	2·26
gas oil	0·86	− 30	200...300	0·15	·
glycerine	1·27 [3]	− 20	290	0·29	2·43
heating oil	0·92	− 5	175...350	0·12	·
hydrochlor acid 10%	1·05	− 14	102	0·50	3·14
hydrochlor acid 40%	1·20	·	·	·	·
hydr. fluoride	0·99	− 92·5	19·5	2·33	2·09
linseed oil	0·96	− 20	316	0·15	·
machine oil	0·91	− 5	380...400	0·126	1·68
mercury	13·6	− 38·9	357	8·4	0·138
methyl alcohol	0·8	− 98	66	·	2·51
nitric acid	1·56 [3]	− 41	86	0·26	1·72
oil of resin	0·94	− 20	150...300	0·15	·
oil of terpent.	0·87	− 10	160	0 10	1·80
perchlor ethyl.	1·62	− 20	119	·	0·905
petroleum	0·80	− 70	150...300	0·159	2·09
" ether	0·67	− 160	40 ... 70	0·14	1·76
sulph.acid conc.	1·84	− 10	338	0·5	1·38
sulph.acid 50%	1·40	·	·	·	·
sulphurus acid	1·49 [4]	− 73	− 10	·	1·34
toluene	0·88	− 94·5	110	0·14	1·59
trichlor ethyl.	1·47	− 86	87	0·16	1·30
water	1 at 4°	0	100	0 58	4 183

1) 1 W/(m K) = 0·8598 kcal/h m K)

2) 1 kJ/(kg K) = 0·2388 kcal/(kg K)

3) at t = 0°C

4) at t = − 20°C

TABLES

Properties of gases

Reference conditions

Density ρ at t = 0°C and p = 1·0132 bar. For perfect gases ρ can be calculated for other pressures and/or temperatures from: $\rho = p/(R \times T)$.

Melting point and boiling point t at p = 1·0132 bar.

Thermal conductivity λ at t = 0°C and p = 1·0132 bar. For other temperatures see Z 15.

Specific heat c_p and c_v at t = 0°C and p = 1·0132 bar. c_p at other temperatures see Z 13.

Substances	density ρ	melting point t	boiling point t	thermal conductivity λ	specific heat c_p	specific heat c_v
	kg/m³	°C	°C	W/(m K) [1]	kJ/(kg K) [2]	
acetylene	1·17	− 83	− 81	0·018	1·616	1·300
air, atmosphere	1·293	− 213	− 192·3	0·02454	1·005	0·718
ammonia	0·77	− 77·9	− 33·4	0·022	2·056	1·568
argon	1·78	− 189·3	− 185·9	0·016	0·52	0·312
blast furn. gas	1·28	− 210	− 170	0·02	1·05	0·75
butane, iso-	2·67	− 145	− 10	·	·	·
butane, n-	2·70	− 135	1	·	·	·
carbon di-oxide	1·97	− 78·2	− 56·6	0·015	0·816	0·627
carbon disulph.	3·40	− 111·5	46·3	0·0069	0·582	0·473
carbon monoxide	1·25	− 205·0	− 191·6	0·023	1·038	0·741
chlorine	3·17	− 100·5	− 34·0	0·0081	0·473	0·36
coal gas	0·58	− 230	− 210	·	2·14	1·59
ethylene	1·26	− 169·3	− 103·7	0·017	1·47	1·173
helium	0·18	− 270·7	− 268·9	0·143	5·20	3·121
hydrochlor acid	1·63	− 111·2	− 84·8	0·013	0·795	0·567
hydrogen	0·09	− 259·2	− 252·8	0·171	14·05	9·934
hydrogen sulph.	1·54	− 85·6	− 60·4	0·013	0·992	0·748
krypton	3·74	− 157·2	− 153·2	0·0088	0·25	0·151
methane	0·72	− 182·5	− 161·5	0·030	2·19	1·672
neon	0·90	− 248·6	− 246·1	0·046	1·03	0·618
nitrogen	1·25	− 210·5	− 195·7	0·024	1·038	0·741
oxigen	1·43	− 218·8	− 182·9	0·024	0·909	0·649
ozone	2·14	− 251	− 112	·	·	·
propane	2·01	− 187·7	− 42·1	0·015	1·549	1·360
sulfur dioxide	2·92	− 75·5	− 10·0	0·0086	0·586	0·456
water vapour [3]	0·77	0·00	100·00	0·016	1·842	1·381
xenon	5·86	− 111·9	− 108·0	0·0051	0·16	0·097

1) 1 W/(m K) = 0·8598 kcal/(h m K)

2) 1 kJ/(kg K) = 0·2388 kcal/(kg K)

3) at t = 100°C

Coefficients of sliding and static friction

material	on material	sliding friction μ: dry	sliding friction μ: on water	sliding friction μ: with lubrication	static friction μ_0: dry	static friction μ_0: on water	static friction μ_0: with lubrication
bronze	bronze	0·20	0·10	0·06			0·11
	cast iron	0·18		0·08			
	steel	0·18		0·07	0·19		
oak	oak ‖	0·50			0·60		
	oak ╫	0·30			0·50		
cast iron	cast iron		0·31	0·10			0·16
	steel	0·18			0·33		
rubber	asphalt	0·50	0·30	0·20			
	concrete	0·60	0·50	0·30			
hemp rope	timber				0·50		
leather belt	oak	0·40			0·50		
	cast iron	0·40	0·40	0·20	0·40	0·50	0·12
steel	oak		0·26	0·08		0·65	0·11
	ice	0·014			0·027		
	steel	0·10		0·10	0·15		0·12

Rolling friction

(for section K 12 and L 9)

material on material	lever arm f of frictional force in mm
rubber on on asphalt	0·010
rubber on concrete	0·015
lignum vitae on lignum vitae	0·050
steel on steel (hard: ball bearing)	0·001
steel on steel (soft)	0·005
elm on lignum vitae	0·080

‖ : movement with grain of both materials
╫ : movement perpendicular to grain of sliding body

Friction factors for flow in pipes $\zeta = f\left(Re, \frac{k}{d}\right)$

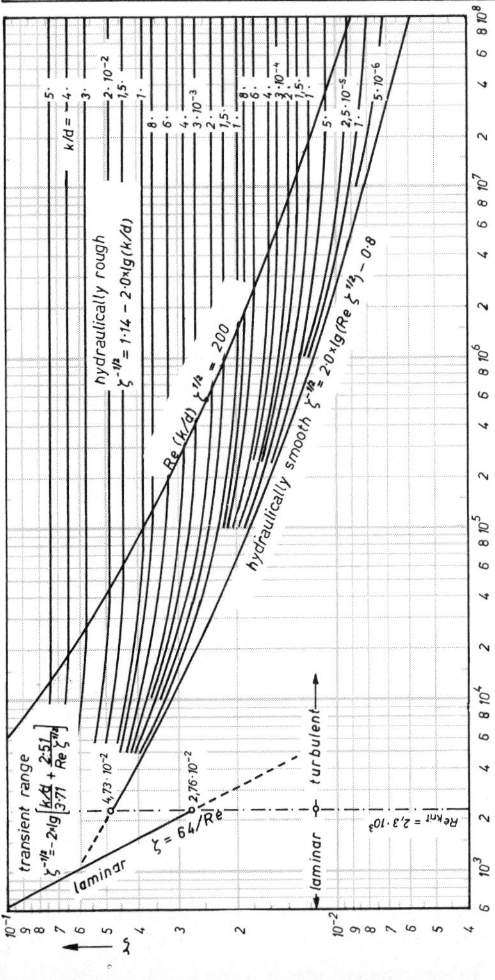

Note: For non-circular pipes k/d is to be replaced by k/d_h.

Roughness k

(according to Richter, Hydraulics of Pipes)

Material and kind of pipe	Condition of pipe	k in mm
new seamless rolled or drawn (commercial) steel pipes	typical rolled finish	0·02...0·06
	pickled	0·03...0·04
	cleanly galvanized (dipping process)	0·07...0·10
	commercial galvanized	0·10...0·16
used steel pipes	uniform corrosion pits	about 0·15
	medium corrosion, light incrustation	0·15...0·4
	medium incrustation	about 1·5
	heavy incrustation	2...4
	cleaned after long use	0·15...0·20
cast iron pipes	new, typical cast finish	0·2 ...0·6
	new, bituminized	0·1 ...0·13
	used, corroded	1 ...1·5
	incrusted	1·5 ...4
	cleaned after several years of use	0·3 ...1·5
	mean value in urban se-werage installations	1·2
	heavily corroded	4·5
pipes folded and riveted of sheet steel	new, folded	about 0·15
	new, depending on kind and quality of riveting	
	light riveting	about 1
	heavy riveting	to 9
	25 years old, heavily incrusted, riveted pipe	12·5

TABLES

Heat values

Latent heat of fusion per unit mass l_f

material	kJ/kg	material	kJ/kg	material	kJ/kg
aluminum	377	glycerine	176	paraffin	147
antimony	164	gold	67	phenol	109
brass	168	ice	335	platinum	113
cadmium	46	iron	205	potassium	59
cast iron	126	lead	23	silver	109
chromium	134	manganese	155	sulfur	38
cobalt	243	mercury	11·7	tin	59
copper	172	naphthaline	151	Wood's alloy	33·5
ethyl ether	113	nickel	234	zinc	117

Latent heat of evaporation per unit mass l_d

at 101·32 kN/m² (= 760 torr)

material	kJ/kg	material	kJ/kg	material	kJ/kg
alcohol	880	hydrogen	503	oxygen	214
ammonia	1410	mercury	281	sulf. dioxide	402
carb. dioxide	595	methyl. chlor.	406	toluene	365
chlorine	293	nitrogen	201	water	2250

Drawing colours of steel and associated temperatures

colouring	t °C	colouring	t °C
pale yellow	200	corn-flower blue	300
straw-yellow	220	light-blue	320
brown	240	bluish gray	350
purple	260	gray	400
violet	280		

Colours of glowing steel and associated temperatures

colouring	t °C	colouring	t °C
dark-red	680	yellowish red	950
dark cherry-red	740	yellow	1000
cherry-red	770	light-yellow	1100
light cherry-red	800	yellowish white	1200
light-red	850	white	1300
light light-red	900		and above

Linear coefficient of expansion α in 1/K

at t = 0...100°C

material	$\alpha/10^{-6}$	material	$\alpha/10^{-6}$	material	$\alpha/10^{-6}$
aluminum	23·8	Germ. silver	18·0	porcelain	4·0
bismuth	13·5	gold	14·2	quartz glass	0·5
brass	18·5	lead	29·0	silver	19·7
bronze	17·5	molybdenum	5·2	steatite	8·5
cadmium	30·0	nickel	13·0	steel, mild	12·0
cast iron	10·5	nickel steel =Invar 36%Ni	1·5	tin	23·0
constantan	15·2			tungsten	4·5
copper	16·5	platinum	9·0	zinc	30·0

Cubic coefficient of expansion β in 1/K

at t = 15°C

material	$\beta/10^{-3}$	material	$\beta/10^{-3}$	material	$\beta/10^{-3}$
alcohol	1·1	glycerine	0·5	petroleum	1·0
benzene	1·0	mercury	0·18	toluene	1·08
ether	1·6	oil of turp.	1·0	water	0·18

Coefficient of heat transfer k in W/(m²K)

(Approx. values, natural convection on both sides)

material	thickness of insulating layer in mm								
	3	10	20	50	100	120	250	380	510
reinforced concr.				4·3	3·7	3·5	2·4		
insul. cement bl. (e. g. thermalite)									
σ_c = 2·45 N/mm²						1·2	0·7	0·5	
σ_c = 4·90 N/mm²						1·6	0·9	0·7	
σ_c = 7·35 N/mm²						1·7	1·0	0·7	
glass	5·8	5·3							
glass-, mineral-wool, hard foam	4·1	2·4	1·5	0·7	0·4				
timber wall			3·8	2·4	1·8	1·7			
chalky sandstone						3·1	2·2	1·7	1·4
gravel concrete				4·1	3·6	3·4	2·3		
slag concrete						2·7	1·7	1·4	1·0
brick						2·9	2·0	1·5	1·3

double or treble glazing	2·6 or 1·9
single window, puttied	5·8
double window, 20 mm spacing, puttied *)	2·9
double window, 120 mm spacing, puttied *)	2·3
tiled roof without/with joint packing	11·6/5·8

*) also for windows with sealed air gaps

TABLES

Heat values

Gas constant R and molecular mass M

material	R J/(kg K)	M kg/kmol	material	R J/(kg K)	M kg/kmol
acetylene	319	26	hydrogen	4124	2
air	287	29	nitrogen	297	28
ammonia	488	17	oxygen	260	32
carbonic acid	189	44	sulphuric acid	130	64
carbon monoxide	297	28	water vapour	462	18

Radiation constant C at 20°C

material	C W/(m² K⁴)	material	C W/(m² K⁴)
silver, polish.	$0 \cdot 17 \times 10^{-8}$	copper, oxid.	$3 \cdot 60 \times 10^{-8}$
aluminum, pol.	$0 \cdot 23 \times 10^{-8}$	water	$3 \cdot 70 \times 10^{-8}$
copper, polish.	$0 \cdot 28 \times 10^{-8}$	timber, planed	$4 \cdot 40 \times 10^{-8}$
brass, polished	$0 \cdot 28 \times 10^{-8}$	porcelain,glaz	$5 \cdot 22 \times 10^{-8}$
zinc, polished	$0 \cdot 28 \times 10^{-8}$	glass	$5 \cdot 30 \times 10^{-8}$
iron, polished	$0 \cdot 34 \times 10^{-8}$	brickwork	$5 \cdot 30 \times 10^{-8}$
tin, polished	$0 \cdot 34 \times 10^{-8}$	soot, smooth	$5 \cdot 30 \times 10^{-8}$
aluminum,unpol.	$0 \cdot 40 \times 10^{-8}$	zinc, unpol.	$5 \cdot 30 \times 10^{-8}$
nickel, polish.	$0 \cdot 40 \times 10^{-8}$	iron, unpol.	$5 \cdot 40 \times 10^{-8}$
brass, unpol.	$1 \cdot 25 \times 10^{-8}$	absolutely	
ice	$3 \cdot 60 \times 10^{-8}$	black surface	$5 \cdot 67 \times 10^{-8}$

Specific heat c for liquids in kJ/(kg K)

t at °C	-25	0	20	50	100	200
ammonia		4·64	4·77			
benzene			1·74			
carbon dioxide			3·64			
machine oil			1·85	1·97	2·18	
mercury			0·139			
sulphur dioxide	1·25	1·355	1·389			
transformer oil			1·89	2·04	2·29	
water		4·216	4·179	4·181	4·212	4·497

Mean specific heat $c_{pm}\big|_0^t$ of various gases in $\frac{kJ}{kg\,k}$ as a function of temperature

t °C	CO	CO_2	H_2	H_2O [1]	N_2 pure	N_2 [2]	O_2	SO_2	air
0	1·039	0·8205	14·38	1·858	1·039	1·026	0·9084	0·607	1·004
100	1·041	0·8689	14·40	1·874	1·041	1·031	0·9218	0·637	1·031
200	1·046	0·9122	14·42	1·894	1·044	1·035	0·9355	0·663	1·013
300	1·054	0·9510	14·45	1·918	1·049	1·041	0·9500	0·687	1·020
400	1·064	0·9852	14·48	1·946	1·057	1·048	0·9646	0·707	1·029
500	1·075	1·016	14·51	1·976	1·066	1·057	0·9791	0·721	1·039
600	1·087	1·043	14·55	2·008	1·076	1·067	0·9926	0·740	1·050
700	1·099	1·067	14·59	2·041	1·087	1·078	1·005	0·754	1·061
800	1·110	1·089	14·64	2·074	1·098	1·088	1·016	0·765	1·072
900	1·121	1·109	14·71	2·108	1·108	1·099	1·026	0·776	1·082
1000	1·131	1·126	14·78	2·142	1·118	1·108	1·035	0·784	1·092
1100	1·141	1·143	14·85	2·175	1·128	1·117	1·043	0·791	1·100
1200	1·150	1·157	14·94	2·208	1·137	1·126	1·051	0·798	1·109
1300	1·158	1·170	15·03	2·240	1·145	1·134	1·058	0·804	1·117
1400	1·166	1·183	15·12	2·271	1·153	1·142	1·065	0·810	1·124
1500	1·173	1·195	15·21	2·302	1·160	1·150	1·071	0·815	1·132
1600	1·180	1·206	15·30	2·331	1·168	1·157	1·077	0·820	1·138
1700	1·186	1·216	15·39	2·359	1·174	1·163	1·083	0·824	1·145
1800	1·193	1·225	15·48	2·386	1·181	1·169	1·089	0·829	1·151
1900	1·198	1·233	15·56	2·412	1·186	1·175	1·094	0·834	1·156
2000	1·204	1·241	15·65	2·437	1·192	1·180	1·099	0·837	1·162
2100	1·209	1·249	15·74	2·461	1·197	1·186	1·104		1·167
2200	1·214	1·256	15·82	2·485	1·202	1·191	1·109		1·172
2300	1·218	1·263	15·91	2·508	1·207	1·195	1·114		1·176
2400	1·222	1·269	15·99	2·530	1·211	1·200	1·118		1·181
2500	1·226	1·275	16·07	2·552	1·215	1·204	1·123		1·185
2600	1·230	1·281	16·14	2·573	1·219	1·207	1·127		1·189
2700	1·234	1·286	16·22	2·594	1·223	1·211	1·131		1·193
2800	1·237	1·292	16·28	2·614	1·227	1·215	1·135		1·196
2900	1·240	1·296	16·35	2·633	1·230	1·218	1·139		1·200
3000	1·243	1·301	16·42	2·652	1·233	1·221	1·143		1·203

[1] at low pressures [2] derived from air

Calculated from figures given in E. Schmidt: Einführung in die Technische Thermodynamik, 11. Auflage, Berlin/Göttingen/Heidelberg: Springer 1975.

Z 14

TABLES

Values for heat transfer, at 1.0132 bar

Dynamic viscosity η of liquids in N s/m² × 10^{-5} *)

t in °C	-25	0	20	50	100	200
acetic acid			121 0	79·2	45·8	
ammonia	21·5	16·9	13·8	10·3		
benzene			64·9	43·6	26·1	11·3
carbon dioxide	13·0	10·1	7·0			
chloroform		70·6	56·3	42·4		
diethylene ether	38·7	29·6	24·3	18·3	11·8	
glycerine		12 10	15 10			
hydrochloric acid	17·0	11·0	8·6	5·4		
hydrofluoric acid	37	25·6				
hydrogen sulfide		15·2	12·6	9·4		
propane	15·8	12·7	10·2	7·1		
sulfur dioxide	47·5	36·8	30·4	23·4		
toluene		77·3	58·6	41·9	26·9	13·3
trichlorethylene		71·0	60·0	44·7		
water		175·7	100·2	54·3	27·8	13·3

Dynamic viscosity η of motor oils in N s/m² × 10^{-1} *)

		-25	0	20	50	100
SAE	10		3·1	0·79	0·20	0·05
	20		7·2	1·70	0·33	0·07
	30		15·3	3·10	0·61	0·10
	40		26·1	4·30	0·72	0·12
	50		38·2	6·30	0·97	0·15

Dynamic viscosity η of gases in N s/m² × 10^{-5} *)

t in °C	-25	0	20	100	200	400
air	1·50	1·72	1·82	2·18	2·58	3·28
ammonia	0·85	0·930	1·00	1·28	1·65	2·34
argon	1·19	2·10	2·35	2·68	3·21	4·10
carbon dioxide		1·37	1·46	1·82	2·22	2·93
chlorine		1·25	1·32	1·68	2·10	2·86
helium	1·75	1·89	1·93	2·28	2·67	3·41
hydrogene chloride		1·31	1·43	1·83	2·30	
hydrogene sulfide		1·17	1·15	1·59	1·99	
krypton		2·33	2·48	3·06	3·74	4·91
neon		2·99	3·10	3·65	4·26	5·32
oxygen	1·75	1·92	2·00	2·43	2·88	3·67
sulfur dioxide	1·06	1·17	1·16	1·63	2·07	2·80
nitrogen	1·50	1·66	1·75	2·09	2·47	3·14
water vapour		0·90	0·97	1·25	1·61	2·33
xenon		2·11	2·35	2·83	3·50	4·73

*) 1 N s/m² = 1 kg/(m s) = 1 Pa s = 1000 cP

Thermal conductivity λ of liquids in W/(m K)

t in °C	-25	0	20	50	100	200
ammonia		0·540	0·494			
benzene			0·153			
carbon dioxide			0·087			
mercury			0·304			
spindle oil			0·144	0·143	0·140	
sulfur dioxide	0·228	0·212	0·199			
transformer oil			0·124	0·122	0·119	
water		0·554	0·598	0·640	0·681	0·665

Thermal conductivity λ of vapours and gases in W/(m K)

t in °C	-25	0	20	100	200	400
air		0·022	0·023	0·029	0·035	0·047
ammonia		0·020	0·024	0·027		
carbon dioxide	0·011	0·013	0·014	0·019	0·026	
hydrogen		0·160	0·150	0·209	0·252	
sulfur oxide		0·008				
water vapour				0·022	0·030	0·050

Prandtl number Pr for liquids

t in °C	-25	0	20	50	100	200
ammonia		2·07	2·12			
benzene			7·33			
carbon dioxide			2·00			
mercury			0·023			
spindle oil			168	71	31	
sulfur dioxide	2·80	2·36	2·14			
transformer oil			481	165	60	
water		13·6	7·03	3·90	1·75	0·94

Prandtl number Pr for vapours and gases

t in °C	-25	0	20	100	200	400
air		0·715	0·713	0·707	0·685	0·685
ammonia		0·92	0·93	0·97		
carbon dioxide		0·80	0·80	0·80	0·81	
hydrogen	0·67	0·67	0·67	0·67	0·67	0·67
sulfur oxide		0·86				
water vapour				1·12	0·97	0·88

TABLES

Strength values in N/mm²

Modulus of elasticity (Young's modulus) of steel:
$E = 210\,000$ N/mm²

Specifications from	Material	Tensile strength R_m	Yield point 0·2 proof stress R_e; $R_{p\,0\cdot2}$	Fatigue strength: tension, compression A $p_{t\,A}$	tension, compression U $p_{t\,U}$	bending A $p_{bt\,A}$	bending U $p_{bt\,U}$	torsion A $p_{q\,A}$	torsion U $p_{q\,U}$	Notes: sizes in mm
ASTM	A 283 Gr. C	340	225	153	225	170	283	99	131	
	A 284 Gr. D	410	265	185	265	205	342	119	154	
	A 572 Gr. 55	470	285	212	285	235	392	136	165	$16<d\leq 40$
	A 572 Gr. 65	570	325	257	325	285	455	165	189	
SAE	1044	670	355	302	355	335	497	194	206	
SAE	1045	697	481	314	481	349	582	202	279	$d\leq 16$
		657	412	296	412	329	548	191	239	$16<d\leq 40$
		618	373	278	373	309	515	179	216	$40<d\leq 100$
SAE	4140	1079	883	486	810	540	900	313	512	$d\leq 16$
		981	765	441	735	491	818	285	444	$16<d\leq 40$
		883	638	397	638	442	737	256	370	$40<d\leq 100$
UNS	K 31820	1226	1030	552	920	613	1022	356	593	$16<d\leq 40$
ASTM A536	60-40-18	420	250	180	250	200	333	116	145	
	80-55-06	600	350	245	350	270	450	155	200	
	100-70-03	700	400	280	400	315	525	180	230	

A : Alternating } see P 2
U : Undulating }

*) Allow a safety factor for the permissible stresses (see P 2 and P 18)

Note: The strength depends on the diameter, especially with heat-treated steel

Allowable bending and torsional stresses; E and G moduli for elastic materials in N/mm²

Material	Modulus of elasticity E	Type of loading *)	p_{bt} A	p_{bt} B	p_{bt} C	Modulus of ridig. G	p_{qt}
Spring Steel		I	1000	500	150		650
SAE 1078	210 000	II	750	350	120	80000	500
hard. and temp.		III	500	250	80		350
Yellow Brass		I	200	100	40		120
ASTM-B134(274)	110 000	II	150	80	30	42000	100
HV 150		III	100	50	20		80
Nickel Silver		I	300	150	50		200
65-18; HV 160	142 000	II	250	120	40	55000	180
ASTM-B122(752)		III	200	100	30		150
Tin Bronze		I	200	100	40		120
CDA-419	110 000	II	150	80	30	42000	100
HV 190		III	100	50	20		80
Phosphor Bronze		I	300	150	50		200
CDA-529	117 000	II	220	110	40	45000	180
HV 190		III	150	80	30		150

A : for simple springs (safety factor ≈ 1·5)
B : for bent and shaped springs (" " ≈ 3)
C : for springs with no hysteresis effect (" " ≈ 10)

For cylindrical helical springs use diagram on page Q 9.

*) For explanation refer to P 1

Characteristic quantities for machining (for turning outside longitudnally)

Material	Strength in N/mm² or hardness	z	$1 - z$	$k_{s\,1.1}$ N/mm²
ASTM - A 572 Grade	520	0·26	0·74	1990
UNS - K 04600	720	0·30	0·70	2260
SAE - 1045	670	0·14	0·86	2220
SAE - 1060	770	0·18	0·82	2130
SAE - 5120	770	0·26	0·74	2100
SAE - 3140	630	0·30	0·70	2260
SAE - 4135	600	0·21	0·79	2240
SAE - 4140	730	0·26	0·74	2500
SAE - 6150	600	0·26	0·74	2220
SAE - L 6 annealed	940	0·24	0·76	1740
SAE - L 6 tempered	ASTM E18-74-HRD54	0·24	0·76	1920
Mehanite A	360	0·26	0·74	1270
Chilled cast iron	ASTM E18-74-HRD60	0·19	0·81	2060
ASTM - A 48 - 40 B	ASTM E18-74-HRD33	0·26	0·74	1160

Specified values apply directly for turning with carbide [tip
Cutting speed v = 90...125 m/min
Chip thickn. h=0·05mm ⩽ h ⩽ 2·5mm | Ratio of slenderness $\varepsilon_s = 4$
Normal side-rake angle $\gamma = 6^\circ$ for steel, 2° for cast iron

Strength values in N/mm^2

Permissible contact pressure p_b in N/mm^2

Bearing pressure of joint bolts – Building construction

Load characteristic	material	p_b	material	p_b
main load	ASTM-	206	ASTM-	304
main and additional load	A 283 Gr. C	235	A 440	343

Journals and bearings, bearing plates (see q 13)

Hydrodynamic lubrication see q 47.

Mixed lubrication, shaft hardened and ground:[1), 2)]

Material		v m/s	p_b
gray cast iron			5
ASTM-B30 Cast-	(836) Lead. Red Brass		8...12
		1	20 [3)]
	(938) Lead. Tin Bronze	0·3 ...1	15 [3)]
sintered iron		<1	6
		3	1
sintered iron with copper		<1	8
		3	3
sintered bronze		<1	12
		3	6
		5	4
tin-bronze graphite (DEVA metal)		<1	20 ... 90 [4)]

Material	v m/s	p_b
Cast Tin Bronze CDA 902 grease lubricat.	<0·03	4...12
quality bearings	<1	60
PA 66 (polyamide)	→ 0	15
dry [5)]	1	0·09
grease lubric. [5)]	1	0·35
HDPE (high-density polyethylene)	→ 0	2... 4
	1	0·02
PTFE (polytetrafluorethylene enclosed)	→ 0	30
	1	0·06
PTFE + lead + bronze (GLACIER-DU)	<0·005	80... 140[4)]
	0·5...5	1

General, non-sliding surfaces: Max. values are possible up to the compressive yield point at the material ($\sigma_{dF} \triangleq R_e$). But normal values for good p_b are lower.

Material	Normal values of p_b under: dead load	undulating load	shock load
bronze	30... 40	20... 30	10...15
cast iron	70... 80	45... 55	20...30
governm. bronze	25... 35	15... 25	8...12
malleable iron	50... 80	30... 55	20...30
steel	80...150	60...100	30...50

1) $(p \times v)_{perm}$ are closely related to heat dissipation, load, bearing pressure, type of lubrication.

2) Sometimes a much higher load capacity with hydrodynamic lubrication is possible.

3) Limited life (wearing parts).

4) Specially developed metals | 5) For shell thickness 1 mm

TABLES

Data for clutches and brakes

Z 19

Properties of friction materials

(Q 15...Q 17)

	Material pairing	Sliding friction coefficient μ_{slide} [4] –	Max. temperature continous °C	Max. temperature transient °C	Contact pressure p_b N/mm²	Thermal capacity per unit area q_b kW/m²
dry	organic friction lining/steel or cast iron					
dry	general [5]	0·2...0·65	150...300 (K...H)	300...600 (K...H)	0·1...10	2·2...23
dry	single plate friction-clutch	0·35...0·4	150...300	400	1	12...23
dry	automobile drum brake	0·2...0·3	250...300	350...450	0·5...1·5 ...2·0	
dry	automobile disc brake	0·3...0·4	400	600	10 (emergency braking)	
dry	cast iron/steel	0·15...0·2	300		0·8...1·4	
dry	sintered bronze/steel	0·05...0·3	400...450	500...600	1	5·5
wet [6]	sintered bronze/steel	0·05...0·1 [1]	180	500...600	3	12...23
wet [6]	steel/steel	0·06...0·1 [2]	200...250		1	3·5...5·5 [3]

[1] $\mu_{static} = (1{\cdot}3...1{\cdot}5)\,\mu_{slide}$

[2] $\mu_{static} = (1{\cdot}8...2{\cdot}0)\,\mu_{slide}$

[3] splash lubrication lower, internal lubrication higher

[4] often: μ_{static} 1·25 μ_{slide}

[5] K = rubber bond; H = synthetic resin bond

[6] running in oil

Strain energy w and yield strength k_f

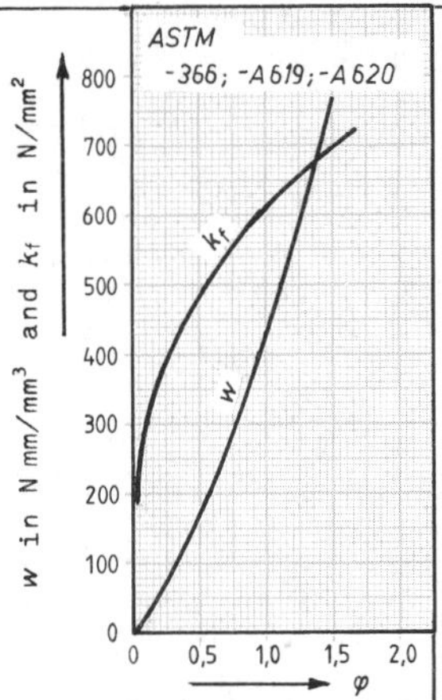

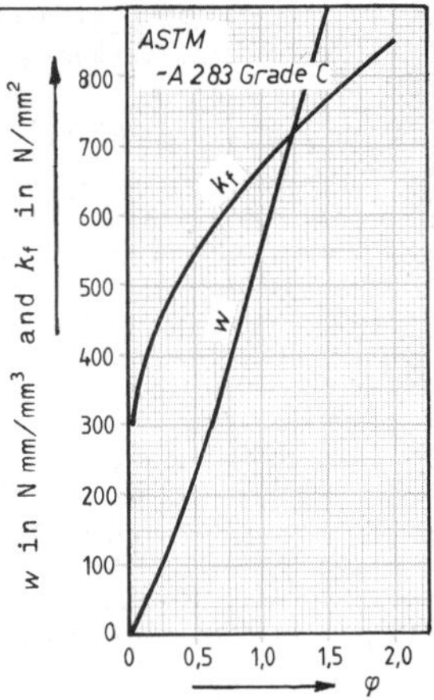

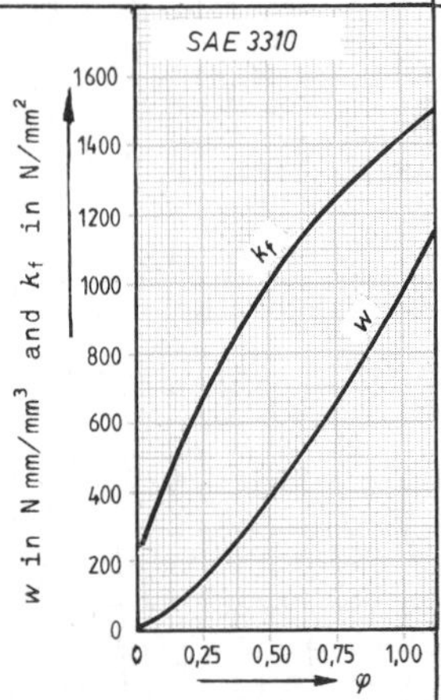

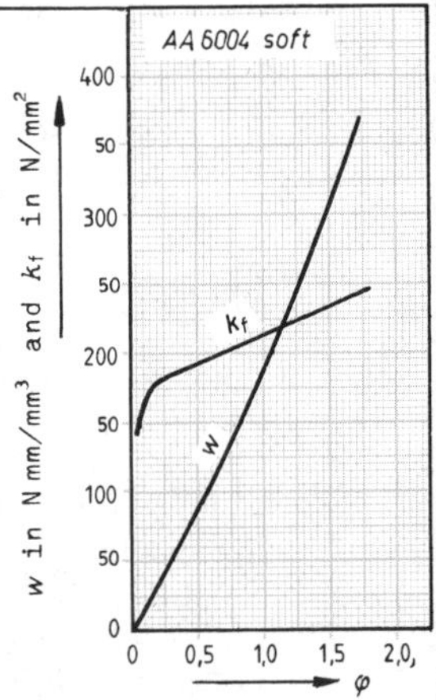

φ : logarithmic deformation ratio | k_f : yield strength
w : strain energy per unit volume
For other materials see VDI 3200

Electrical specific resistance ρ and specific conductance γ of conductors at 20°C

material	ρ $\frac{\Omega\, mm^2}{m}$	γ $\frac{m}{\Omega\, mm^2}$	material	ρ $\frac{\Omega\, mm^2}{m}$	γ $\frac{m}{\Omega\, mm^2}$
aluminum	0·0278	36	iron (pure)	0·10	10
antimony	0·417	2·4	lead	0·208	4·8
brass-58%Cu	0·059	17	magnesium	0·0435	23
brass-63%Cu	0·071	14	manganese	0·423	2·37
cadmium	0·076	13·1	mercury	0·941	1·063
carbon	40	0·025	mild steel	0·13	7·7
cast iron	1	1	nickel	0·087	11·5
chromium-Ni-Fe	0·10	10	nickeline	0·5	2·0
constantan	0·48	2·08	platinum	0·111	9
copper	0·0172	58	silver	0·016	62·5
German silver	0·369	2·71	tin	0·12	8·3
gold	0·0222	45	tungsten	0·059	17
graphite	8·00	0·125	zinc	0·061	16·5

Electrical resistance ρ of insulators

material	ρ Ω cm	material	ρ Ω cm
bakelite	10^{14}	plexiglass	10^{15}
glass	10^{15}	polystyrene	10^{18}
marble	10^{10}	porcelain	10^{14}
mica	10^{17}	pressed amber	10^{18}
paraffin oil	10^{18}	vulcanite	10^{16}
paraffin wax (pure)	10^{18}	water, distilled	10^{7}

Electric temperature coefficient α_{20} at 20°C

material	α_{20} 1/K or 1/°C	material	α_{20} 1/K or 1/°C
aluminum	+ 0·00390	mercury	+ 0·00090
brass	+ 0·00150	mild steel	+ 0·00660
carbon	− 0·00030	nickel	+ 0·00400
constantan	− 0·00003	nickeline	+ 0·00023
copper	+ 0·00380	platinum	+ 0·00390
German silver	+ 0·00070	silver	+ 0·00377
graphite	− 0·00020	tin	+ 0·00420
manganese	± 0·00001	zinc	+ 0·00370

TABLES

Electrical properties

Dielectric constant ε_r

insulant	ε_r	insulant	ε_r
araldite	3·6	paraffin oil	2·2
atmosphere	1	paraffin wax	2·2
bakelite	3·6	petroleum	2·2
casting compound	2·5	phenolic resin	8
castor oil	4·7	plexiglass	3·2
ebonite	2·5	polystyrene	3
glass	5	porcelain	4·4
guttapercha	4	pressed board	4
hard paper (lamin.)	4·5	quartz	4·5
insulation of high voltage cables	4·2	shellac	3·5
		slate	4
insulation of telephone cables	1·5	soft rubber	2·5
		steatite	6
marble	8	sulfur	3·5
mica	6	teflon	2
micanite	5	transf. oil mineral	2·2
nylon	5	transf. oil vegetab.	2·5
oil paper	4	turpentine	2·2
olive oil	3	vulcanised fibres	2·5
paper	2·3	vulcanite	4
paper, impregnated	5	water	80

Electro-motive series
(potential difference with respect to hydrogen electrode)

material	U volt	material	U volt
potassium	−2·93	cadmium	−0·40
calcium	−2·87	cobalt	−0·28
sodium	−2·71	nickel	−0·23
magnesium	−2·37	tin	−0·14
beryllium	−1·85	lead	−0·13
aluminum	−1·66	hydrogen	0·00
manganese	−1·19	copper	+0·34
zinc	−0·76	silver	+0·80
chromium	−0·74	mercury	+0·85
tungsten	−0·58	platinum	+1·20
iron	−0·41	gold	+1·50

Magnetic field strength H and relative permeability μ_r as a function of induction B

induction B		cast iron		steel casting and dynamo sheet steel $p_{Fe10} = 3\cdot6\ \frac{W}{kg}$		alloyed dynamo sheet steel $p_{Fe10} = 1\cdot3\ \frac{W}{kg}$	
		H	μ_r	H	μ_r	H	μ_r
$T = \frac{V\,s}{m^2}$ tesla	[G gauss]	A/m	–	A/m	–	A/m	–
0·1	1 000	440	181	30	2 650	8·5	9 390
0·2	2 000	740	215	60	2 650	25	6 350
0·3	3 000	980	243	80	2 980	40	5 970
0·4	4 000	1 250	254	100	4 180	65	4 900
0·5	5 000	1 650	241	120	3 310	90	4 420
0·6	6 000	2 100	227	140	3 410	125	3 810
0·7	7 000	3 600	154	170	3 280	170	3 280
0·8	8 000	5 300	120	190	3 350	220	2 900
0·9	9 000	7 400	97	230	3 110	280	2 550
1·0	10 000	10 300	77	295	2 690	355	2 240
1·1	11 000	14 000	63	370	2 360	460	1 900
1·2	12 000	19 500	49	520	1 830	660	1 445
1·3	13 000	29 000	36	750	1 380	820	1 260
1·4	14 000	42 000	26	1 250	890	2 250	495
1·5	15 000	65 000	18	2 000	600	4 500	265
1·6	16 000			3 500	363	8 500	150
1·7	17 000			7 900	171	13 100	103
1·8	18 000			12 000	119	21 500	67
1·9	19 000			19 100	79	39 000	39
2·0	20 000			30 500	52	115 000	14
2·1	21 000			50 700	33		
2·2	22 000			130 000	13		
2·3	23 000			218 000	4		

——— practical limit

TABLES
Magnetic properties

Dynamo sheet properties

type			mild sheet steel	alloy sheet steel			
				low	medium	high	
class of sheet			I 3·6	II 3·0	III 2·3	IV 1·5	IV 1·3
panel size mm x mm			1000 x 2000				750x 1500
thickness mm			0·5				0·35
density kg/dm³			7·8	7·75	7·65	7·6	
core losses per unit mass at f = 50Hz W/kg (max.)		p_{Fe10}	3·6	3·0	2·3	1·5	1·3
		p_{Fe15}	8·6	7·2	5·6	3·7	3·3
in-duc-tion (min.)	B_{25}	V s/m² [gauss]	1·53 15 300	1·50 15 000	1·47 14 700	1·43 14 300	
	B_{50}	V s/m² [gauss]	1·63 16 300	1·60 16 000	1·57 15 700	1·55 15 500	
	B_{100}	V s/m² [gauss]	1·73 17 300	1·71 17 100	1·69 16 900	1·65 16 500	
	B_{300}	V s/m² [gauss]	1·98 19 800	1·95 19 500	1·93 19 300	1·85 18 500	

Explanations

B_{25} = 1·53 Vs/m² indicates that a minimum induction of 1·53 V s/m² [or 15 300 gauss] is reached with a field strength of 25 A/cm. Thus a flux length of e.g. 5 cm requires a circulation of 5 x 25 A = 125 A.

p_{Fe10}	describes the core losses per unit mass at f = 50 Hz and an induction of	1·0 V s/m² = [10 000 G]
p_{Fe15}		1·5 V s/m² = [15 000 G]

Guide values for illumination E_v in lx = lm/m²

Type of establishment or location		General lighting only	General and spec. lighting: bench	General and spec. lighting: general
workshops according to work done	rough	100	50	200
	medium	200	100	500
	precise	300	200	1000
	very prec.	500	300	1500
offices	normal	500		
	open	750		
living rooms, lighting	medium	200		
	bright	500		
streets and squares with traffic	light	20		
	medium	50		
	heavy	100		
factory yards with traffic	light	20		
	heavy	50		

Luminous efficacy η

Type of lighting	Colour of illuminated surface: light	medium	dark
direct	0·60	0·45	0·30
indirect	0·35	0·25	0·15

Type of lighting	deep bowl reflector	widespread reflector
street and square lighting	0·45	0·40

Luminous flux Φ_v of lamps

Standard lamps with single coiled filament (at operating voltage)		P_{el}	W	15	25	40	60	75	100
		Φ_v	klm	0·12	0·23	0·43	0·73	0·96	1·39
		P_{el}	W	150	200	300	500	1000	2000
		Φ_v	klm	2·22	3·15	5·0	8·4	18·8	40·0
Fluorescent lamps values for 'Warmwhite' 'Daylight'	tubular diamet.								
	26 mm	P_{el}	W		18	36	58		
		Φ_v	klm		1·45	3·47	5·4		
	38 mm	P_{el}	W	15	20	40	65		
		Φ_v	klm	0·59	1·20	3·1	5·0		
High-pressure lamps filled with mercury vapour		P_{el}	W	125	250	400	700	1000	2000
		Φ_v	klm	6·5	14	24	42	60	125

INDEX